D1432832

THE FORBIDDEN FUEL

A History of Power Alcohol

NEW EDITION

HAL BERNTON
WILLIAM KOVARIK
SCOTT SKLAR

Foreword to the Bison Books edition by R. James Woolsey

Preface to the Bison Books edition by Boyd Griffin

Introduction to the Bison Books edition by Hal Bernton

UNIVERSITY OF NEBRASKA PRESS
LINCOLN AND LONDON

This volume is a reissue, with new material, of the original edition published in 1982 by Boyd Griffin, Inc. The new edition is published under an arrangement with Boyd Griffin.

Library of Congress Cataloging-in-Publication Data
Bernton, Hal.
The forbidden fuel: a history of power alcohol / Hal Bernton, William Kovarik, and Scott Sklar; foreword to the Bison Books edition by R. James Woolsey; preface to the Bison Books edition by Boyd Griffin; introduction to the Bison Books edition by Hal Bernton. — New ed.
p. cm.
Includes bibliographical references and index.
ISBN 978-0-8032-2808-5 (pbk.: alk. paper)
1. Alcohol as fuel. I. Kovarik, Bill. II. Sklar, Scott. III. Title.
TP358.B4 2010
333.79'68—dc22
2009045323

Foreword

This fascinating book, first published just over a quarter of a century ago, was beyond prescient when written and has aged like fine wine.

The story of using alcohols as fuel for vehicles reaches back to the early years of the last century and to Henry Ford's enthusiasm for ethanol's being the principal fuel for the internal combustion engine. Realistically, the amount of fuel required and the dependence on agricultural feedstocks such as corn and sugarcane made ethanol more a candidate for being a major additive to gasoline to improve its octane rather than a total replacement. However, throughout most of the twentieth century, each time the question would arise whether to use ethanol in place of tetraethyl lead or the carcinogenic "aromatics" (benzene, toluene, xylene) as a safer way of enhancing octane, the oil industry pulled out every stop to defeat ethanol's use for any fuel purpose. Ethanol's far more benign chemical nature (compared to tetraethyl lead and the aromatics) was ultimately of no avail. These stories from the 1920s and the 1970s are the first two acts of a struggle still going on today, one in which much the same arguments predominate.

National security was very much on the minds of the participants in the 1970s debate—and that was when we only imported about one-third of our oil compared with the approximately two-thirds we import today. But although a repeat of the Saudi 1973 oil embargo certainly came to be a huge concern of the public, politicians, and policy makers, the conclusion reached at the end of the Carter administration was, sadly, to emphasize government investment in extremely expensive coal-to-liquid synfuels. Three years after the 1982 publication of *The Forbidden Fuel*, the Saudis decided to drop the bottom out of the oil market (prices fell below $10 a barrel in 1986), and the next year the Synfuels Corporation went bankrupt. Yet another chance to move toward biofuels was lost.

Climate change was not on many people's horizon in 1982, but Messrs. Bernton, Kovarik, and Sklar included a thorough tutorial on its importance and the reasons ethanol is superior to fossil fuels in terms of well-to-wheel CO_2 emissions. There are also sophisticated assessments of not only the potential of various feedstocks—particularly the relative ability of different plants to grow on arid land largely unsuitable for crops—but also of the prospects for evolving toward the use of cellulosic feedstocks. This book appreciated and explained over twenty-five years ago the revolutionary nature of such a shift toward the use of waste and agricultural and forest residues to produce alcohol-based fuels.

With extraordinary foresight the authors also advocated "decentralized technologies available for producing renewable alcohol fuels" rather than the "centralized technologies" and large facilities required to produce synthetic fuels from fossil fuels. They even note the complementary nature of decentralized production of alcohol fuels and the decentralized generation of electricity, for example, with biomass being used in both processes.

It is a very fair question: How with this kind of insight in hand in 1982, could we have gone so wrong?

Several factors seem to share responsibility for our misdirection.

First, very low oil prices in the mid-1980s did much to discourage not only synfuels but also embryonic biofuel efforts by insightful scientists and entrepreneurs motivated by the oil crises of the 1970s such as those described in these pages.

Second, oil companies and, to some extent, automobile companies fought hard against anything that would change their nineteenth-century business plans (oil: drill, refine, market; autos: bend steel into cars, power them with oil products).

Third, the views of those who were heavily influenced by the memory of the oil embargo of 1973 and the disruption of 1979 were not yet enhanced by 9/11 or the recent level of concern about climate change.

Finally, and perhaps most importantly, until quite recently batteries were boring. For 150 years the only interesting thing that happened to batteries (almost all lead acid) was that we stopped needing to add water to them. Until a few years ago bright graduates of MIT, Cal Tech, and Stanford were not starting battery companies. But the consumer electronics revolution of the last three decades has changed that. Cell phones, PDAS, and such have been a huge stimulus to innovation in electricity storage. So also has been the increasing fragility of the electricity grid and the consequent demand for uninterrupted power supplies for, for example, communications equipment. As a result the development of lithium-ion batteries for transportation, flow batteries for stationary storage, capacitors, innovative use of compressed air, vehicle-to-grid and vehicle-to-home hookups for plug-in hybrids, and other approaches to electricity storage are multiplying and becoming more efficient by the month.

Some might take the view that the resulting rebirth of electricity in transportation will make alcohol-based fuels unnecessary. I disagree. I believe that the cost of batteries for transportation will remain sufficiently high for some time, and that most vehicles will have an all-electric range of only a few dozen miles on a single charge, covering most daily driving. Vehicles will thus consequently need a backup for longer trips, such as fast charging or battery swapping at special facilities, or rely on liquid fuel in the tank, as with a plug-in hybrid. If the market shifts toward using liquid fuel as the backup to electricity in a substantial number of plug-in hybrid vehicles, the natural evolution would be toward alcohol-based fuels in plug-ins that are also open-standard flexible-fuel vehicles (FFVs) since this system requires the smallest change in the existing infrastructure. These types of FFVs could use ethanol, methanol, biobutanol, or other such fuels along with electricity. The added cost of their being FFVs is only about one hundred dollars per car. New pumps at filling stations? Not a showstopper. We've changed these before, to get the lead out.

The key point is that whether oil prices are driven down intentionally by the Saudis, as in 1986, or, more likely, driven down by recession, as was the case in the spring of 2009, oil products cannot now become cheap enough to undersell electricity.

In plug-in hybrids now on the road, the cost of electricity is in the range of one to four cents per mile. With time-of-day pricing for electricity in your garage at night, such as the pricing offered business customers, those costs would approximately halve. For gasoline to compete with, say, two-cents-per-mile electricity, it would have to sell at solidly under a dollar a gallon, and oil prices would have to stay below thirty dollars per barrel. Because of Saudi and other OPEC commitments to welfare programs and other perceived needs (such as Saudi citizens paying essentially no taxes), this prospect is not in the cards. And if the Saudis, who have the reserve capacity, cannot beat electricity, I think it is far more likely that they would milk as much out of the oil market for as long as they can rather than going partway and trying, say, to drop oil to a price in the forty-dollar-per-barrel range, which would cause severe competitive problems for liquid fuels but would not defeat electricity. Electricity, in a sense, protects biofuel innovation.

So, as the song in the musical *Oklahoma* says, "The farmer and the cowman should be friends," and so should the alcohol fuel innovators and the electricity advocates. There's room for both in a transportation-fuel market that has been freed from the twin yokes of oil's monopoly (96 percent) of transportation fuel and OPEC's domination of oil. Electricity and alcohol fuels should prosper together and end the age of oil.

R. James Woolsey
Director of Central Intelligence (1993–95)
Central Intelligence Agency, United States of America

Preface

Back in 1978 the three authors—Hal Bernton, William Kovarik, and Scott Sklar—shared a common fascination with the rebirth of the power alcohol industry and began a four-year collaboration that resulted in the manuscript for *The Forbidden Fuel*. I met the authors in 1980 and agreed to play the roles of editor and publisher of the eventual book; I had just started a new publishing company, and the subject of alcohol fuels was of great interest to me also. The authors were diligent in their efforts to tell the whole story; they reviewed historical archives and traveled around the country to interview farmers, distillers, and engineers, among others. They even flew down to Brazil to meet with sugarcane growers and automobile executives. In Washington DC, they interviewed politicians, lobbyists, and scientists, who informed them—in 1979—of the greenhouse gas dilemma and the perils of global warming. (As evidence of the authors' diligent research and the book's prescience on the subject of global warming, it was published less than a year after James Hansen—widely considered in the scientific community to be the leading authority on the subject of global warming—and six other NASA atmospheric scientists published a seminal article in *Science* magazine, "Climate Impact of Increasing Atmospheric Carbon Dioxide," that alerted the world to the threat of global warming caused by the burning of fossil fuels.) After much drafting, fact checking, and redrafting—without personal computers, word-processing software, or the World Wide Web, which emerged in the decade that followed—*The Forbidden Fuel* was published in April 1982.

As a history of the agricultural, social, and political forces in the twentieth century that created conditions for the birth and growth of the modern renewable fuels industry, the book is unique and holds up very well today. (That's the way it should be with all good history books: If you want to read a history of World War II, Winston Churchill's books still are a good place to begin.) The book's examination of controversial issues—the food-or-fuel debate, limits of the land, environmental effects, even global warming—probed in detail many of the pressing questions that still spark debate today.

As book publishing goes, *The Forbidden Fuel* was a modest success. It received very good critical reviews; for example, *Choice*, a leading book-review periodical for librarians, wrote in its 1983 review:

An intriguing history of alcohol fuel in America . . . a study of alcohol as an energy source in the contemporary world . . . at its best in those sections concerned with that part of 20th-century Americana dating from Henry Ford, through the Depression years (when American farmers hoping to grow corn for fuel were frustrated in their objectives), into the more recent period of farmers' movements and "modern moon shiners." Particularly good is Chapter 5, which focuses on interest-group politics in Washington during the 1970s, and the American Petroleum Institute's efforts to block development of the gasohol option on a national basis. With its good bibliography, helpful glossary, and valuable appendixes on the economics and technology of alcohol power, the book is recommended to all institutions seeking to cover the wide area of energy options and energy politics.

The book sold over two thousand copies after its publication in 1982. So it's fair to say that, even among people (and librarians) who follow closely the evolving story of biofuels, the book is not well known. This reissued edition, backed by the distribution capabilities of the University of Nebraska Press, with all the Internet book-marketing possibilities in play, should gain a much wider readership.

It is remarkable that in 2009 the authors and I could reunite and collaborate on this reissued edition. (As one of the authors observed, it was kind of like the Blues Brothers getting the band back together again.) All three of the authors took time from their full-time occupations to help with the reissue. Hal Bernton, currently a reporter and journalist for the *Seattle Times*, miraculously found the time and energy to write the new introductory chapter of this edition, which provides an excellent overview of the developments since 1982 and the current, pressing issues facing policy makers and the renewable fuels industry. William Kovarik, a professor at Radford University in Virginia, and Scott Sklar, head of the Stella Group in Washington DC, are both still very much involved in the field of renewable energy and provided vital input for this reissued edition.

The authors and I are deeply grateful to R. James Woolsey for his informative and insightful foreword. Considering the importance of a renewable-fuel capability to energy independence and national security, who better to offer an informed perspective than a former director of the Central Intelligence Agency of the United States. Also, this reissued edition includes a wholly new appendix B, "Economics of Ethanol," thanks to the gracious permission of Chris Hurt, Wally Tyner, and Otto Doering, distinguished agricultural economists at Purdue University who are the authors of the piece, and to the Purdue Extension, the article's original publisher.

Boyd Griffin
Editor and publisher of the original edition of *The Forbidden Fuel*

Introduction

HAL BERNTON

When *The Forbidden Fuel* was published in April 1982, the oil markets were retreating from record highs as recession took hold in the United States. Oil prices would eventually plunge as low as $13.49 a barrel average in 1986,[1] and as the crude markets collapsed so, too, did America's efforts to gain energy independence.

President Jimmy Carter's plan for a "moral equivalent of war" to wean Americans away from petroleum gave way to President Ronald Reagan's call for limited government in a nation where markets, not bureaucrats, would choose the liquid fuels of the future. Ambitious projects to turn coal into gasoline were scaled back as well as efforts to wrest oil from shale rock and tar sands.

The fledgling corn ethanol industry—producing a few hundred million gallons of fuel nurtured by a large federal subsidy—seemed to be another casualty of Reaganomics. But a coalition of farm-belt lobbyists and their Republican allies in Congress helped preserve a fifty-four-cent-per-gallon tax break for gasohol, the 10 percent blend of alcohol and gasoline, which by the early 1980s was sold at thousands of midwestern service stations. As a result, the ethanol industry not only survived but dramatically expanded. By the end of 2008 more than 170 plants stretching from New York to Oregon produced some nine billion gallons of ethanol annually, an amount equivalent to more than 6 percent of the nation's gasoline consumption.[2]

The United States' appetite for oil also increased during the past quarter century as automotive fuel-efficiency standards stagnated, and sport utility vehicles populated the spreading realms of suburbia. Thus despite the growth of the ethanol industry, the nation's oil consumption in the first decade of the twenty-first century climbed to record levels, and the nation's imports of Middle Eastern oil were larger than ever.

As this reissue of *The Forbidden Fuel* is readied for publication in June 2009, a deep worldwide recession and financial crisis have taken hold, and oil markets have demonstrated unprecedented volatility; oil prices soared to record heights of nearly $150 a barrel in July 2008 before plummeting to below $40 a barrel in December of that year, then rebounding to more than $70 per barrel by June 2009.

Exhibit Intro-1
Source: Renewable Fuels Association. Chart © Boyd Griffin & Company, LLC.

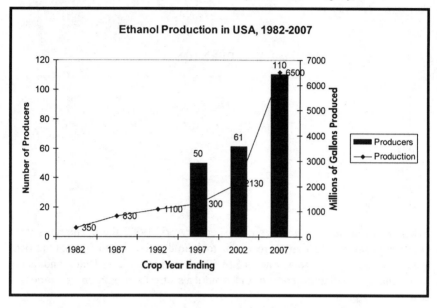

It is easy to imagine a repeat of history. Perhaps Americans struggling to reclaim jobs and financial security once again will balk at shouldering the costs of moving into the postpetroleum era. But there are also plenty of reasons to believe that the United States will chart a different course for the twenty-first century.

First, oil is expected to be an increasingly scarce resource in this new century as the explosive growth of China, India, and other emerging economic powers puts huge new strains on global petroleum markets. A prolonged world recession offers temporary relief from that pressure, but forecasters—citing some forty years of declining success in exploration—predict that oil production will peak somewhere between 2017 and 2023.[3]

Regardless of the global supply situation, in the United States there is bipartisan consensus that national security demands a cut in oil consumption: The largest petroleum producer in the world, Saudi Arabia, continues to be an incubator for Islamic fundamentalism, which produced most of the men who on September 11, 2001, turned civilian airplanes into bombs that claimed the lives of 2,973 people. The second largest global oil producer is a nuclear-armed and resurgent Russia that appears to be entering an uneasy new relationship with the United States.

There is also a new moral imperative to reduce the consumption of fossil fuels as climate change has emerged as one of the defining issues of the twenty-first century, and the action—or inaction—of today's world leaders is expected to have profound consequences for future generations. Global warming was noted in the first edition of *The Forbidden Fuel*, which cited a 1979 report from the U.S. Department of Energy calling the atmospheric buildup of greenhouse gases "the most important environmental issue facing mankind." Yet during the last two decades of the twentieth century, Exxon Mobil and other oil companies helped fund an army of skeptics to discount the science of global warming and in the United States derail serious regulatory efforts. Those countermeasures increased in the aftermath of the 1997 Kyoto Protocol, signed by thirty-seven nations and the European Union to reduce greenhouse gases. "Unless climate change becomes a non-issue, and the Kyoto proposal is defeated, and there are no further initiatives to thwart the threat of climate change, there may be no moment when we can declare victory in our efforts," wrote Joe Walker, an American Petroleum Institute official, in a 1998 memo to officials at Exxon and other oil companies.[4]

The oil industry found an ally in the White House as President George W. Bush took office in 2000, and the United States never ratified the Kyoto treaty. But the issue of global warming did not go away. Scientists reported startling physical changes already attributed to global warming, such as the rapid shrinkage of summer ice in the Arctic, which put polar bears at risk of extinction. They warned that a warming world would endanger millions of people by intensifying droughts, shrinking glaciers and snow packs that are vital sources of freshwater, and raising sea levels with potentially catastrophic effects on some heavily populated lowland areas. Such research earned a Nobel Prize for the scientists serving on the United Nations Intergovernmental Panel on Climate Change and for Al Gore because of his efforts to alert the world to that threat. Scientists also have documented that fossil fuel combustion has increased the acidity levels of the ocean; this process will continue for decades to come and create major new risks for coral, zooplankton, and other creatures that are the underpinnings of the marine ecosystems.

By 2007 the skeptics were in retreat. Even Exxon chairman Rex Tillerson, in a speech that year, conceded "legitimate concerns about the risks of climate change due to rising greenhouse gas emissions resulting from the world's enormous requirements for fossil fuels" and acknowledged that it was prudent to reduce greenhouse gas emissions.[5]

In his 2008 presidential campaign, President Barack Obama pledged to act on global warming by building a clean-energy industry that would end U.S. dependence on Middle Eastern oil and reduce greenhouse gas emissions by regulating carbon dioxide as a pollutant. "We have the opportunity now to create jobs all across the country in all fifty states to repower America, . . . to make us more competitive for decades to come even as we are saving the planet," Obama said in a December 2008 press conference. "We are not going to miss that opportunity."[6]

In his first months in office, as the nation was mired in the worst economic downturn since the Great Depression, with two of the largest companies in the U.S. automotive industry, General Motors and Chrysler, bankrupt, Obama appeared poised to try to follow through on those promises. His plan embraces ethanol and other biofuels as one part of a much wider effort to reduce oil consumption. The plan calls for putting one million new vehicles on the road by 2015 that can run either on electricity for normal commutes or switch to liquid fuel for long distances. These plug-in vehicles as well as hydrogen- and fuel-cell-powered vehicles hold enormous potential, especially when they are linked to wind, solar, and other forms of renewable electricity. Also augmenting their impact would be improvements in mass transit and fuel conservation resulting from the 30 percent increase in automotive fuel efficiency to be in place by 2016 through an agreement that Obama reached with the automotive industry.

Table Intro-1
History of Ethanol Tax Incentive Legislation

1978 Energy Tax Act	Created 40-cent-per-gallon tax exemption for ethanol for gasoline excise tax
1982 Surface Transportation Assistance Act	Increased exemption for 10 percent gasohol to 5 cents per gallon
1984 Tax Reform Act	Increased exemption for 10 percent gasohol to 6 cents per gallon
1990 Omnibus Budget Reconciliation Act	Extended exemption to 2000 but decreased incentive to 5.4 cents per gallon
1998 Transportation Efficiency Act of the Twenty-first Century	Extended exemptions until 2007
2004 Jobs Creation Act	Mechanism of incentive changed to a blender tax credit and was extended to 2010
2005 Energy Policy Act	Mandates for renewable fuel use to be at least 7.5 billion gallons per year by 2012
2007 Energy Independence and Security Act (EISA)	Set new mandates for renewable fuel to reach 36 billion gallons per year by 2022 and stipulated that 16 billion (or 44 percent) come from cellulosic production
2008 Food, Conservation and Energy Act ("2008 Farm Bill")	Established "biorefinery assistance program" offering loan guarantees of up to $250 million to plants producing "advanced" biofuels, defined as those produced from feedstock other than corn

Sources: EIA; Wallace E. Tyner, Purdue University; E85; Whipnet.com; and Alternative Fuels and Advanced Vehicles Data Center (AFDC) of U.S. Department of Energy.

As these new automotive technologies develop, demand for liquid fuels could eventually plummet to less than half of today's annual consumption. Then a significant portion of this much smaller volume of liquid combustibles could come from a new generation of ethanol, biodiesel, and other biofuels produced from wood wastes, trash, grasses, and even algae. General Motors, for example, plans to introduce in 2010 the Chevy Volt, an electric vehicle that could use an 85 percent blend of ethanol and gasoline to maintain the car's primary power source, a lithium-ion battery.

In his own energy plan released on the campaign trail, Obama calls for another major expansion of biofuels to reach sixty billion gallons of "advanced" biofuels by 2030. When Obama moved into the White House, he had the legislative spurs to develop new biofuels already in place: a provision of the 2007 Energy Independence and Security Act (EISA) mandates thirty-six billion barrels of biofuel production by 2022 and requires that sixteen billion gallons, or 44 percent, of that supply be derived from sources other than corn, soybeans, and other traditional feedstocks.

Many oil industry officials are rankled by that mandate and would like to see it weakened. Yet corporate positions have changed dramatically since the early twentieth century, when, as this book documents, there were campaigns to undermine alcohol fuels. Today there is widespread acceptance of 10 percent ethanol blending with gasoline, and many companies are investing in alcohol and other biofuels research.

But the role of biofuels in the nation's energy future remains a subject of intense debate. Critics question the ethics of diverting the starch from corn into producing energy for cars rather than fattening livestock for human consumption. They are also alarmed by the environmental repercussions of a global expansion of agriculture on water, forests, and soil fertility. Many of these concerns have surrounded the ethanol industry since its inception and are explored in the book. In chapter 6, "Agriculture: The Limits of the Land," we note that "the prospect of the United States turning to agriculture to help satisfy its voracious appetite for energy has come to be the single most controversial question surrounding the development of alcohol fuels industry."

In the twenty-seven years since the book's publication, that controversy has exploded into a bitter global dialogue as corn and soybean prices soared along with ethanol and biodiesel plant expansion; there have been street riots from Mexico to Morocco to protest rising food prices, and desperate Haitians have turned to mud pies to sate their hunger. A University of Minnesota report predicts that in the years ahead the global ranks of the malnourished will rise from 850 million people to at least 1.2 billion by 2025 because of "competition for land and water from biofuels."[7]

Some reports have concluded that biofuels already have increased hunger. A 2007 report by the UN Food and Agriculture Organization (FAO) cited critical food shortages in forty countries and singled out biofuels "as putting an upward trend on food prices"; Dr. Jean Ziegler, the UN "special rapporteur" on the right to food, decried using agricultural lands to produce biofuels as a "crime against humanity."[8] He has also insisted that there should be a five-year moratorium on biofuels until new technologies can be advanced.

But the FAO has noted that there are plenty of other issues driving food insecurity, such as lack of rural development, infrastructure, energy cost, and fair-trade standards for agricultural products. The FAO also rejects Ziegler's call for a halt to biofuels development.

"Biofuels present both opportunities and risks," said FAO director-general Jacques Diouf. "Current policies tend to favor producers in some developed countries over producers in most developing countries. The challenge is to reduce or manage the risks while sharing the opportunities more widely."[9] On another occasion Diouf stated, "Bioenergy provides us with a historic chance to fast-forward growth in many of the world's poorest countries, to bring about an agricultural renaissance and to supply modern energy to a third of the world's population."[10]

Ethanol's role in boosting midwestern corn production has also had far-reaching effects on water quality. Vast amounts of nitrogen fertilizers are applied to boost yields, and the excess runoff empties into the Mississippi River, which then dumps this nitrogen-rich load into the Gulf of Mexico. The end result has been a huge dead zone in the Gulf of Mexico that, during some warm months, is roughly the size of New Jersey. Scientists, in a report submitted to the Environmental Protection Agency (EPA), warned that federal policies favoring corn ethanol production would continue to increase corn acreage with "profound implications" for water quality.[11]

U.S. environmentalists opposed to corn ethanol want major policy reforms that would strip away the federal corn ethanol subsidies. They also seek a broader reduction of biofuel mandates, which they fear will increase the global expansion of plantation energy into sensitive tropical forests that are important carbon storehouses and wildlife refuges. "We've seen perverse outcomes with biofuels, like the destruction of some of our richest rain forests in the world," said Carter Roberts, president of the World Wildlife Fund.[12]

President Obama is unlikely to abandon the ethanol industry. He launched his political career in Illinois, the heart of the nation's Corn Belt and home state for Archer Daniels Midlands, one of the nation's largest ethanol producers. The president has been a longtime supporter of corn ethanol, which he labels a "transitional fuel" to the next generation of biofuels.

His cabinet appointments reflect that thinking. For energy secretary, he picked Stephen Chu, a Nobel Prize–winning physicist who has been critical of corn ethanol but is an aggressive advocate for developing biofuels from wood wastes, grasses, and other cellulose materials to combat global warming. As head of the Lawrence Berkeley National Laboratory, Chu launched a biofuels research effort funded largely by $500 million from BP, the oil giant who in ad campaigns of recent years boasts that its initials now stand for "Beyond Petroleum."[13]

For his agriculture secretary, Obama reached into the heart of ethanol country and chose Tom Vilsack, the former governor of Iowa, a state with more than thirty ethanol plants, where opposition to ethanol is akin to political suicide. Vilsack, in his first year in office, found many of these operations in tough financial straits and, like Wall Street banks and Detroit's automobile giants, looking to the federal government for support. But in its first year in office, the Obama administration has also shown a willingness to buck the biofuels industry. In May 2009 the EPA released draft rules concluding that new corn ethanol production and most U.S. biodiesel production did not go far enough in reducing greenhouse gases to be included in the expanded federal mandate to produce thirty-six billion gallons by 2022.

FARM ROOTS OF ETHANOL

Our research for the book unfolded in the late 1970s, a period of widespread anger and sometimes despair among American farmers who struggled with low grain prices. By the winter of 1978, a farm strike movement had staged tractorcade protests along interstate highways and even encircled the U.S. Capitol. As chronicled in the book, some of these farmers began to build small-distillation ethanol plants in an effort to consume some of the surplus crops. It was a prairie populist effort to reclaim at least a small portion of the energy industry that had been lost to oil long ago, and there was hope that new demand could be created to force up the price of crops, especially corn.

But through the close of the twentieth century, markets for the most part remained glutted. Corn prices averaged less than $2.50 a bushel through the 1980s and 1990s even as fertilizers, machinery, and other production costs soared. Most farm-based alcohol plants folded, while in the 1980s and 1990s Archer Daniels Midlands ruled the ethanol market as the dominant supplier of the corn-based fuel.

Some of the disillusioned veterans of the farm alcohol movement veered off into extremism. Colorado wheat farmer Gene Schroder, who, as described in the book, helped organize tractorcade protests and built an impressive early distillery, later gained notoriety for hosting paramilitary bomb-making seminars on his family farm and aligning himself with a militia movement.[14]

The farm alcohol movement also produced Jeff Broin, who, as I write, is the chief executive of POET, a privately held ethanol giant that in 2008 had a stake in some twenty-six plants capable of producing approximately 1.5 billion gallons of fuel annually. Broin is the son of a Minnesota farmer who in the 1970s built a farm distillery that after years of tinkering and engineering reached a capacity of approximately 125,000 gallons of ethanol per year. That plant provided a hands-on education for Broin and his two brothers that led to the large operation managed by POET today. Similar stories describe the backgrounds of other midwestern farmers who started small in the late 1970s and early 1980s and today manage much larger plants ("biorefineries" as the industry leaders call them). "That [Minnesota] plant still operates today and in fact is the only farm-scale ethanol plant from the 1980s that still runs today," Broin says. "The *only* one," he emphasizes.[15]

As Broin and others built ethanol plants in the 1990s, federal subsidies kept the industry alive. But it was ethanol's ability to help reduce air pollution in two ways that provided a huge boost to the industry at the decade's end. First, standard gasoline must have an octane-boosting additive, and ethanol is a clean replacement for toxic leaded gasoline and carcinogenic benzene. Second, federal legislation in 1990 mandated that dozens of cities suffering from high levels of smog-forming ozone or carbon dioxide use special "reformulated gasoline" that reduces pollution by adding oxygen, which reduces the amount of unburned hydrocarbons.

The oil industry served up an additive called MTBE (methyl-tertiary-butyl-ether), but this fossil-fuel cocktail leached from storage tanks and poisoned drinking-water supplies. MTBE was eventually banned by many states, and Congress declined to pass legislation to shield the oil industry from pollution lawsuits. That prompted a huge surge in demand for ethanol as an octane-boosting oxygenate to meet federal standards.

Between 1999 and 2006, U.S. ethanol production more than doubled from 1.47 billion gallons a year to 4.85 billion gallons, and billions of new investment poured into the rural Midwest as construction crews fanned out to install huge cooking-fermentation tanks, boilers, pumps, heat exchangers, and distilling columns that rose like rocket ships from concrete pads set amid corn and soybean fields.

Many plants were largely controlled by farmers. Just outside Steamboat Rock, Iowa, for example, four hundred farmers and a few other local investors raised $34.5 million to erect the Pine Lake Processors distillery, which began operations in 2005. Farmers hoped the plant would serve as a hedge, yielding big returns when corn prices were low and eventually driving up corn prices and making farming more profitable. And the company earned profits of $24 million in its first two years of operation, according to Larry Meints, a farmer who served as Pine Lake's president.[16]

Other ethanol companies became hot Wall Street stock picks. When VeraSun debuted on the New York Stock Exchange in 2006, the stock rose 30 percent on the first day of trading as the company made plans to double production. Pacific Ethanol, building plants in western states, drew $84 million from Microsoft founder Bill Gates.

By February 2007 ethanol was on an epic roll as the industry trade group, the Renewable Fuels Association, held its annual conference at a sprawling Marriott resort at the edge of Tucson, where golf greens are bordered by saguaro cactus. Producers during the prior year (2006) had endured a scintillating mix of cheap corn prices and record high prices for ethanol as oil companies scrambled for the additive to meet federal clean air requirements. Ethanol prices surged to over three dollars per gallon in the summer of 2006. Some producers were able to clear one dollar per gallon or more of profit, while others claimed to have paid off plant construction costs in less than a year. The industry's success was reflected by the list of some two thousand conference participants. There were many midwesterners from farms and agribusinesses who scorned ties and jackets in favor of open-necked collars and slacks. They were joined by oil and automotive industry executives and investment bankers from Wall Street.

Yet the industry's dramatic growth triggered a fierce backlash. In some communities, new plants sparked outrage from citizens who feared their health would be put at risk from plant emissions, and many of the early plants did indeed have air quality as well as water pollution violations. There were also concerns about the consumption of water resources by ethanol plants and increased congestion as caravans of trucks carried grains to the plants.

The food versus fuel debate had also flared anew. "Cars, not people, will claim most of the increase in world grain consumption this year [2006]," wrote Lester Brown, director of the Earth Policy Institute. "The U.S. Department of Agriculture projects that world grain use will grow by 20 million tons in 2006. Of this, 14 million tons will be used to produce fuel for cars in the United States, leaving only 6 million tons to satisfy the world's growing food needs."[17]

Brown had been speaking out against ethanol for decades, as evidenced by the quotations from him in *The Forbidden Fuel*. As ethanol consumed unprecedented percentages of the corn crop, he had powerful new allies in the livestock industry, which feared an increase in the price of grains crucial to feeding cattle, pigs, and chickens. "To say we're wary is an understatement," noted Bill Roenigk, vice president of the National Chicken Council, in 2006.[18]

Ethanol advocates noted that most U.S. corn is fed to livestock and that the ethanol industry yields a by-product of the fermentation process known as distillers dried grain (DDG), which is used as a feed supplement. Each fifty-six-pound bushel of corn yields starch that produces about 2.7 gallons of ethanol in a dry milling process (compared to about 2.5 gallons when *The Forbidden Fuel* was first published in 1982) and some seventeen pounds of DDG.[19] DDG is fed to dairy and beef cattle and in some instances to hogs and poultry.

Ethanol critics were not given the podium at the 2007 Renewable Fuels Association conference in Tucson, but they weren't out of mind. In a kickoff speech, Bob Dinneen, a burly Bostonian who was then the executive director of the association, exhorted the faithful to rally around the ethanol flag. "Who is willing to stand up here and embrace a new vision for biofuels? . . . Who is willing to stand up to the skeptics, firm in . . . the conviction that our farms can meet the commitments to produce both food and fuel? . . . Stand up and believe in your fuel. The state of the ethanol industry is sound. We have met the nattering nabobs of negativity, and we have soared."[20]

Yet not even two years later, beginning in 2008, the industry—along with the rest of the U.S. economy—would make a crash landing that shuttered some ethanol plants, cratered stock values, and shook the faith of many corn farmers. The stage was set for industry woes by a record rise in prices for their key feedstock, corn, which by June 2008 had surged to more than seven dollars a bushel in a market stoked in part by the expanding production of corn-based fuel but also by tight global grain markets responding to growing world demand as more and more people in emerging economies added grain-eating livestock to their diets. (Many analysts also attributed the hot commodity markets of 2008 to financial speculators and commodity traders.) Then, in the second half of 2008, ethanol prices collapsed in tandem with those for oil—which dropped by approximately one hundred dollars per barrel from July to December—and most other commodities, rocked by a world economic slowdown. Corn prices also tumbled from record highs, although they still ended the year at roughly double the prices of 2005.

One of the most spectacular ethanol failures in 2008 involved VeraSun, a company that had made an impressive debut on Wall Street just two years earlier and operated ten ethanol plants. In an effort to reduce risk in what corporate officials bet would be a runaway market, VeraSun contracted to buy high-priced corn from dozens of farmers and then failed to make good on those purchases as ethanol prices tanked below $1.60 a gallon. "This puts people in a terrible bind," said Mark Kuhn, a farmer and state legislator from Charles City, Iowa. "There is so much uncertainty with the contracts, and how they will be dealt with. We're being held hostage."[21] By November 3, 2008, the company's stock had dipped to just 28 cents a share, from a fifty-two-week high of $17.75. VeraSun had ceased trading on the New York Stock Exchange and had filed a petition in U.S. bankruptcy court for financial reorganization, which ended with the company's sale of its ethanol plants to Valero Energy, an oil refiner.

Many other ethanol producers were struggling, and as credit markets seized up, construction on many new plants was halted. Cargill, for example, abandoned plans for a $200-million ethanol plant near Topeka, Kansas.[22] The economic vise also tightened around the necks of farmer-controlled plants such as Pine Lake Processors, discussed earlier, which filed for bankruptcy protection in December 2008 after buying high-priced corn that could not be profitably converted into ethanol. "We were still chewing through $6 corn recently and ethanol was around $1.50," said Larry Meints, the plant's president.[23] Out west Pacific Ethanol by May 2009 had idled three or four plants and had subsidiaries file for reorganization in U.S. bankruptcy court.

Some ethanol producers, however, were seeing opportunity amid the financial carnage. Broin, for example, increased production at POET plants by 30 percent in 2008 and was credited with innovative efforts to improve energy efficiency: building a pipeline to harness landfill methane to help power one plant and further refining the protein-rich DDG that is a by-product of fermentation. At the end of 2008, Broin was seeking to expand even more by buying out distressed ethanol producers.[24]

At the end of the Bush administration, the U.S. Department of Agriculture extended federal loan guarantees to help bail out some of the corn ethanol companies, a proposal that quickly drew opposition from eight livestock organizations that protested the plan in a letter. Congressional critics also went on the attack. "The federal government's ethanol policies have driven up the price of corn," said Representative Jeff Flake, an Arizona Republican who had submitted legislation to repeal ethanol subsidies. "But rather than reforming the policies that have caused a spike in corn prices, the federal government wants to bail out ethanol producers who speculated in the price of corn."[25]

Federal loan guarantees would not be a cure-all. Pine Lake Processors, for example, considered the loan but found that banks still balked at financing the 30 percent of the loan that lacked federal backing, and the farmer-owned plant, though still operating, faces an uncertain future. "It's really tight now," Meints, the president, said. "We hope we can keep control of the plant, but right now we don't know what will happen."[26]

In the years ahead, the corn ethanol industry appears poised for a new era of consolidation and much more challenging economics. There will undoubtedly be a continuing contest in Washington DC between the corn-based ethanol industry, which favors continued subsidy, and a coalition of food-industry lobbyists and environmentalists, who are opposed to such support and are likely to wield much more political clout than in decades past.

ETHANOL IN BRAZIL

Back in 1975 Brazil launched what we describe in the book as "the most ambitious national effort ever made to break the oil industry's hammerlock hold on liquid fuel markets." In a memorable visit to São Paulo and Rio de Janeiro in the late 1970s, we met with engineers, bureaucrats, and sugar producers and executives of Petrobras, Brazil's nationalized oil company responsible for blending ethanol with gasoline. In a 1979 interview Antonio Seabra Moggi, then general manager of Petrobras's research center, scoffed at what he called the "artificial economics" of ethanol but said he could eventually get used to the idea. "Many things we do in life are not so natural as we think, such as drinking Coke. . . . I used to hate Coke. Now I like it, and spend a lot of money on it."[27]

Over the past quarter century, the Brazilian ethanol program has had plenty of setbacks. Some resulted from technology that rushed underperforming ethanol cars into the marketplace. Others reflected economic hard times as oil prices plunged in the late 1980s and stuck motorists with cars that ran on 100 percent ethanol rather than then-abundant gasoline.

But today many Brazilians have followed Moggi's lead and acquired a taste for ethanol. In the first six months of 2008, Brazilian motorists purchased 2.38 billion gallons of ethanol, almost equaling the 2.4 billion gallons of gasoline consumed during the same period. Many Brazilian motorists who purchase ethanol drive flex-fuel cars that were introduced in 2003 and are mandated by the government. These cars can run on 100 percent ethanol or any blend of gasoline and ethanol, and they now make up more than half of all vehicles on the road. Other motorists drive cars that run on blends of 25 percent ethanol and 75 percent gasoline, the minimum ethanol blend mandated by the Brazilian government. (In the United States government policy encouraged production of more than 5 million flex-fuel vehicles by granting manufacturers credits that allowed them to make other cars less fuel efficient.)

In Brazil the pure sugarcane ethanol is priced cheaper than gasoline. To give a good value, that ethanol price must be 30 percent less than gasoline or better to compensate for the roughly 30 percent lower mileage per gallon that the flex-fuel vehicles obtain compared to gasoline-powered vehicles. (Flex-fuel vehicles have low-compression engines to accommodate gasoline. Vehicles with higher-compression engines can take advantage of ethanol's higher octane and run more efficiently, with results that reduce the mileage penalty from 30 percent to 15 or 20 percent fewer miles per gallon.[28] Some researchers believe that the mileage penalty for ethanol can be reduced even further.)

Overall, Brazilian officials say that sugarcane ethanol can be a competitive alternative to gasoline when oil is priced at forty dollars per barrel or higher, and the alcohol fuels industry is likely to keep gaining market share since most new cars sold in Brazil are flex-fuel vehicles. In April 2008 Cosan, one of Brazil's largest ethanol producers, made a big bet on the future by buying 1,500 Exxon-Mobil service stations in hopes of increasing profit margins by eliminating the middle man that separates distilleries from motorists.[29] Cosan is a vertically integrated

Exhibit Intro-2
Sources: Brazilian Sugarcane Industry Association (UNICA) and Ministry of Agriculture, Livestock and Food Supply (MAPA). Chart © Boyd Griffin & Company, LLC.

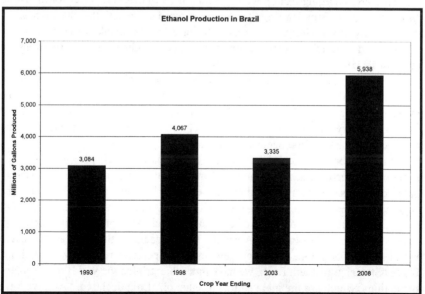

corporation that owns vast sugarcane acreage, distillation plants, and, with the acquisition of the Exxon-Mobil stations, the means of distributing its fuel. This type of field-to-pump control represents a far different model than has emerged in the United States, where even big producers like Archer Daniels Midlands don't grow their own corn or operate service stations.

Brazilian ethanol producers also benefit from cost savings resulting from their feedstock, sugarcane, which typically yields more than a third more ethanol per acre than corn and produces a sugar that can be directly fermented into alcohol. Crop residues typically are burned to provide steam for distillation, eliminating the need for natural-gas or coal boilers typically used by U.S. producers.

As ethanol production has grown, Cosan and other Brazilian producers have ventured into exporting fuel. In the crop year ending in the summer of 2008, Brazil produced nearly six billion gallons of ethanol—second only to the United States—and exported about 20 percent of that fuel to more than a dozen nations, including Japan, Sweden, and the European Union.

The ethanol boom in Brazil in the first decade of the twenty-first century was accompanied by a broader economic resurgence that helped buoy the popularity of President Luis Inácio "Lula" da Silva, a former shoe-shine boy and union organizer who was elected president in 2002 with a campaign pledge to focus his efforts on reducing poverty with a goal of "zero hunger." As Lula assumed office, Brazil was producing far in excess of the calories required to feed its people, but wealth was highly concentrated, and a government national food and nutrition survey found that nearly half of all Brazilians believed they couldn't get all the food they need.[30]

Lula's policies—such as offering cash payments to families who keep their children in school—were credited with helping to improve the plight of eleven million people, and he won reelection in 2006. "It is undeniable that millions of people in this vast country, which occupies almost half the landmass of South America, are emerging from poverty," said Simon Marks, a reporter for the *News Hour*, in a Public Broadcasting System report aired in June 2008. "Statistics show the economy is growing at its fastest rate for 20 years, thanks in large measure to the world's voracious appetite for Brazilian commodities. That is swelling the ranks of Brazil's middle-income earners."[31]

The social gains have eased but not erased the tensions created by devoting vast amounts of prime acreage to fuel crops rather than food in a nation where so many still struggle for the basics of life. Back in 1979 we noted this tension and said that victory over the OPEC oil barrels will be a hollow one if achieved at the expense of the basic food needs of Brazil's people.

Lula has been a vocal defender of the Brazilian biofuels program. He touts initiatives to persuade small farmers in Brazil's poorest regions to produce African palm, cottonseed, sunflower, and castor beans for biodiesel fuels. He credits the sugarcane industry with creating one million jobs and restoring abandoned pasture land to rotating crops that often include food. At a press conference with President Bush in São Paulo in April 2007, Lula declared that he was "obsessed with biofuels" and hoped to forge new, more cooperative relationships with the United States. "President Bush, we have more than tripled the yields of sugarcane plantations, which are the main source of ethanol," Lula declared. "And we have demonstrated that it is possible to increase the production of biofuels without harming the production of food."[32]

Yet the work conditions in the sugarcane jobs have come under harsh scrutiny, with numerous reported abuses of the labor force making headlines in the Brazilian and international press. "They have the clothes on their bodies and nothing else," said Valeria Gardiano, a director of a social-service department in Palmares Paulista, a town whose population increases each year as migrant workers arrive to help cut the cane. "They bring their children with malnutrition, their ill mothers-in-law. We try to reduce the problem. But there is no way we can fix it 100%. It is total exploitation."[33] Ethanol producers are mechanizing more of the harvest, which may reduce worker abuses in the fields but will also strip away hundreds of thousands of jobs for the rural poor.

Lula's willingness to harness Brazilian lands to meet the energy demands of other nations has also stirred protests. During a 2007 visit to Brazil by then-president Bush, Via Campensia, a movement of landless rural workers, organized nonviolent occupations of agribusiness corporations, including one in which nine hundred women showed up at the Cevasa ethanol distillery in São Paulo.[34] Cevasa is one of the largest fuel producers (partially owned by the giant commodities company Cargill). The movement organizers maintained that they opposed not the ethanol industry but the large-scale production for export. They argued that Brazil would be better off turning prime land over to small farmers, who would employ more people per hectare of land, than by expanding large plantations producing sugarcane fuel for export.

The protests of organizations such as Via Campensia stem from longstanding awareness and suspicion in Brazil, and throughout South America, of the far-

Exhibit Intro-3
Source: Renewable Fuels Association. Chart © Boyd Griffin & Company, LLC.

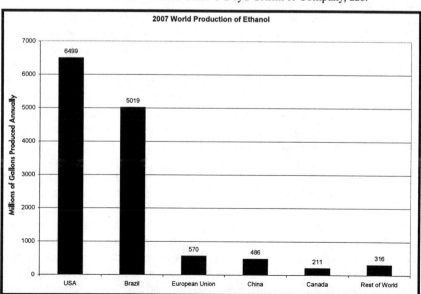

reaching economic power and political influence of U.S.-based large multinational corporations such as Cargill, Archer Daniels Midland, Exxon-Mobil, and Ford Motor Company. Such suspicions were expressed by Pedro Stedile, a spokesman for the Via Campesina movement, who said, "Bush came to Brazil as a messenger boy for the multinational companies, the agribusiness companies, the oil companies and the automobile companies that want to control the biofuels."[35]

GLOBAL ETHANOL

The book contains an appendix about the international history of alcohol fuel programs. It documents how alcohol has been promoted and used as a transport fuel since the beginning of the twentieth century. Since *The Forbidden Fuel* was published in 1982, global production of ethanol, biodiesel, and other biofuels has increased dramatically, especially in the first years of the twenty-first century. At present, more alcohol fuel is in use than at any time in world history. The largest producers are the United States, Brazil, the European Union, and China, but in dozens of countries there is some level of activity as corn, sugarcane, cassava, and other starch and sugar crops are being converted to fuel.

CELLULOSIC ETHANOL: THE HOLY GRAIL?

In chapter 4, "Rebirth of the Power Alcohol Industry," under the heading "Biotechnology and Alcohol Fuel Production," the book explains the essential science and early efforts of manipulating biochemistry to produce what is now commonly called "cellulosic" ethanol. This fuel can be produced from nonfood-plant feedstock such as wood, cornstalks, grasses, paper, and even municipal waste. At the time of this writing in 2009, it is fair to say that the optimistic projections for cellulosic ethanol in 1982 were just that.

The book reports one projection in the early 1980s that "by 1990, nearly 10 billion gallons of alcohol, enough fuel to replace almost 10 percent of current gasoline consumption, could be supplied by municipal solid wastes—more than double the amount available from grain resources." It notes that the Department of Energy's Alcohol Fuels Task Force tagged cellulosic ethanol as a potential game-changing technology that, if commercialized, "could yield more than 50 billion annually by the year 2000 and open up vast new storehouses of nonfood feedstocks." As it turned out, in the year 2000 there were no commercial plants in operation that turned cellulose into ethanol fuel.

Economic and technical challenges have been the main roadblock to commercialization of cellulosic ethanol. The price of oil through the 1980s and much of the 1990s was far below what many had forecast, and that dramatically reduced incentives to invest in cellulosic technology. There were also formidable challenges in producing commercial quantities of fuel. Most of these materials are bulky and expensive to transport to a centralized area. The cellulose is held together with tough molecular bonds that can be very difficult to break down consistently and on a large scale for efficient conversion to fuel. "Cellulosic ethanol may be the Holy Grail," said John Felmy, an economist for the Petroleum Institute who spoke at the Renewable Fuels Association industry conference in Tucson in early 2007. "But remember, they never found the Holy Grail."[36]

As oil markets tightened, however, plenty of companies were willing to join in the search as the federal government emerged as a big booster of the technology. In an extraordinary intervention into fuel markets, the federal mandates passed by Congress in 2007 require that sixteen billion gallons a year of cellulosic fuel ethanol be sold annually by 2022.

To help launch the technology, the U.S. Department of Energy in 2007 tapped six of the most promising cellulosic companies to receive up to $385 million in federal grant money to help finance ethanol plants that would eventually produce 130 million gallons of fuel annually. The six grant recipients agreed to raise another $1.2 billion in private funds.

By the end of 2008, two of those projects had already been put on hold. Iogen, a Canadian cellulose company, withdrew a proposal to build a wheat-to-straw ethanol plant in Idaho, and Alico scrapped a proposal to turn agricultural wastes to ethanol in Florida.

Other grant recipients are still moving ahead in 2009. Broin's company, POET, selected enzyme technology to turn crop residues from the corn into ethanol at an Iowa plant. Abengoa, a Spanish company, planned to use enzymes to convert grasses, wheat straw, and other residues at a new plant in Kansas and had a small plant operating in York, Nebraska. A Colorado-based company, Range Fuels, launched construction of a Georgia plant that would use thermochemical processes to gasify wood wastes and then extract ethanol and methanol. In California, Blue Fire Ethanol won federal backing for an acid hydrolysis process to break down wood and other wastes gathered at a landfill.

On the average cellulosic technologies are expected to produce a cost, or "minimum ethanol selling price" (MSEP), for ethanol of about $2.61 a gallon (2008 real dollars), according to a report from the National Renewable Energy Laboratory in Golden, Colorado. Furthermore, the report states: "It is believed that the technical targets needed to reduce this number to $1.49 by 2012 are still achievable."[37]

Still, there is powerful momentum to push forward and develop stronger enzymes, more digestible cellulose plants, and other innovations. BP's $500-million commitment to the Berkeley National Laboratory and the University of Illinois was noted earlier. Also, General Motors in January 2008 announced an investment in Cosaka, an Illinois company claiming to wield new technology that can undercut competitors and even extract ethanol from old tires. In May 2008, General Motors teamed up with outspoken ethanol proponent Vinod Khosla, who made a fortune in the high-tech industry, to invest in Mascoma, a Massachusetts-based cellulosic ethanol company that works with scientists from MIT. In May 2009, Mascoma announced an engineered microorganism that company officials say could dramatically increase the efficiency of cellulose-to-ethanol and reduce production costs.

There were also signs of increased interest in cellulose ethanol by oil companies; BP, for example, invested in Verenium, a biofuels company that in May 2009 began turning a bagasse—a by-product of sugarcane processing—into ethanol at a 1.4-million-gallon-a-year experimental plant in Louisiana.[38]

Cellulosic ethanol also has strong financial support from the federal government. In 2008 Congress passed a new farm bill authorizing federal loan guarantees to biofuel producers that use feedstocks other than corn. Billions of federal dollars also may be directly invested in new cellulosic ethanol research and development as Obama calls on Energy Secretary Chu and Agriculture Secretary Vilsack to make good on his campaign pledge to create more clean-energy jobs in the United States. A first step came in May 2009, when Chu announced $480 million in spending to test new ethanol technologies.

Given the new industry and federal focus on cellulosic ethanol production, there will likely be at least modest production in the years ahead. But the industry, due to a mix of politics, economics, and continuing technical challenges, could still fall well short of the ambitious mandates set by Congress.

GREENHOUSE GASES AND GLOBAL WARMING: DOES ETHANOL FUEL HELP OR HINDER?

In the world of the future, a "clean-energy job" will be one that helps combat global warming, rather than contributes to the problem. In 1982 we cautioned that alcohol fuels would not be an environmental ally if forests and fields are simply "mined" for biomass feedstock and erosion undermines soil fertility. The book describes the threats of "greenhouse effect" and "global warming" linked to fossil fuels and emissions of carbon dioxide and, on the same page, warns of "an unplanned rush to use forest or other biomass resources [that] could trigger soil erosion and other ecological disruption."

In recent years abundant new research has documented the importance of carbon-rich organic matter in maintaining soil fertility. Washington State University scientists have found that in drier areas of the state's wheat country, it is best to return all the wheat straw to the soil. So in those areas there would be no surplus "waste" wheat straw to use as feedstock for cellulosic ethanol. In the Midwest some corn land could safely yield at least a portion of its cobs, stalks, and husks to an ethanol plant. But others should retain all these residues.

"These crop residues are not being wasted. They are being used quite wisely," notes Wally Wilhelm, a federal plant physiologist in Nebraska who has spent much of the past quarter century studying how crop residues return organic materials to the soil. "To wholeheartedly say that every wheat straw and cornstalk that is out there in November is now fair game—that is not the way we need to do it."[39]

There is also intense debate about how to tally the greenhouse gas contributions of cellulosic ethanol and other biofuels. This is a key issue in the United States, Europe, and elsewhere as nations step up the worldwide effort to combat global warming. In the years ahead biofuels deemed to be reducing these emissions are likely to receive substantial government support, while those deemed to be contributing to the problem are likely to face carbon taxes that would slow or halt their development.

In an April 2007 report, the EPA, citing the work of Argonne National Laboratory in Illinois, gave cellulosic ethanol the highest ranking of fourteen different transportation fuels. That report credited cellulosic ethanol with a 90.9 percent reduction in carbon dioxide emissions, compared to gasoline. Biodiesel was rated as offering a 67.6 percent drop in those emissions, while corn ethanol ranked the lowest of the biofuels at a 21.8 percent cutback.[40] That 2007 report estimated the carbon dioxide emissions generated by growing, processing, distributing, and combusting these biofuels. Thus, for example, the corn ethanol tally included the emissions from natural gas used to make fertilizer, diesel fuels used to harvest the crop, boiler fuels used to distill the alcohol, and transportation fuels to get the ethanol to market.

But in February 2008 those calculations came under assault when *Science* magazine published the results of two studies that sought to tally the indirect impacts of biofuels as energy cropping expanded demand for agricultural and forestry crops and triggered the conversion of tropical forests, savannahs, and other grasslands to agriculture. One study, coauthored by Ted Searchinger of Princeton, projected the effects of land conversion as an expanding U.S. ethanol industry consumed more of the nation's corn crop. It found that corn ethanol, rather than fighting global warming, would contribute to global warming.[41]

The other study, authored by Joe Fargione of the Nature Conservancy and four University of Minnesota professors, took a broader look at how biofuels spurred land conversions and contributed to global warming. The study concluded that large amounts of carbon dioxide would be released as the acreage was burned in preparation for cultivation and later as organic carbon, stored in the soil, decomposed. According to the study this land conversion creates a "carbon debt" that could take decades or even centuries to repay. The authors concluded that during that payback period the total effect of biofuels production from these converted lands would have much greater greenhouse gas emissions than fossil fuels.[42] The authors found that the size of the carbon debt would depend on the type of land conversion: According to their assumptions and calculations, converting grasslands in the midwestern United States to corn production would require about 48 years to repay the carbon debt, while converting Brazilian grasslands to soybean biodiesel could require 37 years of debt to repay. The worst-case example involved the conversion of carbon-rich peat soils in tropical forests to palm biodiesel plantations. That debt could require up to 420 years to repay. "Most people don't realize that globally there is almost three times as much carbon in the plants and soils as there is in the air," Fargione said. "Our natural ecosystems provide an incredibly valuable service of carbon storage and climate stabilization if they are left intact."[43]

This was more than a theoretical exercise. In the Amazon large tracts of forest have been razed for soybean production; rising demand for soybeans has been triggered in part by biodiesel production and reduction in U.S. soybean acreage as American farmers switch more acres to corn to help meet ethanol demand. In Indonesia in recent years peat forests have been burned and cleared for palm oil plantations to produce biodiesel, cooking oil, and cosmetics. In Sumatra's Riau Province, for example, 2.1 million hectares were claimed by palm oil plantations in 2008, compared to 400,000 hectares in 1998.[44]

The two studies were published at a time when large areas of tropical forests were being cleared, at least in part due to the demand for biofuels. The clearing of the forests for palm oil plantations, timber, and farming is destroying important habitat for many species, including the orangutan, which a UN report predicted could be extinct in the wild by 2022 if the current pace of deforestation continues. "They change the forest and say it's for energy sustainability, but they're killing other creatures," noted Ichlas Al Zaqie, of the Los Angeles–based Orangutan Foundation International.[45]

The *Science* article reporting the two studies described here was published as the EPA grappled with a challenging new responsibility to delve into biofuels greenhouse gas emissions. The 2007 Energy Act provides an ambitious mandate for the production of thirty-six billion gallons of U.S. biofuels by 2022. In drawing up rules to administer the act, the EPA was asked to calculate all the carbon costs—including land conversion—attributed to biofuels. Then the agency was supposed to ensure that all new biofuels would attain at least a 20 percent reduction in greenhouse gas emissions compared to petroleum fuels, and two-thirds of the new biofuels subject to the mandate were supposed to reduce those emissions by at least 50 percent.

Biofuels industry officials and their congressional allies worried that the *Science* studies would prompt the EPA to develop stringent new biofuels standards that might discourage production. They noted that many scientists were skeptical of some of the methodology used in the *Science* studies. There was common ground in acknowledging a carbon debt resulting from the destruction of an Indonesian forest. But some scientists argued that Searchinger's study, for example, had grossly overestimated the future expansion of the corn ethanol industry, assuming a thirty-billion-gallon-a-year level rather than the fifteen-billion level mandated by Congress. This assumption then led Searchinger to overestimate the global impacts on land conversion for food and fuel production and thus overstate the greenhouse gas emissions, according to scientists from the Department of Energy and Argonne National Laboratory.[46]

Scientists who support efforts to calculate the land conversion costs also believe that the models used to predict those changes were still too crude to set policy. Industry officials seized on those concerns to lobby the EPA to put the carbon analysis on hold. "There is nothing that requires the EPA to plunge into the unscientific quagmire of questionable numbers, instead of waiting until calculations of defensible numbers, bottomed on appropriate scientific methodology," wrote James Greenwood, president of the Biotechnology Industry Organization, in an October 23, 2008, letter to then–EPA administrator Stephen Johnson. Six U.S. senators, including Senator Ken Salazar, who would later be appointed as Obama's interior secretary, also weighed in and told the EPA that the federal legislation forced the EPA to consider not all the international effects of land conversion but only the effects that biofuel production would have on U.S. land conversion.

There was also heat generated by environmentalists. In a letter to Johnson, representatives of the Environmental Working Group, the Friends of Earth, and the Clean Air Task Force argued that it would be irresponsible to delay development of the greenhouse gas standards for ethanol and other biofuels. They charged that "biofuel developers continue to be less interested in examining the fuels' net social and environmental impacts than they are in maintaining federal support in the form of mandates and tax breaks."[47]

Lobbied on all sides, the Bush administration's EPA balked at releasing an initial draft of the new biofuels standards. Thus it was up to the Obama administration to craft these standards and other rules that will likely play a crucial role in guiding development of the biofuels industry. And in May 2009, EPA analysts—drawing on the work of Fargione and other critics—concluded that using corn, soybeans, and other crops for fuel in the United States would spur major land conversions that would cause significant releases of carbon dioxide. Thus, even when assessed over a one-hundred-year time frame, these standards did not reach the 20 percent greenhouse gas reduction threshold required by the 2007 legislation.

The EPA, in its proposed rules, allowed a grandfather exception for ethanol production from existing plants. Based on the new analysis, the EPA also said that the fuel produced from new corn ethanol plants would be excluded from the federal mandates. The EPA was tougher on biodiesel, excluding production from existing canola oil– and soybean-based biodiesel plants.

The proposed rules triggered a backlash from the biofuels industry officials, who said the climate models were not developed enough to yield an accurate assessment. They quickly lobbied Congress to override the EPA rules through legislation, introduced by late spring of 2009, which would essentially void the analysis. "The clearing of lands in other countries for domestic food production is more appropriately part of the lifecycle of the food product, not part of the lifecycle of fuel production," testified Larry Dinneen, at a May 21, 2009, House Agriculture Committee hearing.[48]

But environmentalists and their allies were not about to retreat on this issue, and they geared up for a big political fight to defend the EPA's efforts to track the lifecycle emissions of biofuels. "Without full and accurate accounting, it would be unjustifiable to go forward with the RFS [Renewable Fuels Mandate]," wrote the leaders of nineteen environmental, conservation, and science groups in a June 10, 2009, letter to three House committee chairmen.

Clearly, there will be much more debate about how to tally the carbon credits. But there is also some consensus on strategies that will produce ethanol and other biofuels and combat global warming. Even the authors of the critical studies that were the basis of the *Science* magazine article concede that some energy crops grown on marginal or abandoned farmlands could yield biofuels that would produce a net reduction in greenhouses gases. "The most efficient use of biofuels will be to produce them in ways that don't require the conversion of natural ecosystems," said the Nature Conservancy's Fargione. "And to use them close to home."[49]

One option would be to plant a new generation of perennial energy crops, which require less cultivation and produce less erosion, on marginal lands. In the book we note that the Jerusalem artichoke might hold promise. But in the United States the twenty-first-century focus is on grasses such as miscanthus or switch grass, a native of the prairies that once was one of the four major grasses that fed the buffalo herds. Switch grass can be grown in the same field for eight to fifteen years in a row. These grass stands may reach ten feet in height and on some lands have the potential to yield more than one thousand gallons of cellulose ethanol per acre. Miscanthus, according to some estimates, could yield as much as two thousand gallons of ethanol per acre.[50]

Like most of the cellulose crops, these grasses have drawbacks. The huge bulk of the bales makes them expensive to transport to a distillery, and they could be difficult to store over long periods of time. But both switch grass and miscanthus require far less pesticide and fuel-intensive tractor work than corn. Their roots may extend ten feet deep, helping hold the soil in place. These roots also store large amounts of carbon and thus help slow the release of carbon dioxide from the soil over long periods of time.

Switch grass and miscanthus could be grown in many areas of the Midwest and the South. Farther west there is interest in poplar plantations, which with the aid of irrigation already spread over vast acreages in arid areas of eastern Washington and eastern Oregon. These are a kind of desert forest—impressive swaths of trees where once there was only sagebrush and tumbleweed—and thus help fix carbon from the air. Some of the trees are turned into lumber, and if cellulosic technology develops, wastes could be made into ethanol.

Jatropha is another plant with significant potential for biofuels production that reduces greenhouse gases yet produces seeds with 25 to 40 percent oils. Jatropha can be grown for decades on marginal, nearly barren lands with very little water and without the use of pesticides. It can help control erosion, and the by-product from the oil is a seed cake that might be used as fertilizer. All these attributes have sparked tremendous interest among researchers, who have just begun to try to improve the genetics of the genus of more than 170 different varieties that range from shrubs to trees and vary widely in oil content. The shrub has also been embraced by many policy makers in developing nations eager for a shrub that can grow on poor lands unsuited for food production and help stimulate rural economies. An October 2008 study by the International Food Policy Research Institute found that in Mozambique, jatropha could be a powerful tool for boosting the incomes of small farmers, creating far more employment and income than sugarcane. In recent years farmers from Mali to Brazil to India (where millions of potential sites for jatropha plantations have been identified) have started to plant this crop as their governments push biodiesel programs. And in an important test in 2008, Air New Zealand filled up a Boeing 740 with a blend of 50 percent jatropha fuel and 50 percent jet fuel and took off on a flight. "We undertook a range of tests on the ground and in-flight with the jatropha biofuel performing well through both the fuel system and engine, just as laboratory tests proved it would," says Captain Dave Morgan, Air New Zealand's chief pilot.[51]

But many questions remain about developing jatropha crops. Taking a scruffy shrub and cultivating it on large plantations might make the plant more susceptible to disease, and there are also concerns that a biofuels boom could prompt farmers to grow jatropha on prime lands—where yields would be highest—and thus create a conflict with food production.

ALGAE AS FEEDSTOCK FOR ETHANOL AND BIODIESEL

In the final chapter of the book, we note that aquaculture is a new frontier for liquid production and cite the potential of algae. In this new century there has been an explosion of research into algae as a feedstock for ethanol, biodiesel, and other feedstocks. Algae holds the promise of producing transportation fuels that do not require the large-scale use of finite resources such as freshwater and fossil fuels and could consume—rather than emit—the global warming gas of carbon dioxide as it produces fuel.

Current research usually involves growing the algae in either ponds or enclosed containers, then harvesting and processing the inherent oil and carbohydrates, an expensive, complex process for producing fuels—biodiesel or ethanol—for aviation, trucking fleets, or diesel automobiles. The investment per acre would be much higher than in traditional farming and by some estimates push the price to nine dollars to sixteen dollars per gallon.[52] But some of the potential yield projections are extraordinary if cheaper ways can be found to produce the algae fuels. For biodiesel, for example, algae could yield from 5,000 to 10,000 gallons per acre compared to 120 to 150 gallons per acre for canola and rapeseed oils and 175 to 250 gallons per acre for jatropha.[53]

Algae's status as a hot energy prospect is reflected in the formation of the Algal Biofuels Consortium, which includes the Sandia National Laboratories, Cargill, Honeywell, and several universities. The key is finding a way to reliably produce these fuels on a large scale at a competitive price—a huge biological and technical challenge that the consortium hopes to help surmount.

Substantial venture capital already has been invested in algae. This has resulted in a plethora of new companies launching different strategies to exploit the science and technology and build successful companies. Some will surely flounder, and there is bound to be some hype amid all the press releases. But some serious investors—ranging from oil companies to Bill Gates—have been willing to take the plunge.

Much of the early focus has been on algae producing oil for biodiesel. Solazyme, for example, is a start-up company based in South San Francisco that uses a microbial fermentation process to produce oil "quickly and efficiently" without the aid of sunlight. That oil has produced a diesel fuel that was tested in a 2008 Jeep Liberty and met American Society for Testing and Materials diesel specifications.[54]

Algae also holds potential for big yields of ethanol. One cynaobacterial strain developed by David Nobles and Malcolm Brown, researchers at the University of Texas, could theoretically produce more than 11,000 gallons of ethanol per acre. Algenol Biofuels, a Florida company, is focused on ethanol, hoping to make a successful leap from an algae research facility in its home state to a commercial plant in the Sonoran Desert in Mexico within the next couple of years.

Algae could help generate a new wave of what some call "third-generation" biofuels that are readily compatible with the current distribution systems set up for petroleum. "We feel pretty confident that it will be possible to engineer organisms to produce . . . biopetrol," says Chris Somerville, director of Energy Biosciences Institute at the University of California at Berkeley, the research effort funded largely by BP.[55]

San Diego–based Sapphire Energy hopes to have a ten-thousand-barrel-a-day facility producing this green crude (similar to a light grade of petroleum oil) in operation between 2012 and 2015. By early 2009 the company had raised more than $100 million in venture capital, some of which came from Cascade Investments, which handles ventures for Gates. According to a company spokesman, the Sapphire process has developed an organism that can thrive in saltwater and with the aid of sun and carbon dioxide can produce the green crude oil. To be economical Sapphire believes the algae will have to grow in open ponds rather than closed capsules, and the goal is to create a sturdy organism that will thrive under hostile saltwater conditions and thus dominate this ecosystem over competing organisms that would thwart the focus on producing the crude oil.

As the world moves toward tougher regulation of greenhouse gases, algae's appeal is likely to increase as utility companies look for ways to reduce emissions. In November 2006 Massachusetts-based Greenfuel Technologies Corporation and Arizona Public Service Company announced that carbon dioxide from a natural-gas power plant had been recycled into algae production that produced both biodiesel and ethanol at a 1,040-megawatt plant in Arlington, Arizona. The algae production exceeded expectation and if replicated on a large scale would yield 37 times more fuel per acre than corn ethanol and 140 times more fuel per acre than soybeans. In terms of carbon dioxide reductions, Greenfuel officials estimate that 80 percent of the daytime production of that greenhouse emission from a natural-gas plant could be consumed in an algae-to-fuel plant.[56]

"The reason algae is so interesting is that it can directly convert CO_2 into biomass very quickly, more efficiently than anything else we know of," says Rodney Andrews, director of the University of Kentucky's Center for Applied Energy Research.[57] But Andrews estimates that it would take about 5,000 to 6,000 acres of algae ponds to handle the carbon dioxide emissions of a 500-megawatt coal plant. That might be feasible in a remote area of the desert but would be a huge hurdle in plants located closer to populated areas where real estate is at a premium.

In our final chapter of the book, one section is headed with the question "How much alcohol—how soon?" This is a question many are still asking in this new century, when the debates about energy and agriculture policy in Washington are much sharper, more capital is being invested in new "biorefineries," and more research is being conducted on a wide range of new technologies—funded by both private investors and government grants—than ever before. In the same chapter, we also report estimates from scientists who postulate that commercialization of cellulosic ethanol technologies could result in production levels "as high as 70 to 100 billion gallons of fuel a year." Today, in 2009, most estimates would fall in a lower range, and the targets set for 2022 by recent mandates are at a significantly lower level; however, as noted at the beginning of this introduction, none other than President Obama, when he was a candidate in 2008, suggested production of 60 billion gallons per year as a reasonable goal. Given that ethanol provides about a third less energy per gallon as gasoline, this would be enough to supplant approximately one-third of the nation's current gasoline consumption of about 140 billion gallons per year even if no technology was employed to utilize ethanol's higher octane rating.

Whatever the realistic possibilities are, we still believe in 2009, just as we suggested in the book back in 1982, that the most promising path to follow in the twenty-first century is one that limits development of ethanol and other biofuels to production processes that are environmentally sound and do not increase world hunger. Today, there is a much louder outcry to define a global policy for sustainable biofuels and develop a plan that would strive to improve the plight of the world's poor. "In the absence of such a plan we run the risk of producing diametrically opposite effects: deeper poverty and greater environmental damage," observes Jacques Diouf, director general of the FAO.[58]

The Worldwatch Institute, an American think tank, has tried to flesh out a sustainable biofuels policy. Their suggestions include a focus on cellulosic biofuels, local control of biorefineries, stricter oversight of biofuel plant emissions, and support for perennial crops and other farming practices that protect soil fertility.

ELECTRIC CARS: A POTENTIAL GAME CHANGER?

The expansion of a sustainable biofuels industry should be linked to a major expansion of conservation and solar, wind, and other renewable energy resources without adding to the risks of global warming. Today, U.S., Japanese, and other automobile companies appear to be entering a promising new era that will launch a wide range of alternative-fuel vehicles. In an epic transformation their vehicles will rely less and less on the overall consumption of liquid fuels, ethanol included.

Some cars will offer consumers an option to use electricity and alternative fuels. The most high-profile example is the Chevy Volt, which General Motors is launching in 2010 in a bid to help restore its reputation and shattered finances. This car is a plug-in electric vehicle with an onboard generator that can be fueled with gasoline of E85 (a blend of 85 percent ethanol and 15 percent gasoline). The electricity that powers the car is provided by a large battery that uses lithium-ion chemistry, a technology used to power cell phones and laptop computers. The battery can be charged from a home electric outlet, run forty miles, and then be recharged by the generator.

Within the next five years, consumers will have many more options for all-electric commuter cars. Virtually every major car company has announced plans for such cars. Mitsubishi in the spring of 2009 introduced an all-electric four-door car, the MiEV, that will travel up to one hundred miles per hour between charges, and Ford unveiled its "Project M," an all-electric prototype designed for commuter travel. "Frankly, I think it's a gamble not to do it," William C. Ford Jr., Ford's executive chairman, said in an interview, speaking of Ford's commitment to electric-car technology. "It's clear that society is headed down this road."[59]

In a *Science* article that added a new wrinkle to the biofuels debate, J. E. Campbell of the University of California, Merced, argued that burning biomass to generate bioelectricity for such cars would be a more efficient route than turning these plant materials into ethanol.[60]

In contrast to plans for electric cars, there appears to be far less consensus within the automobile industry about the future of biofuels. When *The Forbidden Fuel* was published in 1982, Volkswagen and other foreign automakers were on the front lines of ethanol technology as they introduced vehicles in Brazil. Today, the U.S. automotive industry appears to be more bullish about ethanol than their foreign competitors. They tout the potential of ethanol just as Henry Ford did in an earlier era of the U.S. automotive industry. "Live Green, go yellow," proclaimed General Motors in 2006 as it launched U.S. sales of its flex-fuel vehicles, which can run on blends containing as much as 85 percent ethanol. But in early 2009, it is unclear just how many U.S. automotive companies will survive. As this is being written, General Motors has filed for reorganization in U.S. bankruptcy court; following bailout loans, the U.S. government is now GM's largest stockholder. Chrysler in the spring of 2009 sealed a deal with Fiat of Italy to help it survive.

The global automotive giant Toyota has taken a more skeptical approach to ethanol and other biofuels than its American competitors have. At a two-day "sustainability" seminar for journalists in the fall of 2008 in Portland, Oregon, Toyota officials brought in Dr. Jan Kreider, a University of Colorado professor who noted in his presentation the "pervasive biofuel concerns" that included huge logistical, land, investment, and water requirements in an era when global warming is expected to bring new droughts and exacerbate shortages. Overall, Kreider concluded, there is "too much hype, too little attention to engineering and economic basics—more dreamland than reality just now."[61]

HITTING A BLEND WALL

Ethanol production nonetheless continued to climb through 2008 even as the industry went through a severe readjustment. The 2008 output in the United States was expected to top nine billion gallons, about a quarter of the way to the thirty-six-billion-gallon federal mandate for the year 2022. But that mandate makes no sense if U.S. motorists don't have vehicles that can consume all that fuel. All the cars on the road in 2009 can handle 10 percent ethanol blends, but the vast majority can't use the E85 blends of 85 percent ethanol and 15 percent gasoline. So the ethanol industry is wary of a "blend wall" that could be hit in the next few years as production capacity reaches twelve to fourteen billion gallons a year, enough ethanol to meet virtually the total demand for E10 in the United States.

Once a blend wall is reached, the industry's future in the United States would likely depend on a Brazil-style mix of government and market incentives to increase the use of ethanol. One option would be to try to increase the ethanol blend percentage to 15 percent for use in nonflexible-fuel vehicles, and the industry has petitioned the EPA for approval of this new blend standard. Still, the new blend could risk a consumer backlash if these fuels caused performance, corrosion, or other problems in automobiles not designed to handle these blends, as well as potential problems with small engines that power such things as lawn mowers and weed eaters. Another option would be to use government tax incentives and mandates to expand production of flex-fuel vehicles (as Brazil did) and also intervene when necessary in the markets to ensure that high-blend (up to E85) fuels for these vehicles are the cheapest option at the gas pumps. Such measures are economically and technically feasible but would likely face fierce political resistance from environmentalists and the oil industry, especially if they give yet another boost to corn ethanol production and reduce the amount of oil-refined product contained in a gallon of fuel.

Critics are calling for a pullback from the biofuel mandates as the United States develops a broader transportation fuel policy that is tightly linked to the reduction of greenhouses gases and sustainable energy production. "Biofuel mandates and subsidies do nothing to address the increased risk of soil degradation, water pollution and habitat loss as we ramp up production to supply crops for food, feed and fuel," wrote Craig Cox, the Environmental Working Group's Midwest vice president, in a column published in the *Des Moines Register*.[62]

To curb greenhouse gases, some call for an outright tax on carbon production. But as of June 2009, the nation appears headed for a cap-and-trade system,
with pending congressional legislation calling for carbon emitters to buy credits
to offset their production. There are still plenty of questions about how effective
this measure would be, and how the various biofuels would be treated under this
system.

Much of the twenty-first-century debate over biofuels has its roots in the
twentieth-century alcohol fuels movement chronicled in our book. It documents
the tumultuous history of the early pioneers—farmers, scientists, inventors, indus-
trialists, and others in the United States and Brazil who helped lay the foundation
for this global industry—and offers an inside account of the early lobbying efforts
in Washington that helped shape the rules and policies governing this industry in
the United States. It also provides an early overview of the environmental and
ethical controversies that still loom large in this new century.

In the years ahead there will be plenty more wrangling—and no shortage of
new controversies—to flesh out a new political framework that will guide twenty-
first-century energy policies, including those specifically dealing with ethanol and
other biofuels. But as oil production wanes and temperatures rise, it appears certain that we will move by century's end into the postpetroleum world. Alcohol and
other biofuels already are helping the world to head in that direction.

NOTES

To keep informed about ongoing issues and future books by the authors on this topic, readers can visit http://www.forbiddenfuel.com.

1. Energy Information Administration.

2. Renewable Fuels Association, Historic U.S. Ethanol Fuel Production and Ethanol
Industry Overview.

3. Peter Wells, oil industry analyst, presentation at Toyota Sustainability Mobility
Seminar, Portland, Oregon, September 22, 2008.

4. Joe Walker, public relations representative for the American Petroleum Institute,
undated memo, 1997.

5. Rex Tillerson, chairman and CEO, Exxon Mobil Corporation, remarks before the
Royal Institute for International Affairs, London, England, June 21, 2007.

6. Press conference of President-elect Barack Obama, December 15, 2008, as reported at http://www.calcars.org, December 17, 2008.

7. As reported at http://www.foodfirst.org/en/node/2074, Web site of Food First:
Institute for Food & Development Policy.

8. Grant Ferrett, "Biofuels 'Crime against Humanity,'" BBC News, October 27, 2007,
reporting on UN report of Jean Ziegler as released by UN News Centre, October 26, 2007,
http://www.un.org.

9. Statement by Jacques Diouf, director general of FAO, at the launch of SOFA FAO 2008,
Rome, October 7, 2008.

10. *Financial Times (London)*, August 15, 2007, Asia Edition 1.

11. Science Advisory Board (SAB) Hypoxia Panel Draft Advisory Report submitted to
the Environmental Protection Agency, August 30, 2007, p. 3

12. Kent Garber, "Obama under Pressure over Role of Ethanol in Energy Policy,"
U.S. News & World Report, November 21, 2008.

13. Timothy Gardner, "New Ethanol to Face Crunch Time under Chu DOE," Reuters,
December 11, 2008.

14. William Ritz, "Farm Militants Study Bomb-Making: U.S. Probing Guerilla 'Seminars,'" *Denver Post*, February 13, 1983.

15. Author's interview with Jeff Broin, July 2006.

16. Author's interview with Larry Meints, January 2, 2008.

17. Lester R. Brown, "Plan B Updates: Supermarkets and Service Stations Now Competing for Grain," Earth Policy Institute, July 13, 2006, http://www.earth-policy.org.

18. Author's interview with Bill Roegnik, vice president of the National Chicken Council, 2006.

19. *Amber Waves*, U.S. Department of Agriculture publication, April 2006.

20. Author's notes of speech by Bob Dinneen at Renewable Fuels Association conference in Tucson, Arizona, February 20, 2007. The full text of Dinneen's speech can be found in the 2007 media archive of the Renewable Fuels Association at http://www.ethanolrfa.org/media/press/rfa/2008.

21. Matt Andrejczak, "Ethanol Bust Squeezes U.S. Corn Farmers: Bankrupt Verasun Leaves Farmers in Limbo as Prices Sink," *MarketWatch*, December 18, 2008.

22. John Berger, "The Ethanol Bust," *Fortune*, February 28, 2006.

23. Author's interview with Larry Meints, January 2, 2008.

24. Peter Harriman, "POET Seeks to Buy as Verasun Weighs Sale," *Argus Leader*, November 26, 2008.

25. Representative Jeff Flake, press release, October 22, 2008.

26. Author's interview with Larry Meints, January 2, 2008.

27. Author's interview with Antonio Seabra Moggi at Petrobras Research Center, Rio de Janeiro, Brazil, July 1979.

28. Author's interview with Terry Alger, manager of the Advanced Combustion Emissions section of the Southwest Research Institute (which has modified internal combustion engines to run on pure ethanol), January 2009. Also "Efficiency Improvements Associated with Ethanol-Fueled Spark-Ignition Engines," SWRI, http://www.swri.edu/4org/d03/engres/spkeng/sprkign/pbeffimp.htm.

29. "Cosan Signs Contract for the Acquisition of Esso Brasileira de Petroleo Ltda.," Cosan press release, April 24, 2008.

30. Marc J. Cohen, "Health and Nutrition Implications of Food Insecurity and Related Polices, Brazil Case Study," International Food Policy Research Institute, http://www.dse.unifi.it/sviluppo/doc/Cohen_%202_%20Brazil.doc.

31. Simon Marks, "Brazil's Economic Boom Marred by Social Inequalities," *The News Hour*, PBS, June 9, 2008.

32. "President Bush and President Lula of Brazil Discuss Biofuel Technology at Petrobas Facility in São Paulo, Brazil," transcript of remarks provided by White House, March 9, 2007.

33. Tom Phillips, "Brazil's Ethanol Slaves," *Guardian*, March 9, 2007.

34. Isabella Kenfield and Roger Burbach, "Militant Brazilian Opposition to Bush-Lula Ethanol Accords," *América Latina en Movimiento*, March 21, 2007.

35. Kenfield and Burbach, "Militant Brazilian Opposition."

36. Author's notes from Renewable Fuels Association conference in Tucson, Arizona, February 20, 2007.

37. D. Humbird and A. Aden, "Biochemical Production of Ethanol from Corn Stover: 2008 State of Technology Model," National Renewable Energy Laboratory Technical Report (NREL/TP-510-46214), http://www.nrel.gov/docs/fy09osti/46214.pdf.

38. Kris Bevill, "Verenium's Cellulosic Plant Opens," *Ethanol Producer Magazine*, June 2008.

39. Hal Bernton, "Leftovers from Harvest are Potential Fuel Source," *Seattle Times*, October 16, 2006.

40. "Greenhouse Gas Impacts of Expanded Renewable and Alternative Fuel Uses," Environmental Protection Agency, April 2007, http://www.epa.gov/otaq/renew ablefuels/420f07035.htm.

41. Timothy Searchinger, Ralph Heimlich, R. A. Houghton, Fengxia Dong, Amani Elobeid, Jacinto Fabiosa, Simla Tokgoz, Dermot Hayes, and Tun-Hsiang Yu, "Use of U.S. Croplands for Biofuels Increases Greenhouse Gases Through Emissions from Land-Use Change," *Science*, February 29, 2008.

42. Joseph Fargione, Jason Hill, David Tilman, Stephen Polasky, and Peter Hawthorne, "Land Clearing and the Biofuel Carbon Debt," *Science*, February 7, 2008.

43. Joseph Fargione, "The True Cost of Biofuels," interview by the Nature Conservancy, http://www.nature.org/tncscience/archives.

44. Fargione, "True Cost of Biofuels."

45. Paul Watson, "Biofuel Boom Backfiring," *Los Angeles Times*, October 19, 2008.

46. Letter from Michael Wang, Center for Transportation Research, Argonne, to *Science*, March 14, 2008.

47. Letter from Jonathan Lewis, Clean Air Task Force; Sandra Schubert, Environmental Working Group; and Kate McMahon, Friends of the Earth, to EPA administrator Stephen Johnson, October 31, 2008.

48. Testimony of Bob Dinneen, president and CEO, Renewable Fuels Association, before Committee on Agriculture, U.S. House of Representatives, May 21, 2009.

49. Interview by Nature.org with Joseph Fargione, a regional scientist for the Nature Conservancy, as published by the Nature Conservancy, January 1, 2008.

50. Chris Somerville, "The Argument for Biofuels," American Society for Cell Biology, May 2007, http://media.hhmi.org/ibio/somerville/somerville_powerpoint_pt1.pdf. See also Susanne Retka Schill, "Miscanthus versus Switchgrass," *Ethanol Producer Magazine*, October 2007, http://www.ethanolproducer.com/article.jsp?article_id=3334.

51. "Air NZ Biofuel Test Flight a Success," Air New Zealand press statement, TVNZ, December 30, 2008.

52. Nigel Quinn, Energy Biosciences Institute, *EBInsider*, November 2008.

53. Algae 2020 Biofuels Market Survey and Commercialization Outlook, from Emerging Markets Online, http://www.emerging-markets.com/index.html.

54. Solazyme press statement, June 11, 2008.

55. Vaughn Scully, "Ethanol from Woodchips—Via Termites?" *Business Week*, March 26, 2008.

56. "Growth Rates of Emission-Fed Algae Show Viability of New Biomass Crops," Arizona Public Service Company press release, September 26, 2007. See also "Biofuels from Algae: Scotland's First Minister Announces BioMara Funding," Scottish Association for Marine Science, http://www.sams.ac.uk.

57. Jim Warren, "UK Scientists Turning CO_2, Algae into Oil," *Lexington Herald-Leader*, December 7, 2008.

58. Jacques Diouf, "Biofuels Should Benefit the Poor, not the Rich," *Financial Times (London)*, August 15, 2007.

59. *New York Times*, January 11, 2009.

60. J. E. Campbell, D. B. Lobell, and C. B. Field, "Greater Transportation Energy and GHG Offsets from Bioelectricity than Ethanol," *Science*, May 7, 2007.

61. Jan F. Kreider and Peter S. Curtiss, "Comprehensive Life Cycle Analysis of Future, Liquid Fuels for Light Vehicles," http://www.evworld.com.

62. "Revamp Policy: Stress Sustainability," *Des Moines Register*, November 1, 2008, http://www.desmoinesregister.com.

To our families and friends who made this book possible.

And to William Hale, who, in an era when oil was king, refused to keep pace with his companions. He heard the beat of a different drummer, and, following Thoreau's counsel, stepped to its music, however measured or faraway.

Contents

List of Illustrations

Acknowledgments

Research for this book began in the summer of 1977 and proved to be a challenging but thoroughly rewarding task. It involved hundreds of interviews with scientists, farmers, politicians, corporate executives, and energy officials, as well as extensive travel both in Brazil and the United States. We owe a deep debt of gratitude to Senator Jacob Javits, columnist Jack Anderson, Eric Easton of Business Publishers, Inc., and Randy Son of the Washington Small Farm Network for allowing us the time and freedom required to complete this book even when this meant taking off time from our day-to-day work responsibilities.

We received helpful comments on various parts of the manuscript from a number of people: August W. Giebelhaus, Richard Archer, and William Scheller on historical aspects; Paul Middaugh, Cliff Bradley, Ken Runnion, Gene Schroder, and Albert Hubbard on the intricacies of both large and small-scale ethanol production; Donald Hertzmark on the economics involved; Pamela Shaw of the World Watch Institute on the agricultural and environmental material; Carl Duisberg and George Hawrylyshyn on Brazil, Richard Pefley, Louis Browning, Eugene Ecklund, and Roger Lippman on "Alcohol In Engines"; and Paul Bente of the BioEnergy Council on future technologies.

Carol Camelio gave unstintingly of her time and energy to aid in improving the overall structure and flow of the material presented in this book, as well as editing an invaluable resource document for the authors, the *Bioenergy Directory*.

The authors also profited from the counsel of numerous other individuals including Roger Blobaum, Lester Brown, Spence Johnson, Tom Allen, Dan Morgan, Allen Frank, Ginny Earnshaw, David Gitlitz, John Tinker, Charles Walters, Doug Campbell, Tom Lukens, George Spring, Carolyn Spillers, Jim, Sally and Charlie Crouchet, Marilyn Herman, and Gene Kahn.

Strong support from the Washington office of the National Center for Appropriate Technology helped facilitate communications between three authors who were often scattered about the country from Washington state to South Carolina. The tedious job of typing and xeroxing the rough manuscript was greatly aided by Kay Stoner, Pam Terwilliger, Linda Cheffet, Linda Fannon, and Estelle Buccino.

Although the comments of many of the aforementioned individuals helped to improve the final manuscript of this book, it should be noted that all errors of fact and interpretation are solely the responsibility of the authors.

Finally, the support of our families was crucial to the successful completion of this book. Linda Burton Kovarik deserves special mention for persevering through many of the travels associated with the book and for first suggesting its title, while Georgia and David Kovarik provided a welcome retreat for the writers at a crucial point in the manuscript. Constance and Horace Bernton were an unfailing source of strength and encouragement throughout the four-year research effort as were Rhea Bernton and Ray Tashof. Also, we are grateful to Blanche White, who was coaxed out of a well-deserved retirement to index this book.

"They say we have foreign oil.
Well, how are we going to get it in case of war?
It is in Venezuela . . . It is out in the East,
in Persia, and it is in Russia.
Do you think that is much defense
for your children?"

—Francis Garvan, *Chemical Foundation*
President, 1936

Chapter 1
Power Alcohol Comes of Age

When gasohol first went on sale at a few midwestern service station pumps in the winter of 1978, the national media treated its arrival as little more than an amusing diversion from the serious dimensions of the nation's energy crisis. Editorial cartoonists and newspaper columnists had a field day satirizing the notion that a fuel-thirsty nation was finally seeking solace in the intoxicating embrace of alcohol. Drunken cars were pictured lurching away from alcohol fuel pumps and farmers equipped with Snuffy Smith-like contraptions were hailed as the new "prairie Arabs." For most reporters, it took a startling leap of faith to believe that the same basic fermentation process that moonshiners utilized to produce white lightning could also yield a high-quality motor fuel.

Yet, the concept of using alcohol as a fuel is actually as old as the internal-combustion engine. During the 1890s, alcohol powered engines were used in farm machinery, train locomotives, and automobiles in both Europe and the United States. Indeed there was a lively debate between early alcohol advocates and gasoline advocates over which energy source would constitute the "fuel of the future." A major campaign was launched during the 1930s to introduce "power alcohol" blends (similar to the modern gasohol) at midwestern service stations. However, this effort, despite the support of such illustrious figures as Henry Ford and Alexander Graham Bell, eventually ended in failure, the victim of a glut of cheap western oil and a propaganda campaign directed against it by the American Petroleum Institute.

This debate over the "fuel of the future" was revived in the late 1970s as the sharp upward spiral of oil prices triggered an international resurgence of interest in power alcohol. In the United States, as gas lines formed around beleaguered service stations, grass roots support for power alcohol coalesced into what one federal official termed a "populist movement" which drew the bulk of its support from midwestern farmers who hoped to build on-farm and cooperative alcohol plants to process surplus grains. But it also included other groups: alternative-energy advocates who built small "backyard" stills to help make independent homesteaders more fuel self-sufficient; black civil rights groups eager to use the new energy industry as a tool for rural development in the south; and urban community activists seeking to establish municipal fuel plants to process garbage and food wastes into alcohol.

1

Alcohol, as a fuel, has also achieved a growing acceptance in the corporate boardrooms of America as large midwestern grain processors began to produce alcohol from the waste streams of corn processing plants. Oil companies which vigorously fought the promotions of Agrol in the 1930s have found a new use for alcohol in the 1980s as an octane booster in premium unleaded fuels. Chevron, an oil company which, even in the 1970s, still lobbied against federal alcohol fuels legislation, began test-marketing gasohol and decided to help finance the construction of a Kentucky fuel distillery. Texaco, hoping to cash in on the public relations value of gasohol, hired comedian Bob Hope to head an advertising campaign touting the merits of this blend of 10 percent alcohol and 90 percent gasoline. In Brazil, a Texaco subsidiary offers 100 percent alcohol to motorists driving a new generation of vehicles developed by Volkswagen, Ford, General Motors, and other international automobile manufacturers.

Between 1978 and 1980 a broad array of federal and state policy measures were passed in the United States to encourage the development of a national alcohol fuels industry which would provide at least a partial alternative to imported oil. By 1980, in the brief span of seven years, the cost of a barrel of OPEC crude oil had jumped from $3.00 to $32.00 a barrel. The shockwaves which rippled from this more than 1,000 percent price increase revolutionized the economics of energy production and touched off an international search for alternative liquid fuels.

The impacts of the OPEC oil price hikes were starkly evident to motorists each time they pulled up to a gas pump to fill their tank. The broader repercussions of the price hikes were felt by the nation as a whole as skyrocketing costs of petroleum-based products fueled an inflationary spiral which slowed economic growth to a crawl and threw millions out of work. Economists began to speak of a "social cost" for imported oil that dwarfed its actual selling price. One study published by the Harvard Business School put this "social cost" at between $65 and $100 a barrel.[1] When viewed from this perspective, ethanol produced from grain was suddenly competitive with imported oil. Equally important in federal policy considerations was the fact that the money spent on a gallon of corn alcohol remained in the United States to help stimulate the nation's embattled economy while most of the money spent on a gallon of imported gasoline wound up in the coffers of oil exporting nations.

However, the ethanol industry that was born in the oil-short decade of the seventies was forced to adapt to a new set of realities as the 1980s ushered in a period of oil abundance largely unforeseen by energy analysts in Houston, New York, and Washington. The surplus of oil on the world market in the winter of 1981 resulted from a combination of factors, the most important of which was a sharp drop in oil consumption in the industrial world which forced OPEC to cut oil production from a high of 31 million barrels per day in 1978 to 24.5 million barrels per day by the fall of 1981.[2] A 6 percent drop in oil consumption occurred in the United States during that period which was achieved primarily by modest conservation efforts, a boom in the sale of fuel-efficient cars, and a shift in U.S. industry away from the use of petroleum fuels to coal.[3]

News media in the United States quickly labeled the oil surplus a glut. However, as Daniel Yergin, a coauthor of the Harvard *Energy Future* report, pointed out in the editorial pages of the *New York Times*, the nation has heard talk of a glut several times before, most recently in 1978. The following year the overthrow of the Shah of Iran sent international oil markets into a state of panic. The glut quickly vanished and oil prices rose an astounding 170 percent.[4]

Energy analysts cautiously predict that the oil glut of 1981 will prove to be of a more enduring nature than previous gluts. They point not only to the declining oil consumption in the industrial world but also to increased production by non-OPEC oil fields in the North Sea and Mexico to justify their conclusions. Yet all such predictions are tinged with uncertainty. A new political crisis in the troubled Middle East resulting from another Arab-Israeli conflict or possibly the overthrow of the pro-western Saudi Arabian monarchy could soon tighten world oil supplies and launch yet another round of OPEC price increases. The distinct possibility of an abrupt supply cut-off of Persian Gulf oil is implicitly recognized in the Reagan administration decision to continue to stockpile oil in a national "strategic petroleum reserve."

While international oil production may continue to keep pace with world demand during the 1980s, U.S. oil production from conventional oil reserves appears to have entered a period of stagnation. Domestic oil producers could supply only 8,597,000 of the 17,006,000 barrels a day of oil consumed in the United States in 1980.[5] By 1990, the U.S. oil industry will be hard pressed to equal 1980 production levels—much less to register any impressive increases. Earl Hayes, a former chief scientist at the U.S. Bureau of Mines, summed up the consensus of many petroleum geologists when he stated, ". . . there is no longer much argument with the conclusion that U.S. resources of conventional oil will be seriously depleted by the year 2000."[6] And Daniel Yergin and Robert Stobaugh, in their book, *Energy Future*, warned:

> Americans should not delude themselves into thinking that there is some huge hidden reservoir of domestic oil that will free them from the heavy cost of imported oil. To the extent that any solution at all exists to the problems posed by the peaking of U.S. oil production and the growth of imports, it will be found in energy sources other than oil.[7]

During the early 1970s, the conventional wisdom, both in the oil industry and the federal energy bureaucracy, was that synthetic gasoline fuels, derived from the nation's immense reserves of coal, oil shale, and tar sands, would be the first replacements for gasoline to be offered to motorists. However, nearly a decade later, not a single synthetic gasoline fuel is commercially available while some 10,000 service stations market gasohol.[8]

The grain alcohol commonly known as ethanol that is mixed with gasoline to produce the gasohol blend is produced during the biological growth cycle of yeast organisms in sugar-containing solutions. Once this fermentation process is complete, the ethanol is separated out from the rest of the "beer" solution, usually via a distillation step.

With two carbon chains in its molecular structure, ethanol is chemically the same substance as drinking alcohol and thus has often been hailed in the press as the "moonshine fuel." However, fuel ethanol almost always contains toxic impurities such as fusel oils which are routinely removed from the more refined, beverage-grade ethanol. The federal government, in an effort to insure that ethanol is not illegally diverted from fuel to beverage markets, requires that it be thoroughly denatured with additional toxic substances such as gasoline before being sold to motorists.

Any substance containing high concentrations of sugars or starches (which are broken down into simple sugars with the aid of enzymes) can be readily converted to fuel ethanol. The range of feedstocks which alcohol producers have experimented with includes cheese whey, food cannery wastes, stale bakery breads, and moldy potatoes, as well as grain crops and sugarcane. New technologies undergoing intensive research and development may soon commercialize processes to convert tough cellulosic fibers contained in such products as wood, paper garbage, and corn stalks into the simple sugars required for fermentation.

A second alcohol prominently considered as a liquid fuel is methanol, which can be produced either from renewable cellulose materials or from fossil fuels such as coal and natural gas. Methanol was first produced as a minor byproduct of charcoal manufacturing. In this process, wood gases were combined with oxygen to produce a mixture of carbon dioxide and hydrogen which was then pressurized into a crude methanol. Today most of the 1.5 billion gallons of methanol produced each year are synthesized from natural gas for use in the chemical industry.[9]

Methanol, like ethanol, can be blended with gasoline or used alone as a pure fuel. However, it has about 15 percent fewer units of energy per gallon than ethanol, and thus methanol fuel does not yield as many miles per gallon. It also requires more extensive modifications in automobiles if it is to perform satisfactorily as either a blended or pure motor fuel. If mistakenly imbibed as an intoxicant, methanol can cause serious internal injuries as well as blindness.

The commercial fuel market for methanol remains limited to the relatively minor demand for it as an octane booster. Only a few private vehicle fleets in California have switched over from gasoline to methanol. The fleet owners, such as California's Bank of America, are primarily attracted to methanol's potential to be produced from coal and some natural gas for less than 50 cents a gallon and from wood for as little as 67 cents a gallon.[10]

There are numerous other, more complex forms of alcohol, such as proponol, pentanol, and butanol, often labeled "higher alcohols." This terminology stems from the fact that they contain much longer, more complicated carbon chains in their molecular structure than either ethanol or methanol. These alcohols, produced both from renewable and fossil fuel feedstocks, are useful in small amounts both as octane boosters and as stabilizers which help keep methanol-gasoline mixtures from undergoing phase separation (separation of alcohol from gasoline when water is present in the blend). They are also under study as possible replacements for diesel fuel. (See Appendix A, page 215.)

Both methanol and the higher alcohols appear capable of playing a significant role in the nation's energy future. Methanol has long been used to power racing cars in the United States and will be increasingly used as a blended fuel with gasoline in some European nations. Since methanol can be produced at a lower cost per unit of energy than ethanol, its long-term prospects may well rival, if not surpass, that of ethanol. Many major oil and chemical companies are carefully examining the potential for methanol to serve not only as an automotive fuel but also to power electrical turbines, and to serve as a feedstock for the petrochemical industry. However, construction costs to build the large, complex plants required to efficiently process coal, wood, and natural gas feedstocks into methanol are much higher than those for ethanol plants, and the plants take more time to complete. Moreover, ethanol, in the form of gasohol, is the only alcohol to date to gain widespread consumer acceptance. Thus, in the short term, most of the expansion of the alcohol fuels industry will come from increased ethanol production, with hundreds of both small-scale and large-scale plants being built during the 1980s. The first methanol plants producing fuel from coal and wood feedstocks are not expected to begin operation until 1990 at the earliest.

Synthetic gasoline and diesel fuels derived from coal, oil shale, and tar sands appear, like methanol, unlikely to make significant contributions to easing the nation's energy dependence on foreign oil during the 1980s. Their development has proceeded at a much slower pace than Richard Nixon anticipated when he launched his ill-fated Project Energy Independence in 1973 with the aim of eliminating the need for foreign oil by 1980. A plan to revive synthetic fuels development, which President Jimmy Carter pushed through a recalcitrant Democratic Congress at the end of his administration, is unlikely to live up to advance billings. It faces strong opposition from environmentalists who fear that a major push to develop synfuels would have serious adverse impacts on air, water, and land quality.

Without federal aid in the form of loan and price guarantees, to date not a single major oil or chemical industry has committed itself to building a commercial-scale synthetic oil shale or coal plant. While many backers of ethanol plants have also sought similar forms of federal aid, others have managed to finance plant construction entirely with private capital.

When the first synthetic fuel plants finally begin commercial production, it is doubtful that they will offer synthetic gasoline fuels for less than the pump price of alcohol fuels. Ethanol production costs from feedstocks such as paper garbage, grains, and cheese whey generally fall between $10.20 and $17.80 per million Btu of energy. Synthetic gasoline fuel production costs are estimated at a minimum of $12.50 and a maximum of over $17.80 per million Btu of energy.[11]

The years ahead appear certain to bring major technological advancements which may dramatically reduce ethanol production costs. New yeast strains developed by genetic engineers already can increase dramatically the efficiency of the grain fermentation process, while molecular sieve devices perfected by chemical engineers eventually eliminate the need for much of the energy-intensive distillation process. Waste heat and carbon dioxide, two major byproducts of ethanol production, have begun to be piped into greenhouses to help economically produce bumper crops of fresh vegetables, even during the harshest of winters. Renewable feedstocks of the future may be harvested from arid lands that currently contribute little to the nation's agricultural bounty.

Some power-alcohol advocates point to the new advances in the still embryonic fuel industry as proof that ethanol can somehow manage singlehandedly to solve the nation's energy woes. Caught up in the excitement of the "gasohol movement," they tend to forget that the production of alcohol fuels is not a religion but rather a technology. And like most technologies, it can be abused by those who choose to ignore potential problems in favor of short-term production goals.

Already, in Brazil, a singleminded dependence on sugarcane as a feedstock for the nation's rapidly expanding ethanol industry has diverted prime agricultural lands and scarce economic resources into the production of energy crops rather than staple food crops. Critics of the Brazilian alcohol fuels program sharply question the wisdom of such a policy at a time of growing hunger among the nation's poor, who can scarcely afford the luxury of owning an automobile, much less of buying a gallon of ethanol fuel. In the United States, the initial reliance of the ethanol industry on corn as its primary feedstock has sparked similar concerns that a "food versus fuel" conflict might eventually occur in the United States. The questions involved in the food-versus-fuel controversy are complex, since alcohol production utilizes only the starch portion of grains, leaving their essential proteins and oils intact for use as either human foods or livestock feed. Gasohol proponents, claiming that the world is faced with a food shortage that is primarily due to a dearth of proteins, see the diversion of surplus grain starches into ethanol production as a perfectly ethical step. Others disagree; convinced that the world is short on starches as well as proteins, they fear the impact of a major U.S. ethanol industry on world grain reserves. Additional questions have been raised over the wisdom of relying on agricultural practices that require large inputs of fossil fuels and often result in serious soil erosion in order to build the foundations of a renewable-fuels industry.

The important issues raised by the emergence of the alcohol fuels industry in the oil-short decade of the 1970s are not ignored in the pages ahead. The chapters that follow chronicle not only power alcohol's promising history to date, but also the problems it may present in the future. This book is written with the hope that once these problems are honestly confronted, power alcohol will help guide the nation safely into the postpetroleum future.

Chapter 2
The Pioneers of Gasohol

Henry Ford never felt comfortable in the tumultuous world of the twentieth century that his horseless carriage helped to create. He was raised on a farm outside of Dearborn, Michigan in an era when memories of the pioneers' trek westward in covered wagons were still fresh in the minds of his parents' generation. To him, modern cities were "the most unlovely and artificial sights this planet affords." Morever, he had a deep distrust for the giants of the oil industry who made their fortunes fueling his creation. This distrust stemmed in large part from his philosophical differences with men such as John D. Rockefeller, president of the powerful Standard Oil Trust, about the direction industrialization should follow in the United States during the twentieth century. Rockefeller viewed the increasing concentration of manufacturers in the urban centers and their increasing reliance on petroleum fuels as inevitable results of the industrial revolution. Ford predicted that the nation's cities would eventually wither away as the industrial revolution, in its final phase, made the amenities of urban living available in rural areas. He urged industry to decentralize and build small factories that would make a greater use of surplus agricultural products. Once this happened, he believed, the United States would see "as great a development of farming as we have had in the past twenty years in manufacturing." And if Henry Ford could have his way, alcohol would help fuel this new rural renaissance.[1]

Despite the vast amount of economic power that Ford wielded, his efforts to speak out on the great issues of his day were often unsuccessful and were subject to frequent ridicule in the press. One of the worst blows to his reputation came from a bumbling effort to keep the United States out of World War I. In January, 1915 he boldly declared that he would do everything in his power to prevent America from getting involved in this "murderous, wasteful war." Later that year—when the war had settled into a stalemate between German troops on one side and French and British on the other—Ford insisted that he would have the troops home by Christmas.

To accomplish this ambitious goal, he chartered an ocean liner and filled it with a bizarre cross section of American society, ranging from religious fanatics to the governor of South Dakota. The voyage across the Atlantic of this so-called "peace ship" was stormy, and the group's fragile harmony was destroyed by internal quarrels when the vessel reached Europe. The eastern press had a field day reporting Ford's inept attempt at peacemaking, and the Washington political establishment had a good laugh at his expense.

Ford's efforts to establish a farm-based alcohol fuels industry in the United States were somewhat more successful than his campaign to keep the country out of World War I. The power-alcohol movement that he helped launch was the beginning of an alcohol fuels industry which, at its peak in the 1930s, supplied over 2,000 midwestern service stations. Additionally, at a critical stage in World War II, a greatly expanded industrial-alcohol industry provided crucially needed synthetic rubber to the armed forces.

The power-alcohol movement, however, was never able to demonstrate that alcohol fuels could compete with the inexpensive petroleum fuels of the era, and it floundered at the end of World War II. Nevertheless, the movement left behind a rich body of information about alcohol production and related engine technology, on which future efforts would draw.

THE NINETEENTH CENTURY

When considering a portable, powerful fuel, it was quite natural that Henry Ford, along with other automotive inventors, should evaluate the variety of lamp oils common during the 1800s. The cheapest of these was camphine, a mixture of ethyl alcohol, turpentine, and camphor. Other possibilities included whale oil and lard oil. At 54 cents per gallon in 1860, alcohol was one-half the price of lard oil and one-third that of whale oil. At the time, the market for alcohol-based solvents and lamp oil exceeded 25 million gallons per year.[2] Probably, it would have declined with the advent of kerosene, a new and less expensive fuel distilled from petroleum, which had just been introduced to the market.

Instead, the market for alcohol went from boom to bust overnight in 1861 as Congress passed a sales tax of $2.08 per gallon to help pay for the Civil War. An empty Treasury and rampant inflation required increased taxes on many commodities, but it must have seemed unfair to the distillers that kerosene and lard oil were taxed at only 10 cents per gallon. The Pennsylvania oil boom that would earn John D. Rockefeller his millions was on, and the dawn of the petroleum age was at hand.

The idea of taxing alcohol to discourage its use as an intoxicant had been imposed by George Washington in 1791 and had been violently resisted by backwoods farmers, who depended on whiskey sales for much of their income. This so-called "Whiskey Rebellion" was crushed, but eleven years later, in 1802, public sentiment led Thomas Jefferson to repeal the unpopular tax. Repeated efforts during the late 1800s to repeal the Civil War alcohol tax failed, however, and as a result alcohol was too expensive to be considered as a fuel.

Unburdened by heavy taxes, alcohol fuels fared much better in Europe than in the United States during the second half of the nineteenth century. Most European governments, with few domestic oil reserves, were eager to encourage the development of a fuel that could be readily distilled from domestic farm products. The German government worked closely with its distilling industry to pass legislation providing substantial financial incentives to produce alcohol and to create new markets for its industrial use. Protected by tariffs from competition with imported petroleum fuels, Germany's alcohol production went from 10 million gallons a year in 1887 to over 29.5 million gallons by 1902,[3] thereby pleasing German potato farmers who benefited from this new market for their crops.

Some of Europe's earliest internal-combustion engines—developed by Nicholas Otto and other nineteenth-century inventors—were designed to run on alcohol. By the 1880s, alcohol engines were beginning to replace steam engines in light machinery, pumps, and grain threshers. Manufacturing companies in Germany, England, and France marketed these engines on the European continent and even exported some to Egypt and South Africa. By 1906, Deutz-Gas engine works of Cologne Deutz, Germany reported that over a third of its locomotives were powered by alcohol.[4] The European alcohol engines were advertised as safer than steam engines (since they did not give off fire-triggering sparks) and less polluting than gasoline engines. One turn-of-the-century survey found that 87 percent of German farmers considered alcohol engines to be equal or superior to steam engines in performance.[5] Numerous races between alcohol-powered and gasoline-powered automobiles were held in Europe during the first decade of the twentieth century, and a lively debate raged over which fuel delivered the best performance. Such races led to a national commission on alcohol in France which recommended tax incentives. In Greece, revenues from petroleum taxes fell so rapidly due to alcohol substitution that a tax was imposed on alcohol fuels as well. Also, the *New York Times* reported "agitation" for alcohol fuels by English farmers.[6]

This widespread use of alcohol engines in Europe was noted with great interest by American farmers, who were suffering from large grain surpluses as vast new tracts of virgin prairie were plowed under to produce bumper first-year crops. Low grain prices on the glutted feed markets gave farmers scant compensation for their hard work. So, to absorb the huge crop surpluses, many farmers looked to the increasingly large market for liquid fuels created by the widespread acceptance of Henry Ford's horseless carriage.

Before alcohol could be considered as a serious alternative to such power sources as gasoline, kerosene, electricity, steam, and coal, Congress had to be persuaded to remove the stiff $2.08 per gallon federal tax placed on alcohol in 1861. The farm lobby found a strong ally in the White House in Theodore Roosevelt, the outspoken Rough Rider elected to the presidency in 1904 (he had succeeded the assassinated McKinley in 1901) after he bitterly attacked the Standard Oil Trust's powerful influence over the U.S. economy.

In 1906, a Roosevelt-backed bill to repeal the alcohol sales tax was introduced in Congress and received widespread support from farm-belt representatives. One of these was Missouri's Champ Clark, who, in an enthusiastic speech on the merits of alcohol, declared that, "Buckle [Henry Thomas Buckle, a nineteenth-century English historian] . . . says that the three most pertinent factors in modern civilization were the invention of gunpowder, the invention of moveable type, and the use of steam. Had he lived in our day, he would have classed electricity as the fourth. And if the hopes of promoters of untaxed, denatured alcohol find fruition in fact, then it will rank as the fifth. Alcohol can be made from cornstalks . . . in quantities to supply not America alone, but the world, with cheaper and better . . . fuel."[7] The bill repealing the $2.08 tax of 1861 was signed into law a few months after Clark's speech.

Amendments that specifically exempted farm stills from government controls passed with the repeal of the tax, and triumphant farmbelt senators, like North Dakota's Hansbrough, proclaimed that "every farmer could have a still" to supply heat, light, and power at low prices. "Advocates look forward with hope to a big change in the farmer's life," the *New York Times* reported. "If the law accomplishes what is hoped, it will . . . make a revolution on the farm," the paper said.[8] Despite lofty rhetoric that obscured a difficult economic picture, the bill kindled interest in alcohol fuels among farmers.

A flood of inquiries spurred the government to investigate alcohol fuels technology. Tests by the Bureau of Mines showed that high-compression engines specifically adapted to alcohol delivered the same power, gallon for gallon, as gasoline in typical low-compression engines.[9] The Agriculture Department issued several bulletins on the use of alcohol and production techniques.[10] Most of the reports were accompanied by caveats on the economics of alcohol compared to gasoline, however. Even without the tax, the Department noted, alcohol sold for a minimum of 30 cents per gallon, while gasoline sold for a minimum of 10 cents per gallon, and kerosene for 8 cents per gallon.*

Despite the cost differential, alcohol fuels first went on sale in Peoria, Illinois at 32 cents per gallon, and demand reportedly far exceeded supply. Enthusiasm was evident in a test run between New York and Boston immediately after the tax repeal. Three Maxwell automobiles—one fueled with alcohol, one with gasoline, and one with kerosene—made a grueling seventeen-hour drive over deeply rutted, snowy roads during the last week of January, 1907. A committee supervising the tests reported better hill-climbing ability and less stalling with the alcohol-fueled auto, but the low-compression engine consumed 40 percent more fuel than the one using gasoline.[12]

Henry Ford supported the "farmer's fuel" by equipping his early Model T engines with adjustable carburetors to be adapted to run on alcohol. Other manufacturers, such as the Olds Gas Power Company, offered a simple mixer attachment for alcohol and found that "under actual (operating) conditions . . . the fuel consumption per horse-power is about the same, pound for pound, whether

---

*As a point of interest, the *New York Times* reported gasoline prices at 20 cents and kerosene at 18 cents in 1907.[11]

using alcohol or gasoline."[13] The Hart-Parr Company, an engine manufacturer based in Charles City, Iowa, commented in 1907, "We have watched with great interest, and added our efforts to help bring about the free use of alcohol for power purposes. . . . our engine is so constructed that alcohol can be used with very little change. . . . In enclosed situations . . . (such as mining operations) where the odors from kerosene fuels are considerable, it is quite likely that alcohol will take the place . . . of petroleum fuels."[14]

However, despite hopes that alcohol prices would decline, the new industrial alcohol market grew to only about 10 million gallons per year by the start of World War I. The handful of distilleries did not possess the capital to compete with the Standard Oil Trust and find a niche in the growing fuel market.

In the early 1900s, as demand for the more cheaply priced gasoline soared, the oil industry began building an extensive system of pipelines to transport crude oil from the newly discovered oil fields in Oklahoma to refineries which shifted their production emphasis from kerosene to gasoline. Gasoline sales soon topped kerosene sales for the Standard Oil Trust, as hundreds of horse stables were converted into gasoline stations where motorists could fill their newly installed tanks. The western oil boom, along with the automobile, was capturing the imagination of the American public. Eastern newspapers sent their reporters to the central plains states where they filed vivid accounts of the instant wealth created by the Oklahoma oil strikes. A new American folk hero—the free-wheeling, independent oil prospector—was born, and new cities grew alongside the oil fields. Supplied with an abundance of oil, motorists could see little reason to support the development of a costly alcohol fuels industry.

The outbreak of World War I in 1914 triggered a major, although short-lived, expansion of the ethyl alcohol industry in the U.S. Alcohol was in demand not only as a fuel but also as a feedstock in the production of gunpowder and the deadly mustard gas—a weapon widely used in trench warfare. To meet this surge in demand, industrial alcohol production in the United States increased from over 10 million gallons in 1914 to over 50 million gallons by the war's end in 1918.[15]

After the war, interest in alcohol fuels was kept alive within the U.S. scientific community by several gloomy predictions of an impending oil shortage. In April, 1921, T. A. Boyd, a General Motors research scientist, speaking at a meeting of the American Chemical Society, warned, "For operating the motors of the country, an enormous and ever increasing amount of liquid fuel is required. This amount of fuel is so large that not only has great activity in the production of crude oil been necessary to meet the demand, but the reserves of crude oil are also being rapidly depleted. The yearly production of petroleum has become so large that exhaustion of reserves in the United States threatens to occur within a few years." Boyd went on to propose alcohol as ". . . the most direct route which we know for converting energy from its source, the sun, into a material that is suitable for use as fuel in an internal combustion motor."[16]

With such a speech, Boyd was merely reaffirming an earlier assessment of alcohol's potential made in 1917 by Alexander Graham Bell, inventor of the telephone, who called it "a wonderfully clean-burning fuel . . . that can be produced from farm crops, agricultural wastes, and even garbage."[17]

Despite fears within the scientific community that U. S. petroleum reserves were running dry, the power alcohol movement made little headway in the 1920s. The only major experiment with alcohol fuels utilization occurred in 1922–23 when Standard Oil of New Jersey test-marketed a blend of 25 percent alcohol and 75 percent gasoline in the Baltimore, Maryland area.*

PROHIBITION

National Prohibition, generally agreed to be the greatest social experiment of the twentieth century, was adopted in the United States after the thirty-sixth state ratified the Eighteenth Amendment to the Constitution, January 16, 1919, which became effective as law one year later, January 16, 1920.

The Eighteenth Amendment read as follows:

Section 1.   After one year from the ratification of this article the manufacture, sale, or transportation of intoxicating liquors within, the importation thereof into, or the exportation thereof from the United States and all territory subject to the jurisdiction thereof for beverage purposes is hereby prohibited.

In outlawing the consumption of any alcoholic beverage—even in the privacy of one's home—the Eighteenth Amendment also strictly forbade the manufacture of any intoxicating liquors. The fact that ethanol was the active ingredient in any "intoxicating liquors" threw a large shadow over any merits it may have had as a fuel. Enforcement of Prohibition, even with the police powers of the Volstead Act of 1919, was a major problem, and a reduction in the manufacture of alcohol was one of the key objectives in the enforcement strategy.

Prohibition represented a victory for the Anti-Saloon League, an organization founded in Oberlin, Ohio in 1893. When the Eighteenth Amendment was ratified, the League was the foremost temperance organization in the country, having already led the efforts that had resulted in statewide prohibition for thirty-three states.

On the eve of Prohibition, the supporters of the Anti-Saloon League gathered by the thousands in churches to greet the arrival of the "dry" era. One of these victory celebrations was described by Thomas Coffey, author of *The Long Thirst: Prohibition in America, 1919–1933*:

Prohibition represented the triumph of America's towns and rural districts over the sinful cities. The biggest of the [Anti-Saloon League's] celebrations took place in the nation's biggest "small town," Washington, D.C., where the First Congregational Church was jammed with God-fearing, respectable, and prominent people, many of whom had been born in smaller communities. Under ordinary circum-

---

*Standard Oil had access to cheap surplus alcohol from the World War I munitions industry.

stances, these people would have considered the hour of the meeting—midnight—almost indecent, especially since the meeting had been advertised as a "party." They were not the kind of people who indulged in midnight parties. But this was no ordinary reunion. On this night, John Barleycorn was condemned to die. Demon Rum, already reeling back to the nether reaches from whence he came, was on strict notice never to return.[18]

Much to the chagrin of the politically powerful Anti-Saloon League, neither John Barleycorn nor Demon Rum was banished from the United States by Prohibition. They were simply driven underground as thousands of clandestine speakeasies, amply stocked with bootleg liquor, began serving the needs of "wet" America.

Both John D. Rockefeller and Henry Ford were originally staunch supporters of the Anti-Saloon League. Rockefeller, whose family eventually contributed more than $1 million to the League, hoped that Prohibition would result in a more sober and productive American work force. Indeed, many thought that it did, including no less a personage than Herbert Hoover, who, speaking as Secretary of Commerce in 1925, said: ". . . while our productivity should have increased about 15 percent due to increase in population, the actual increase has been from 25 percent to 30 percent. . . . There is no question . . . prohibition is making America more productive."[19] Of course, there were other economic forces at work in the 1920s besides a sober labor force. The Rockefeller family eventually abandoned the ranks of the "drys" in the summer of 1932, when John D. Rockefeller, Jr. publicly announced that ". . . many of our best citizens, piqued at what they regard as an infringement of their private rights, have openly and unabashedly disregarded the Eighteenth Amendment."[20]

Ford's support of the Anti-Saloon League, however, remained steadfast until Prohibition's repeal in 1933. It was a natural outgrowth of his strict, midwestern, Protestant upbringing, which frowned upon any kind of alcoholic consumption. He apparently saw no contradiction between his staunch support of farm-based alcohol fuel production and his opposition to the consumption of beverage alcohol.

In the light of Prohibition, however, most politicians in the nation's capital were reluctant to discuss, much less to propose, any legislation favoring alcohol fuel production. Not only would new production further complicate the already massive enforcement problems of the Eighteenth Amendment, but, considering the huge demand for bootleg liquor, it was feared that farm alcohol producers would be more likely to sell their product on the illegal beverage market than on the legal fuel market.

The fourteen years of national Prohibition came to an end on December 5, 1933 when the Eighteenth Amendment was repealed by the ratification of the Twenty-first Amendment. Thus ended the greatest social experiment of modern times and perhaps the most controversial, divisive question the United States had faced since slavery. Needless to say, it was not a prosperous period for alcohol fuels.

CHEMURGY AND POWER ALCOHOL

Although alcohol manufacture during Prohibition was largely an underground industry, the idea of alcohol as an alternative fuel source remained alive. In 1926, at the height of the Prohibition era, the *Dearborn Independent*, a paper owned by Henry Ford, published a lengthy article by William Jay Hale, an Ohio-born chemist, urging the nation to launch a major alcohol fuel development program. The article, "Farming Must Become A Chemical Industry," explained the basic ideology of a curious blend of science, economics, and philosophy which Hale labeled chemurgy—from the Egyptian word *chem* (from which "chemistry" is derived) and the Greek word *ergon* which meant "to work." Thus the literal meaning of chemurgy was "to put chemistry to work."

Of course, this concept was not entirely new. At the turn of the century, George Washington Carver, the famous black agronomist of Tuskegee Institute's agriculture department, invented numerous chemical products from peanuts, cotton, sweet potatoes, and soybeans. Ford, Hale, and other chemurgists recognized him as "a chemurgist before the word was invented."[21]

A scion of the Dow family, Hale had attended Miami University of Ohio and Harvard, and had received formal, postgraduate training in chemistry in Berlin. He was concerned that the American public did not have an appreciation for chemical research. At the time, Germany was the leader in the field, and the United States was running a distant fourth or fifth to European nations. Hale was fascinated by agriculture and its prospects for producing energy and chemicals for the industrial world. His article, based on several years of research, was a call for America to wake up to the potentials of chemical research and the agricultural industry.

While most Americans were enjoying the prosperity of the "Roaring Twenties," Hale was grimly warning of an impending collapse of the nation's depressed agricultural economy. To avert this collapse, Hale believed new markets should be developed to absorb all of the farmer's surplus crops. With the aid of the chemists' magic touch, these crops would be transformed into a variety of raw materials that could help fill the needs of industrial America. Soybeans could be made into plastics (a concept that fascinated Henry Ford and led him to develop a soybean automobile body), milk into a silklike fabric, coffee beans into dyes, cornstalks into paper, and starch-laden grains into alcohol.

As many oil-industry critics later pointed out, there was nothing revolutionary in Hale's chemurgic ideology; he was simply restating and embellishing many of the concepts originally put forth by Ford and other, earlier advocates. But Hale did this with a verve that created unprecedented, broad support for the power-alcohol movement, causing alarm in the oil industry. When the power-alcohol movement reached its zenith during the Great Depression, leaders of the oil industry genuinely feared that the federal government would follow Hale's counsel and mandate the use of alcohol-gasoline fuel blends, what we now call gasohol. Bills were introduced in a number of state legislatures. For example, in 1933 a bill was introduced in Iowa to require that 10 percent alcohol be blended in all gasoline sold in the state.

Hale was an active spokesman for the chemurgy movement, writing and speaking extensively in the late twenties and throughout the thirties. In 1934, he published a book entitled *The Farm Chemurgic* that contained his views favoring the production of farm alcohol.

Hale was joined in his efforts to legitimize the chemurgist proposals by two other men who made important contributions to the power-alcohol movement of the 1930s, Francis P. Garvan, president of the Chemical Foundation, and Dr. Leo Christensen, a professor of chemistry at Iowa State University. Both shared Hale's vision of a worldwide, agrichemical revolution from which alcohol would emerge as a renewable, alternative fuel.

Garvan, an influential New York lawyer, had managed in 1919 to convince Woodrow Wilson and Congress that access to German patents was needed to build up a chemical industry in the U.S. World War I had demonstrated how dependent America and the allies were for such common items as iodine, sulfite, and other chemical necessities. The Chemical Foundation, a private organization, was established to fill this need. Through the Foundation, patents were licensed from the advanced German chemical industry by the then underdeveloped U.S. chemical industry, with fees paid by the benefiting U.S. corporations. As president of the Foundation, Garvan was a vigorous promoter of U.S. self-sufficiency in the production of industrial chemicals, and during the 1920s he lobbied actively to protect the chemical industry from foreign competition. In this capacity, Garvan became an active supporter of Hale's chemurgy movement, and throughout the thirties he funneled over $600,000 of the Chemical Foundation's funds into the power-alcohol movement.

Christensen was the lead author of a report, "Power Alcohol and Farm Relief," that resulted from conferences at Iowa State University in the early thirties. Delighted to find chemurgist views coming from another credible source, Hale wasted little time in meeting Christensen and introducing him to Garvan. Garvan was so impressed with Christensen and his report that in 1934 he published "Power Alcohol and Farm Relief" in a series sponsored by the Chemical Foundation. Throughout the thirties and World War II, Christensen was a leading advocate of chemurgy and power alcohol.

THE GREAT DEPRESSION

The decade of the 1930s was the period of the greatest sustained economic depression that the United States has ever known. From the fall of 1929—when roughly $25 billion was lost in three months in the stock market—to December, 1933, real national income fell by 36 percent. Within four years of the 1929 crash, the American economy was operating at less than 50 percent of its capacity. For many people, all lifetime savings were lost. Factories closed. Banks closed. Twenty-five percent of the work force became unemployed. Foreign trade fell. Buying power was paralyzed. And farmers were driven from the land because they could not sell their products for enough money to begin to cover the costs of production.[22]

In fact, U.S. farmers had been suffering from their own depression through-out the twenties. The last prosperity they had known was during World War I when European demand for American farm products drove up prices and assured high profits. To meet the rising demand, U.S. agriculture drastically increased production. New, efficient machinery, such as the tractor, allowed cultivation of millions of acres previously grazed upon by horses and mules. New fertilizers were used, producing unprecedented yields per acre. Then in 1920, with the war ended, demand—especially from Europe—dropped drastically, creating large surpluses and falling prices. With their new equipment, fertilizer, expanded acreage, and improved technology, farmers continued to produce more than the markets could absorb, and prices continued to fall.

During the twenties, there were several attempts to rescue the distressed farming community, but with little success. The Agricultural Credits Act of 1923 extended federal credit to farmers but did nothing to reduce the supply or stimulate prices. In 1929, the Agricultural Marketing Act was passed, establishing a Federal Farm Board and a fund of $500 million to purchase products at a guaranteed price. The availability of this fund actually stimulated production, adding still more excess supply. After a loss of $200 million by the Farm Board, this scheme was finally abandoned.

The years between 1929 and 1932 were the bleakest in U.S. agricultural history. In that period the ratio of prices paid by the farmers to prices received for their products was the lowest on record. A bushel of wheat that sold for over $3.00 during World War I brought less than 25 cents; a wagon of oats would barely pay for a new pair of shoes; thousands of acres of Missouri corn were left to rot in the fields; California oranges ripened on the trees and fell to the ground. In 1932, 10 percent of all farmers went bankrupt and had to leave their land. Others, slightly more fortunate, fell back on the land to survive—planting large gardens, hunting wild game, and storing whatever they could for the long winter.[23]

The Great Depression also wrecked the economies of the nation's cities, as millions of industrial workers were forced into breadlines. It abruptly ended a century-long trend of urban growth: in 1930, 17,000 people left the cities, and by 1934 the figure had grown to 533,000. Clarence Berger, a Harvard sociologist, speculated at the time that "It may be that this reversal of city-country migration will mark the apogee of the urbanization that has characterized American social and economic life during the past hundred years."[24] This outmigration was not—as Henry Ford had predicted—due to any new measure of prosperity in rural America but was due instead to the bleakness of urban life.

John Steinbeck summed up the tragic irony of the period in *The Grapes of Wrath*:

The decay spreads over the State, and the sweet smell is a great sorrow on the land. Men who can graft the trees and make the seed fertile and big can find no way to let the hungry people eat their produce. Men who have created new fruits in the world cannot create a system whereby their fruits may be eaten. And the failure hangs over the State like a great sorrow. . . .

The people come with nets to fish for potatoes in the river, and the guards hold them back; they come in rattling cars to get the dumped oranges but the kerosene is sprayed. And they stand still and watch the potatoes float by, listen to the screaming pigs being killed in a ditch and covered with quicklime, watch the mountains of oranges slop down to a putrefying ooze.[25]

It was a terrifying time for the nation's farmers. Their confidence in the economic policies of the past decade shattered, they looked desperately for ways out of the crisis, and many embraced the power-alcohol movement created by the chemurgists. With prices for their crops so low, it suddenly made sense to convert a bushel of corn worth 7 cents into 2.5 gallons of alcohol, rather than barter it for a gallon of gasoline worth 5 cents.

In 1933, the combination of the farmers' desperation and the optimistic claims of the chemurgists produced a flurry of lobbying activity in midwestern state legislatures, calling for alcohol-gasoline blends and related tax breaks. Although only a few of these efforts were successful, the oil industry was sufficiently concerned to mount a propaganda campaign deriding alcohol fuels.

OIL INDUSTRY BACKLASH

The first signs of an oil industry backlash to the power-alcohol movement followed the 1933 proposal in Iowa to require mandatory blending of ethyl alcohol in gasoline. The Iowa Petroleum Council printed a pamphlet headlined "the alcohol-gas scheme outrages common sense." It warned: "Compulsory or subsidized dilution of gasoline with 10% alcohol made from corn would result in a devastating raid on the pocketbooks of Iowa motorists." The pamphlet was illustrated with a pig labeled "cost of alcohol gasoline" feeding from a trough filled with Iowa taxpayers' dollars.

Many of the initial attacks on the power-alcohol movement came from new oil companies that had been formed by the dissolution of the Standard Oil Trust. In 1911, the Supreme Court ruled that the existence of the trust represented an illegal restraint of trade under the statutes of the Sherman Anti-Trust Act of 1890. It ordered the parent company to dispose of its stock holdings by distributing them to its corporate stockholders. This ruling came after a decade of effort begun under Theodore Roosevelt. The ruling, however, did little to restore competition within the oil industry. Carl Solberg, in *Oil Power*, noted that, "Although Standard's trust was broken up into some 27 separate companies, the very small group of officers who controlled them received proportionate shares of stock in all 27 companies . . . and Rockefeller continued to be the dominant shareholder."[26]

Spokesmen for the oil industry denounced the chemurgist proposals in sarcastic and derogatory articles in their trade publications and persuaded motorist magazines such as *Traffic World* to do the same. The antialcohol campaign was strengthened by the involvement of the American Petroleum Institute (API), the oil industry's Washington-based trade association. The API reprinted for mass distribution a March, 1933 *Business Week* article entitled "Farmers Would Make

[WHO WOULD PAY-

COST OF ALCOHOL GASOLINE

60% OF IOWA MOTOR-FUEL-BILL

40% OF IOWA MOTOR-FUEL BILL

IOWAN'S POCKETBOOK

IOWA MOTORIST

IOWA FARMER

FOR CORN-ALCOHOL?]

Illustration in pamphlet circulated in 1933 by the Iowa Petroleum Council

Motorists Pay for Farm Relief.'' The article warned that "corn state farmers, bankers, and merchants take it [power alcohol] very seriously indeed. Spontaneous support is considerable. Legislatures, congressmen, and editors are being bombarded with letters extolling the virtues of the mixture.''[27]

According to August W. Giebelhaus, a history professor at the Georgia Institute of Technology, the API in April, 1933 circulated a high-priority memo to the select members of its Industries Committee, spelling out plans to intensify the antialcohol campaign. The memo described the situation as "critical" and urged all committee members to contact key individuals and organizations in their areas, urging them to oppose legislative measures favoring alcohol.[28] Following distribution of the memo, the National Petroleum Association released a report asserting that the general use of alcohol blends would enable bootleggers to obtain a steady supply of alcohol by simply using water to separate it from the gasoline. "To force the use of alcohol in motor fuel would be to make every filling station and gasoline pump a potential speakeasy,'' the report charged.[29]

Radio, rapidly establishing itself as the nation's most powerful means for mass communication, was also used by the oil industry to wage its antialcohol campaign. Standard Oil of New Jersey, a decade after it had marketed an alcohol blend in the Baltimore area, sponsored a radio spot claiming that national blending legislation would "make alcoholics out of America's 22 million motor cars.''[30] Another spot sponsored by the American Petroleum Institute featured John Simpson, the president of the powerful National Farm Bureau, declaring that enacting blending legislation "would be like passing a bill for the cotton farmers in which all the clothing had to be made out of cotton." Simpson's talk was labeled "Quack Remedies.''[31]

AAA TESTS

The oil industry was aided in its attempts to discredit alcohol fuels by the results of a test performed in 1933 by the American Automobile Association (AAA). The stated purpose of the AAA test was to determine the technical performance of ethanol-gasoline blends. Test observers included Secretary of Agriculture Henry Wallace, a young Illinois congressman named Everett Dirksen, and the chemurgist, Dr. Leo Christensen, then teaching at Iowa State University. At noon on a sweltering June day, the test observers gathered a few miles south of Alexandria, Virginia to watch five conventional stock cars, fueled by alcohol-gasoline blends, run over a ten-mile circuital course. The AAA, in its official report of the event, proudly noted that "White-uniformed men with armbands checked stop watches with a master watch to assure accuracy. Large white scales glistened in the sun as U.S. Bureau of Standards officials weighed the fuel for the test containers. The click of shutters could be heard as photographers from news syndicates snapped pictures to capture the event for the print media."

The test findings proved to be potent ammunition for the oil industry. The final report indicated that the power-alcohol blends gave poorer mileage and less power and "separated from gasoline after coming into contact with moisture from the atmosphere."[32]

Dozens of newspapers across the nation carried the AAA results. The general consensus was that power-alcohol blends were technically inferior to gasoline. The *Grand Rapids Press* commented that "road and laboratory tests of alcohol-gasoline blends recently completed by the American Automobile Association in cooperation with the government have produced results that should definitely set at rest all agitation for compulsory blending of the two fuels for motor use."

The *Louisville Times* noted, "the American Automobile Association, representing millions of automobilists, gives a new meaning to the old adage 'alcohol and gasoline do not mix.' " And the *Buffalo News* warned that, "in view of the information which the A.A.A. makes available, motorists generally should be alert to oppose legislation to require a blend of alcohol."[33]

But the AAA results were soon embroiled in controversy as Leo Christensen charged that several members of the official test committee had refused to approve the final report released to the press. "The main trouble with the tests was that they were made in extreme heat and with unusually volatile gasoline," Christensen maintained. While the AAA tests showed a 2.75 percent decrease in fuel economy with the power-alcohol blend, Christensen reported laboratory tests indicating a 4.85 percent increase in fuel efficiency.[34] Three years later the significance of the AAA tests was further clouded when Dr. Oscar C. Bridgeman, a government scientist with the National Bureau of Standards who supervised the testing procedures, told a meeting of the American Chemical Society, "If the need or desire for such [alcohol] fuels should arise in this country, sufficient technical information is available to insure that blends containing ethyl alcohol could be utilized satisfactorily as motor fuels, provided full advantage could be taken of the available technical information. For the most part, therefore, the problem is whether or not it is economically feasible."[35]

In 1933, the U.S. Department of Agriculture released a report encouraging to chemurgists stating, "The Depression has intensified interest in the possibility of developing new uses for farm products. The development of the internal-combustion engine has displaced horse and mule power and reduced the feed requirements for power animals to the extent of the production of about 35,000,000 acres. Finding an additional outlet for some part of this production, for which the demand has disappeared, would contribute to an improvement of the agricultural situation."[36]

In the two years that followed this report, however, the grain surpluses depressing the crop markets dried up as the great American Breadbasket was abruptly transformed into the great American Dust Bowl. The bountiful productivity of American agriculture was revealed to rest on a quite fragile soil foundation—a foundation easily eroded by wind, rain, and the works of man. A fierce drought in 1934 baked the Great Plains, transforming marginal farm lands into swirling clouds of dust that swept across the country, darkening the sky as far away as New York. For one of the few times in American history, Eastern Seaboard cities were forced to import grain from abroad as farmers faced near total crop failures in many areas of the country, especially in the prairie states of the Dust Bowl.

FORD SPONSORS CONFERENCE

Against this stark backdrop, in May, 1935 a major chemurgic conference, planned by Garvan's Chemical Foundation, was sponsored by Henry Ford, the American Farm Federation, the National Grange, and the National Agricultural Conference. It was held in Ford's hometown of Dearborn, Michigan in a hall designed to look like a replica of Independence Hall in Philadelphia.

Heated debates broke out during the two days of the conference, which ended on an inconclusive note. Oil industry representatives and their allies in the farm community repeated their arguments that alcohol fuels were uneconomic substitutes for gasoline, while the chemurgists insisted that, if the federal government supported an intensive research and development program in alcohol production technology, the nation would soon pull out of the Great Depression. Between the two warring groups there appeared little room for agreement.[37]

The chemurgist proposals at the conference received a cool press reception, with the *New York Times* reporting, "Drs. L. M. Christensen and William Hale read lyrics in which alcohol was glorified as a miracle-worker which would permanently place the farmer beyond the pale of distress. . . . Both Drs. Hale and Christensen talked of alcohol at 10¢ a gallon—an impossible price even if corn sold for only half of what it brings now." The editorial concluded that Hale was guilty of indulging in "Jules Verne dreaming."[38]

Despite pressure from the oil industry to back away from the chemurgists, Ford's support for Hale, Christensen, and Garvan never faltered, and he agreed to host a second conference at Dearborn in 1936. In order to regain momentum after the first conference, the chemurgists needed new ammunition.

Shortly after the first chemurgic conference adjourned, Garvan compiled an extensive dossier of reports, pamphlets, and advertisements documenting the performance of alcohol fuels abroad. When the second Dearborn conference was convened in May, 1936, Garvan was ready to use the oil industry's own material in rebutting its attempts to portray power alcohol as an inferior fuel.

"We have been fed volumes to the effect that power alcohol is not a practical fuel," Garvan cried after he stepped up to the podium. "Were they [the oil industry] quite sincere? I think you can judge." He then handed out copies of a booklet printed by Standard Oil of New Jersey's British subsidiary, Cleveland Discol, for British motorists. "If you take your little pamphlet," Garvan continued, "you will find that all these worries have been settled for us. All this chemical research has been done for us, and all the testing. The Standard Oil Company of New Jersey has gone over to England, and in its delightful international aspect of life has joined hands with the English Distillers Company and they together have produced, in their own words, 'the most perfect motor fuel the world has ever known,' and they have stations of it all over England."[39]

The pamphlet boasted that Cleveland Discol's mixture of alcohol, gasoline,

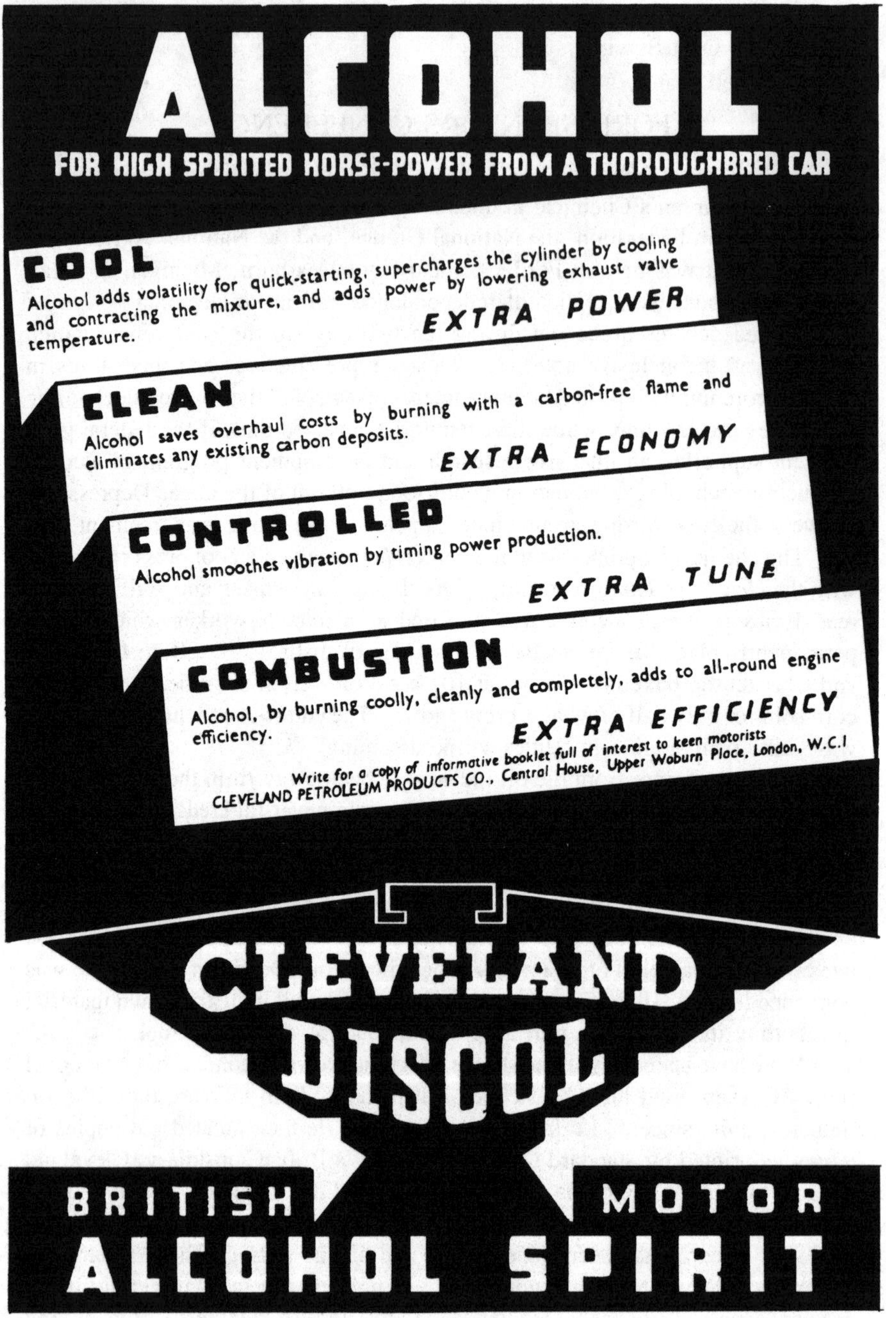

1936 alcohol advertisement in England

and coal-derived benzol provided, "extra power, extra economy, and extra efficiency," possessing the highest octane rating of any motor fuel in England. In the United States, Standard Oil of New Jersey chemists had claimed that alcohol blends suffered from the troublesome "phase separation" problem, in which the alcohol, water, and gasoline in the fuel separated into three layers, causing engines to stall out. But in England, this phase separation problem had apparently been resolved. "It is possible," the pamphlet stated, "to pour almost a pint of water into a car tank containing 10 gallons of Discol without the slightest trouble—in fact in some circumstances with better running."

"When the oil people come before you and raise objections," Garvan told his audience, "I want you to keep their little English book before you."

In addition to Cleveland Discol, Garvan cited a host of other power-alcohol blends promoted in foreign markets by oil companies. Cities Service marketed in England a "Koolmotor Alcohol blend" which it advertised as giving "to every motorist the advantages of super racing petrol—at no additional cost." Texaco, Ethyl, and Esso also marketed foreign ethanol blends during the thirties.

It should be pointed out, however, that these oil companies were promoting alcohol for reasons other than the fuel's merits. Oil was scarce in the countries cited by Garvan and a number of European governments were subsidizing alcohol programs. A comparison could be made to a similar situation in Brazil today.[40]

Garvan pointed out that demand for the alcohol fuels outside the United States was so great that U.S. farm equipment and auto manufacturers had begun producing specially modified alcohol-powered vehicles. International Harvester took out advertisements in the Philippine press promoting "International motor trucks powered with engines especially designed to burn alcohol. These facts have been proved: they are absolutely dependable; they are more economical; they are free from carbon; there is no loss of power." And Chrysler shipped Detroit-made cars equipped with different sized carburetor jets and manifold modifications to New Zealand.

But neither Garvan nor the other chemurgist leaders could hope to convince an increasingly skeptical U.S. motoring public of the merits of power alcohol with words alone. The public and the press had tired of the endless debates and elaborately staged Dearborn conferences. If the chemurgists hoped to maintain their credibility, they would have to begin to transform their rhetoric into reality.

At the end of the second Dearborn conference in 1936, Francis Garvan announced that the Chemical Foundation would loan the Bailor Manufacturing Company $116,000 to convert a brewery in the small farming town of Atchison, Kansas into an experimental fuel-grade distillery. To help ensure the success of the project, Leo Christenson personally took charge of the new operation. For the next year and a half, he experimented with a variety of farm products including wheat, corn, sweet potatoes, rye, and barley. Grain was hauled from nearby farms to the plant, where it was cleaned, ground, broken down into simple sugars, fermented, and finally distilled into 190-proof alcohol. Local farmers were paid 56 cents a bushel for their corn, 14 cents per bushel over the average price at the time. The farmers were also encouraged to grow new energy crops such as sweet sorghum and jerusalem artichokes. The distillery returned some of the protein-rich byproducts of the distillation process to the farmers and sold some of it to livestock-feed dealers. William Buffum, Treasurer of the Chemical Foundation at the time, estimated that the Agrol plant had the potential to increase farm income by $2 million per year in a 50-mile radius around the Atchison plant.

In the latter part of 1937, the Chemical Foundation bought outright the Atchison distillery from the Bailor Manufacturing Company and formed the Atchison Agrol Company, with plans to produce 10,000 gallons per day of commercial alcohol. Christensen worked feverishly for eighteen months to upgrade the distillery facilities and develop a cost-effective production process. Then he turned his efforts to devising a successful marketing strategy for the new Agrol blends. A vigorous advertising campaign was launched, urging motorists to "Ask for Agrol by name. Be persistent. It will pay for your trouble even if you have to ask at several stations. When you buy Agrol you help yourself by getting a superior fuel." Three blends were marketed by the Agrol plant. The highest-grade fuel, Agrol 15, had an antiknock rating above that of premium gasoline. Agrol 10 was suitable for use in high-compression engines, and Agrol 5 was the power-alcohol equivalent of regular gasoline.[41]

"The major companies," reported *Business Week* in December, 1937, "are still opposed to the idea [of marketing Agrol blends] but the Agrol boys don't care, they say the midwest farmers are apt to favor anything the big oil companies oppose."[42]

The Atchison distillery used independent oil distributors (jobbers) and farm cooperatives to market the fuels (a similar strategy would be followed by the first gasohol producers in the late 1970s to circumvent the majors). The Agrol blends, ranging from 6 to 12.5 percent ethanol, cost at least 2 cents more than most gasoline brands, but most of the early distributors initially agreed to absorb half the extra cost of the fuels. Thus the motorists usually paid only a penny a gallon more for the first Agrol blends.

The concerted sales campaign launched by the Chemical Foundation appeared to pay off. By the spring of 1938, the Agrol blends were being sold at over 2,000 service stations in eight midwestern states and at a few stations as far east as Maryland. Some 15 million gallons of alcohol-gasoline blends were sold, and sales increased 1,500 percent in just a few months. Customers reported better mileage, cooler engines, and increased power. Briefly, it appeared that the chemurgists would confound their critics and turn the Atchison venture into a financial as well as technical success.

Nervous oil industry officials quickly sought to develop a strategy to discourage the spread of the Agrol blends. The oil industry, already plagued with crude oil surpluses from newly discovered western fields, could ill afford any reduction in its share of the motor fuel market. But it would have to proceed cautiously with any plans to combat the Agrol blends. *Business Week* observed, "Standard of Indiana realizes that there is potential dynamite in having itself spotlighted as opposing an effort designed to aid the farmers' market. So the company's rebuttal carefully explains its attitude on alky-gas. It explains that Indiana Standard had told its dealers they can try out any motor fuel but that Standard's insignia must be removed from pumps serving blends. The field men of Standard are to regard the alky-gas campaign as a social or political movement and not to interfere."[43]

But all the oil industry field men may not have taken such gentlemanly attitudes toward their new competitor. Christensen soon received reports of a brigade of "traveling experts" who showed up at Agrol distributors' offices and performed a simple test to prove the unreliability of the fuel. A small glass test tube was filled with alcohol and gasoline and then shaken. The two fuels immediately separated into two layers, one of gasoline and one of alcohol and water. The tests worked like a charm because the "expert" always made sure to wash the test tube out with water before the experiment. A few drops of water left in the tube then caused the alcohol and gasoline to separate. The water drops would have this dramatic effect because water will not mix with gasoline, although it blends easily with alcohol. The "experts" insisted that if this happened on the road a car's engine would stall.[44]

It was indeed possible for motorists to experience phase separation with the Agrol blends, but it is unlikely that this happened with anything like the magnitude claimed by the mysterious "experts." When alcohol was distilled up to the anhydrous 200-proof level, phase separation was rarely a problem. Even when some water was present in the blending alcohol, the use of additives such as benzene by the Agrol Company often prevented phase separation from occurring. In the modern gasohol industry, phase separation has rarely been a major problem hampering engine performance.

In the wake of such intrigue and suspicion, power alcohol sales began to drop off. For example, in Sioux City, Iowa sales fell from 2,500 gallons a month to less than 1,000. As the novelty of the new fuels faded, many distributors stopped absorbing half of the added costs, forcing motorists to pay at least 2 cents a gallon more for Agrol blends. Despite the chemurgists' earlier predictions of 15-cent-per-gallon alcohol, the Atchison distillery could not manage to produce it for less than 25 cents and still pay a fair market value for their crops. This 25-cent-a-gallon production cost was more than five times the cost of refining gasoline. Moreover, the distillery, plagued by equipment breakdowns, seldom operated at full capacity. Normal commercial efficiency for industrial alcohol plants of that era was close to 93 percent, while the Atchison distillery managed only 48 percent in its first year of operation, 63 percent in its second year, and 71 percent in its final year.

In November, 1938 the Atchison distillery shut down operations. The Chemical Foundation had lost over $600,000 in less than three years and failed to demonstrate that alcohol fuels could be profitably produced from midwestern grain. The federal government, supportive of the chemurgist efforts during the early 1930s, turned a cold shoulder to additional requests for financial assistance. The Federal Surplus Commodities Crop Corporation refused to supply the Agrol plant with distressed corn (which could be made available at below market

prices), and the U.S. Department of Agriculture released a lengthy report recommending against any federal financial incentives to help maintain the farm-
based alcohol fuels industry.[45]

DAMAGING RUMOR

The Chemical Foundation, struggling without the leadership of Francis Garvan,
who died of pneumonia in 1937, tried to continue supplying Agrol distributors
by purchasing alcohol from distillers of blackstrap molasses (a mixture of crude
sugar and fiber which is left over from the sugarcane refining process). But
rumors asserting that imported blackstrap was being used instead of domestic
supplies were exploited by petroleum marketers, who spread the word that Agrol
was made from foreign products. Agrol Company executives angrily denied the
charges and produced affidavits from the U.S. Industrial Alcohol Company, the
principal supplier of the blending stock, indicating that the product was indeed
derived from Louisiana blackstrap. However, the charges still proved extremely
damaging to sales, and many independent distributors abruptly refused to continue to carry the Agrol blends. An angry Christensen told a U.S. Senate subcommittee in 1939:

> Disaster struck the Agrol plant in the form of a false and I think malicious rumor
> which spread all over the territory which it was serving, to the effect that the
> alcohol it was making was made from blackstrap molasses, and the impression
> was that it was imported.
>
> Since most of the distributors to which the Agrol Company sold alcohol were
> farmer cooperative stations, and the stations were certainly interested in the Amer-
> ican farmer, you can easily visualize what happened. Sales dropped very, very
> rapidly, and all of the capital of the Atchison Agrol Company was used up in
> overcoming the damage which resulted from that rumor.[46]

By the close of the decade, the power-alcohol movement in the United
States was on the verge of collapse. Nine legislative bills introduced to Congress
in 1939 promoting alcohol fuels never even reached the full House or Senate for
a vote. As farmers left behind the dark years of the Great Depression and grain
markets improved, the political base for the movement began to dissolve. Oil
industry spokesmen, such as the American Petroleum Institute's Conger Reynolds, chided legislators for having been seduced by what Reynolds termed the
"fantasies of dream chemists"; and even President Franklin D. Roosevelt's
Secretary of Agriculture, Henry Wallace, once a cautious supporter of the power-
alcohol movement, criticized the chemurgists' "cavalier" attitude towards the
difficulties in maintaining long-term soil fertility. In October, 1939 he publicly
attacked Hale for "his assumption that an inexhaustible amount of farm products
can be produced from the soil." Wallace, sobered by the devastation wrought
by wind erosion during the Dust Bowl years, warned that "the danger from his
[Hale's] erroneous assumption lies in this—that if the soil is to be exploited to
produce farm products for industry, very soon there won't be any soil left to

draw upon, and industry and agriculture will both be left holding the bag."[47]

William Hale remained uncowed by his critics and continued to lead the thinning ranks of the power-alcohol movement. In 1939 he was once again called to Washington to testify before Congress. After reviewing the dismal health of the alcohol fuels industry in the United States, he concluded, "We are the boobs of the world."[48]

EUROPE AND BRAZIL

At the time of Hale's 1939 testimony, power alcohol had become a fact of life throughout much of the industrialized world. Some forty nations, including Germany, France, Brazil, and New Zealand, had either enacted legislation making alcohol blending programs mandatory or had offered subsidies to make the fuels more cost competitive with imported petroleum. In Europe, power-alcohol consumption rose from 59,000 tons in 1930 to 540,000 tons in 1937, constituting over 4 percent of all liquid fuels consumed on the continent.[49]

Despite their widespread use, not all of the European alcohol blends were readily accepted by motorists. Some awkward blends of low-proof ethanol caused considerable problems for motorists. In France, where ethanol was distilled from the remains of the grape harvest, motorists such as Arthur Bar complained of "very bad efficiency" with the blend; Bar noted that "during hot weather starting is more than difficult, and the motor stops almost at once if you don't immediately step on the accelerator."[50]

"France is the one country in the world in which power alcohol has not worked well," conceded Leo Christensen, who blamed the program's failure on a French law requiring blends of up to 50 percent ethanol with gasoline. The fluctuating levels caused constant changes in the octane rating of the fuel, each necessitating a carburetor adjustment to ensure the proper air-to-fuel ratio. "It has been an unsound law; power alcohol has not been unsound," Christensen said after reviewing the French blending legislation.

The French power-alcohol program during the 1930s, like many of its European counterparts, was not primarily geared towards meeting the needs of the nation's motoring public. A much more important objective of the French government was to ensure the economic survival of a large domestic alcohol industry in case war should once again erupt in Europe. Alcohol would, as in World War I, be needed as an alternative fuel (since most European nations relied on imported oil) and as an important ingredient in a wartime munitions industry. It would also be an important feedstock in the manufacture of synthetic rubber—a product which was being developed during the period between the two wars.

Nowhere was the military emphasis on the alcohol blending programs more pronounced than in Germany, where the Nazi government incorporated it into an ambitious plan to achieve full self-sufficiency by the end of the decade. Hitler was convinced that Germany would have emerged from World War I victorious if the nation had been able to obtain sufficient supplies of oil, and he vowed that his new Nazi war machine would never be hobbled by empty fuel tanks.

After Hitler came to power in 1933, the Nazi government decreed the mandatory blending of 10 percent alcohol in all imported fuels. Ethanol production, largely from potatoes, rose from 278,000 barrels in 1932 to 1,423,000 barrels in 1935, and the methyl alcohol industry expanded to the point that it was producing 70,000 tons of fuel in 1937. Ethyl and methyl alcohol fuels were mixed with gasoline and sold as a blended fuel called *Kraftspirits*.[51]

The accelerated development of the German alcohol fuels industry was only one facet of a wide-ranging Nazi effort to create a massive new alternative-fuels industry unprecedented anywhere in the world. The backbone of this industry was the production of synthetic gasoline from the abundant coal deposits of the Ruhr Valley. The production processes of advanced hydrogenation and synthesis were developed by I. G. Farben, the leading German chemical manufacturer. These new synthetic gasoline fuels were supplemented by solid wood fuels, benzol liquids derived from coal, and several different forms of compressed natural and methane gases. By 1937, one year before the Nazi invasion of Austria, these substitute fuels accounted for 54.5 percent of Germany's total light-motor-fuel consumption.[52]

As the flames of a new war spread across oil-starved Europe, liquid fuels were in short supply. In Yugoslavia, the situation was so desperate by 1942 that the Yugoslav General Mihailovitch set up a system of gasoline barter in return for Italian prisoners. According to a report published in the *New York Times*, the Yugoslavs would agree to release "one Italian soldier or non-commissioned officer for one can of gasoline, one Italian officer up to the rank of colonel for four cans of gasoline and one Italian colonel for fifty cans of gasoline."[53]

Throughout much of Europe, liquid fuels were reserved for military uses. Civilians were forced to utilize less-refined fuels, such as wood, in cumbersome vehicles called "gasogens"—which were a cross between an automobile and a wood stove. Over 200,000 of these vehicles were on the roads in Germany, forcing service stations to stockpile chipped wood.[54]

Much of the alcohol produced during World War II was diverted into high-priority munitions, medicines, and synthetic rubber industries. But a limited amount remained in use as a motor fuel. Grape-derived alcohol continued to power many vehicles in occupied France, and Switzerland, which struggled to remain neutral throughout the conflict, turned to methyl alcohol fuel when its oil supplies were cut off. Japan enacted laws requiring all Japanese oil companies to maintain emergency reserves of alcohol equivalent to at least 20 percent of a six-month supply of gasoline.[55]

With World War II, new life was breathed into the ailing power-alcohol movement in the United States. The desperate need for a feedstock in the production of synthetic rubber triggered a dramatic rebirth of the farm-based alcohol industry the chemurgists fought to establish in the 1930s.

Prior to Pearl Harbor, the nation had been almost totally dependent on imported natural rubber to meet its military and industrial needs. Most of the United States supply of this vital commodity was cut off in 1941 after the

Japanese army invaded the Dutch East Indies. Suddenly the United States found itself severely short of the rubber needed to put boots on soldiers, tires on tanks, and gaskets on engines. Twenty days after the Japanese attacked Pearl Harbor, U.S. military experts were warning of a 100,000-ton rubber shortage, an estimate that would quadruple in a year's time. A major rubber conservation program was implemented immediately. Speed limits were reduced. Old tires were recycled. Even used rubber bath mats were collected and transformed into shock absorbers for guns, and new automobiles and trucks were distributed without spare tires.[56]

President Roosevelt established a Rubber Reserve Corporation in 1940 to stockpile surplus rubber and stimulate the development of a new synthetic rubber industry. But Jesse Jones, who headed the Rubber Reserve Corporation, made little progress toward accomplishing his important task in the years prior to the Japanese attack on Pearl Harbor. Bernard Baruch, Dean of Harvard's Law School, was appointed by Roosevelt to investigate the rubber shortage; he concluded, "When we were drawn into the war, our rubber stockpile was sufficient to meet just one year's normal demand. By this time the only solution lay in rapid development of synthetic rubber. But the synthetic rubber program then existed largely on paper, and its development was hampered by conflict over whether grain alcohol or petroleum should be used as the basic raw material." The contestants in this conflict were essentially the same as those who fought over power alcohol in the 1930s—the chemurgists and the oil industry.[57]

The synthetic rubber was to be produced from what was known as a Buna-S process. Two products—styrene, a derivative of alcohol, and butadiene, which could be manufactured from either alcohol or petroleum—were combined in this process. The debate was over what would be the source of the butadiene: alcohol or petroleum.[58]

Early in 1942 Leo Christensen journeyed to Washington to persuade the federal government to assist financially in the commercialization of a process developed by midwestern chemists to produce 8 to 10 pounds of synthetic rubber from a bushel's worth of grain alcohol. Christensen asserted that this butadiene process, based on the dehydration of alcohol molecules, was less expensive than similar petroleum-based processes and could be carried out in plants constructed in six to nine months. But Christensen later testified to a Senate subcommittee that he found officials of the Rubber Reserve Corporation to be unenthusiastic about alcohol-based, synthetic rubber processes.[59]

In March, 1942, William Hale appeared before the same Senate subcommittee and warned that "the war cannot be won without full chemurgic enterprise" in the establishment of the new synthetic rubber industry. He repeated Christensen's statements that the chemurgists could make synthetic rubber from grain alcohol for less than the oil companies could make it from petroleum. "The oil companies have always claimed that they can make anything on earth cheaper and better than any other organization known," he lectured the senators. "That is their hallucination. They cannot make cheap rubber!"

The Washington bureaucracy, perhaps remembering similar claims about the chemurgists' ability to make alcohol fuels at a competitive price, remained unmoved by Hale's passionate outburst. And Sen. Guy Gillette, a chemurgist sympathizer who led a congressional investigation into the rubber shortage, reported that the chemurgists "were met . . . in the government with a cold shoulder, turned down, and sometimes hardly with respect."

The U.S. oil industry had no trouble commanding respect in wartime Washington. Dozens of oil executives were asked to take sensitive federal positions, and those remaining outside the government were closely consulted by those in power. Harold Ickes, the federal Petroleum Administrator, would later praise the "exhibitions of patriotism and good faith the oil industry has undertaken for the monumental task of providing fuel for 185,000 bombers and fighters, 18 million tons of merchant ships, 120,000 tanks, and hundreds of thousands of trucks the U.S. military would need."[60]

STANDARD OIL PACT WITH GERMANS EXPOSED

It was in this atmosphere of trust that the Rubber Reserve Corporation awarded $650 million to twenty-five different companies—most of them subsidiaries of Standard Oil—to produce 700,000 tons of petroleum-based synthetic rubber. The rubber would be produced through a process patented by Standard.

But just as the contracts were awarded, Attorney General Thurmond Arnold announced that Standard had been involved in a secret, illegal pact with the German chemical company, I. G. Farben. The pact—which officials of both companies called a "full marriage"—had been signed in 1929. It was prompted by Standard's fear that the new German process for turning coal into liquid fuel would make inroads on their worldwide markets. Standard agreed to stay out of the world chemical business and Farben agreed to stay out of the world fuel industry. The complex Standard-Farben marriage was renewed in October, 1939, when officials met secretly in the Netherlands to agree that the marriage would "outlast the war." Standard gave Farben data on tetraethyl lead octane boosters, and Farben gave Standard the general outlines for a synthetic rubber process. But Standard was unable to obtain Farben know-how in the actual manufacturing of the oil based "Butyl" rubber, and paid little attention to developing it.[61]

When the war broke out, an executive with a Standard subsidiary, Richard Dearborn, was appointed to a key post with the Rubber Reserve. But muckraking columnist Drew Pearson revealed that Dearborn had helped negotiate the October, 1939 agreement with Farben, "continuing Hitler's monopoly on synthetic rubber—a monopoly which is charged with preventing American development of synthetic rubber." Other ties between the Rubber Reserve and Standard Oil, indicating a deep bias toward Standard's process, also were revealed.[62]

When war broke out, Standard Oil was completely unprepared for synthetic rubber production. In March, 1942, Assistant Attorney General Thurmond Arnold charged:

Not only was the production of synthetic rubber in this country absolutely stifled by Standard's adherence to the restrictions imposed on them by I. G. Farben—which they always loyally preserved—but after 1939, when Standard received permission [from Farben] to enter into negotiations with rubber companies, Standard proceeded to further retard the development of synthetic rubber because of its natural monopolistic desire to keep complete domination over this industry.*[63]

When Standard and other companies using the I. G. Farben process to produce butadiene for Butyl and Buna-S processes (known as the "Jersey patent pool") received the lucrative federal contracts in April, 1942, they were ill-prepared to launch the crash program needed to avert a disastrous rubber shortage. "By June of 1942," wrote war historian Charles Wilson, "the techniques had scarcely begun to assume workable form. . . . Public money was being spent by the millions [on synthetic rubber processes] which might well be obsolete before going into production."[64]

Senator Gillette's subcommittee continued turning up evidence that neither of the two oil-based processes would produce enough rubber, and a bill to give funding to other processes passed Congress. Especially promising was the butadiene process, developed in Russia and Poland during the 1920s and thirties. Instead of using oil, butadiene was produced from grain alcohol. Two refugee Poles, Waclaw Szukiewicz and M. M. Rosten, provided the committee with data showing the alcohol process, as compared to the oil-based process, would cost one-tenth as much to set up, would use one-fifth the crucial steel, and would take only six to eight months per plant, instead of eighteen.**[65]

Drastic measures were taken to ensure some kind of rubber production. Sugar was rationed and diverted to alcohol production. The nation's whiskey distilleries were ordered to convert to industrial alcohol, and the old Atchison Agrol plant was put back into operation. Federal funds were appropriated for construction of three midwestern alcohol plants in Omaha, Nebraska, Kansas City, Missouri, and Muscatine, Iowa, as well as wood-pulp ethanol plants in Oregon and Washington.

By October, 1943, 87 percent of the 230,000-ton yearly quota of alcohol-based synthetic rubber was being delivered on schedule while only 30 percent of the 460,000-ton yearly production quota of the oil-based processes was available. In December, Donald Nelson, an official with the federal Rubber Reserve Corporation, testified in Gillette's subcommittee:

*For example, Standard sued a major tire manufacturer for patent infringement in October, 1941, after cabling Berlin for advice. Eventually, some thirty out of forty members of the War Production Board's Petroleum Industry War Council were indicted by the Justice Department for restraint of trade in synthetic rubber and other war materials. By December, 1942, Standard Oil chairman Walter Teagle and two other top executives had resigned in disgrace.

**Szukiewicz and an associate were kept under secrecy wraps and not allowed to demonstrate the process between their arrival in the U.S. in 1941 and February, 1942. When they were released, they went to work in a Philadelphia alcohol plant, and they were able to begin delivering the intermediate chemical for Buna-S within three months. It is likely that they, like many other scientists, were smuggled into the U.S. by a British spy ring headquartered in New York City. The British Security Coordination had declared Standard Oil a "hostile and dangerous element of the enemy," and had decided to introduce an alternative to the Jersey patent pool to a U.S. war industry desperate for rubber, since they did not trust Standard Oil.[66]

Alcohol requirements have increased because petroleum [processes] on which the rubber program was based have fallen so seriously behind expectations. . . . Had the War Production Board not taken steps to produce alcohol, the position of the rubber program would be truly desperate. The Rubber Director is now counting on pushing alcohol facilities up to 150 percent to 175 percent of rated capacity. In short, that part of the rubber raw material program [alcohol] which was counted on to supply less than one-third must now be counted on to supply two-thirds to three-quarters of the whole burden through 1944.[67]

Oil-based synthetic rubber plants scheduled for completion in July, 1943, were still not operating when Allied armies launched their Normandy invasion of Nazi-occupied France in June, 1944. In contrast, the grain-alcohol plant in Omaha was ready to begin full production in February, 1944, only thirteen months after the first federal construction funds were awarded. Twice the size of the Agrol facility, the Omaha plant had a yearly capacity of 7.5 million gallons of alcohol and 25 million pounds of byproduct livestock feed. It employed over 1,000 people who worked around the clock, seven days a week. The plant, which Leo Christensen helped design, was successful beyond all expectations. Although it was originally thought to have a maximum daily processing capacity of 20,000 bushels of grain, it succeeded in processing as many as 30,000 bushels of grain daily.[68]

By the end of 1944, the United States was producing almost 600 million gallons of alcohol, nearly four times the nation's 1942 level.[69] The alcohol was made from grain, molasses, and, in Bellingham, Washington, from paper-pulp byproducts. About half was used for synthetic rubber; the rest went into other vital materials: smokeless gunpowder, medicines, and power boosters for aviation and submarine fuels. In China, U.S. army troops used alcohol to fuel Jeeps, Land-Rovers, and generators. James Kerwin, who as a G.I. was stationed there, recalls that this alcohol once inadvertently doubled as a cocktail on a lonely December evening.

We were celebrating Christmas, but ran short of alcohol. We sent the houseboy with an empty bottle and $100 (Chinese). He came back after an hour, and we sent him out for reinforcements. He was gone for forty minutes, and with every following trip, he came back sooner. When his trip was down to ten minutes, some sober G.I. trailed him and found that he was selling us our own fuel alcohol. . . . It was not denatured, and it was potable and potent![70]

The future of the farm-based U.S. alcohol industry remained uncertain as World War II entered its final year. The long delayed petroleum-based synthetic rubber plants finally geared up to full production and, much to the chemurgists' dismay, produced a less expensive product than that from grain alcohol.

It should be noted that the success of the alcohol-based synthetic rubber program was not because of its superiority to a petroleum-based process, in terms of its economics or technical characteristics. It was merely a very good substitute for the petroleum processes when petroleum feedstocks were in great demand for such needs as a high-octane aviation fuel in the war effort.[71]

The alcohol-based process produced a pound of rubber for about 21 cents on average, while the oil-based process produced a pound of rubber for about 11 cents.

The oil industry lobbyists urged the federal government to abandon the grain distilleries as soon as the war ended. But the chemurgists' congressional ally, Sen. Guy Gillette, argued in a September, 1944 report to Congress that "Maximum production of synthetic rubber from grain should be maintained. It rather startles one," he remarked, "to be immediately forced to consider the abandonment of plants that we already have built and already own."

But after the war, grain prices rose rapidly, and surpluses dwindled as the U.S. government sent food shipments overseas to help Europe recover from the devastation of the war. The grain-based synthetic rubber became too expensive compared to that based on petroleum, and imported natural rubber was once again available. The government refused to give the alcohol-based plants any postwar contracts that might have kept them financially solvent. The ownership of the three major alcohol plants shifted in rapid succession from the Defense Plant Production Board to the Reconstruction Finance Commission to the U.S. Department of Agriculture in the summer of 1948. The Secretary of Agriculture informed a disillusioned Senator Gillette that his agency could no longer justify spending government money to keep the plants open, and they were soon turned over to the General Services Administration for sale to private industry.*

This policy was criticized by columnist Drew Pearson in November, 1950, in his widely syndicated column.

> For four years the American people were forced to walk as a result of the rubber and tire shortage. . . . now though we are much better off when it comes to synthetic rubber, we would have to have rubber rationing all over again—if war came. . . . Alcohol-based synthetic rubber plants were built at a cost of millions of dollars after Japan cut off rubber from Malaya and Indonesia. Now . . . the administration is insisting on selling these rubber factories for about 15 cents on the dollar.[72]

The grain-alcohol distilleries were turned over to private spirit distillers. The Kansas City plant and the Muscatine plant were put to use producing beverage alcohol, but the Omaha plant eventually deteriorated into a gutted shell that still sits forlornly in the middle of an industrial section of the city. By 1949, less than 10 percent of U.S. industrial alcohol was made from grain; most of it came from a new process developed by the oil industry which used natural gas as its principal feedstock.

It took World War II to transform—however briefly—the dreams of Henry

---

*Within two years, the price of synthetic rubber rose from 15 cents to 60 cents.

Ford and the chemurgists into reality. The destruction of the wartime industrial alcohol industry coincided with the death of Henry Ford, one of the power-alcohol movement's most important leaders, in 1947. Leo Christensen sought to carry on the work the chemurgists had begun in the early 1930s by joining a new chemurgy program created at the University of Nebraska. However, Christensen resigned in 1945, charging that the opposition of large chemical manufacturers was hampering his research.[73]

Interest in alcohol fuels continued sporadically during the 1950s as large grain surpluses once again became a headache for the federal government. A seventy-member commission appointed by President Dwight Eisenhower to find new uses for farm crops concluded in 1958, "The Commission has not found any encouragement for believing that, in the present state of knowledge and under current economic conditions, the use of industrial alcohol for motor fuels can be justified."[74]

It would take another world upheaval in the form of OPEC and its more than 1,000-percent price escalations to rekindle interest in alcohol fuels.

Chapter 3
The Return of the Farm Alcohol Movement

In October, 1973 the thirteen ministers of the Organization of Petroleum Exporting Countries (OPEC) met around a horseshoe-shaped conference table at their headquarters in Vienna, Austria. As news film rolled and cameras flashed, they announced a rise in world oil prices from $5.12 to $11.65 a barrel. This action, swiftly followed by the Arab OPEC members' oil boycott of the United States in retaliation for its support of Israel in the 1973 war, rudely shocked a complacent American public into an awareness of the nation's increasingly dangerous reliance on foreign oil. As gas lines formed around beleaguered service stations, the nation plunged into its deepest recession of the postwar era.

The three-month embargo was only the first flexing of the mighty economic and political muscle of the Arab-dominated OPEC cartel, which would lift world oil prices to over $35 a barrel in the 1980s. OPEC's decision put an abrupt end to a century of cheap oil prices in the United States, and this inadvertently set the stage for a rebirth of the power alcohol movement.

No Americans were more profoundly affected by the tumultuous economic events of 1973 than the nation's farmers, whose dependence on oil products—fertilizers, pesticides, and fuel—had risen dramatically since the end of World War II. It was natural, then, that the leadership of a movement to create an alcohol fuels industry first arose from the nation's farming community and not from the ranks of academia, industry, or government. These farmers sought a more decentralized, smaller-scaled industry than the one envisioned by the chemurgists. While William Hale and Leo Christensen and, later, the mainstream gasohol movement of the 1970s concentrated their efforts on promoting alcohol-gasoline blends (gasohol) on a national scale, the farm alcohol movement began with the more modest goal of simply making farms less dependent on OPEC. They drew their initial inspiration not from the chemurgists' sophisticated 10,000-gallon-a-day Agrol distillery, but from cruder 10- to 500-gallon-a-day moonshine stills that flourished in rural America during Prohibition.

TEN BUSHELS OF CORN FOR A BARREL OF OIL

Farmers' use of oil-based fuels, fertilizers, and pesticides became a heavy burden in the years that followed the first OPEC price rises of 1973. Energy production costs doubled in three years, while crop prices dropped off sharply from the record high grain prices of 1973 when wheat reached $4.41 per bushel, pushed up by worldwide crop failures and a massive purchase of U.S. wheat by the Soviet Union.

Predicting an unprecedented worldwide demand for U.S. grain, then Secretary of Agriculture Earl Butz urged farmers in 1974 to plant "from fence row to fence row." That year's massive harvest, which greatly exceeded demand, helped send prices down to pre-1973 levels. By 1978, the price of wheat had dropped to less than $2 per bushel. While international grain traders profited from the wild price fluctuations of the 1970s, many farmers found themselves caught in a disturbing cycle. In order to make a profit, farmers needed to achieve maximum productivity from each acre of land. This in turn led to an ever more intensive use of petrochemicals and fuel. But as farmers produced more and more grain, surpluses built up and prices dropped. And the more energy they used, the more money farmers needed to borrow to finance next year's crop.

All the irony of the situation was bitterly summed up by Ellen Estes, the wife of a Texas wheat farmer, who said, "We helped to create our own problem. We were asked to produce to feed a starving world, and we did."

Thousands of farmers were driven to bankruptcy in the closing years of the decade, their narrow profit margins having been increasingly squeezed by rising energy costs. In June, 1971 a single bushel of corn would purchase a barrel of OPEC oil, but by June, 1979 it took ten bushels to buy that same barrel. A fundamental shift in economic power—from U.S. farmers to OPEC oil producers—had begun.[1]

With a sense of frustration sometimes bordering on desperation, a few farmers began to experiment with making their own alcohol fuels. Many were aware of Brazil's far-reaching program to make alcohol fuel from sugarcane, which, like grain, had also hit rock-bottom prices. Some remembered the power-alcohol movement of the 1930s.

The first farm-fuel producers achieved varying degrees of success in their early efforts. Regardless of their actual technical accomplishments, they often became instant media celebrities, the subjects of a barrage of newspaper, television, and radio reports. These overnight folk heroes, with their colorful distilling contraptions and moonshine-guzzling tractors, provided an almost comic relief to an American public that had grown weary of hearing the nation's energy options framed in terms of multibillion dollar federal programs to develop nuclear power plants and synthetic fuels from coal and shale oil. Millions of Americans were intrigued by the idea that the fuel for their automobiles could be produced not only in the complex refineries of the oil industry, but also in stills built by farmers.

By the end of the decade, the movement for small-scale alcohol fuels production had spread across rural America with the speed of a wind-whipped prairie fire, sparking a rebirth in the ancient art of moonshining unequaled since the Prohibition era. The chemurgists' books detailing alcohol production processes were rediscovered and rushed back into print. Dozens of new "how-to" books, many of which were misleading and inaccurate, were rushed to press by small publishers. Thousands of people made lengthy pilgrimages to visit the sites of the first farm distilleries, sometimes paying as much as $100 for the privilege of simply touring the premises.

The oil industry, meanwhile, launched a new campaign to convince the American public that alcohol fuels were not an attractive energy alternative, even in the new era of spiraling oil prices and Arab embargos. The American Petroleum Institute released its summary of a 1976 report by six industry scientists who concluded that technical problems and costs for alcohol would prohibit short-term use. Over the long run, they said, "efforts to promote alcohols as synthetic fuels will prove to be misguided unless overall costs to the consumer compare favorably to those of gasoline derived from coal and shale."

The federal energy establishment mirrored the oil industry's viewpoint. The 1977 National Energy Plan developed by the Carter administration failed even to mention alcohol as a potential source of liquid fuel, while a major report on fuel alternatives asserted that processes favored by the oil industry would dominate.

AN UNLIKELY COLLABORATION: ALBERT TURNER AND ALBERT HUBBARD

One of America's first small-scale alcohol fuel plants was the product of a joint effort in Alabama between Albert Turner, a black civil rights worker, and Albert Hubbard, a retired white moonshiner. The success of the plant, built during the summer of 1977, could be measured more by the tremendous amount of favorable press coverage it received than by any new technical achievements. Even if the rickety accumulation of rusted trash bins and crudely welded, second-hand pipes failed to produce high-proof alcohol, it at least helped counteract the effects of the oil industry's new antialcohol campaign.

Albert Turner was the director of the Southwest Alabama Farmers' Cooperative Association (SWAFCA), an organization of about 2,000 black farmers created during the turbulent period of racial strife that swept through the South in the 1960s. In 1963, Turner left his job as a bricklayer in the small town of Marion to join a voter-registration drive sponsored by Martin Luther King, Jr. Turner worked as an organizer, spending long hours on the road arranging the marches, demonstrations, and prayer meetings that thrust the civil rights movement onto the front pages of newspapers across the country.

One of the cruelest reactions to the civil rights movement was the manner in which thousands of black tenant farmers who registered to vote were unceremoniously thrown off land they had occupied for generations. The cotton fields these farmers had toiled in were fenced off and planted in stands of southern pine, while the farmers often ended up in makeshift tent cities. Most of the older of these dispossessed farmers struggled to remain in Alabama, while many of their sons and daughters left for northern cities.

As credit and supplies for these farmers became increasingly hard to obtain from the white establishment, Turner and other civil rights leaders fought to organize community stores, credit unions, and small cooperatives to meet their most basic needs. In the winter of 1967, representatives from ten Alabama counties met to form SWAFCA. Some 800 farm families joined the new organization, naming Turner as one of the cooperative's three original directors. Despite vigorous protests from the Alabama governor's office, the federal Office of Economic Opportunity gave the co-op $400,000 to start its operations. Black farmers were urged to abandon the old system of one-crop cotton cultivation and shift to a more diversified planting system of vegetable cultivation that was better suited to smaller tracts of land.

SWAFCA was resisted by the white community at every step of the way. It was forced to market produce outside the state—a task not easily accomplished. One shipment of thirteen truckloads of cucumbers was repeatedly halted by Alabama state troopers for spot searches. When the cucumbers finally reached the market, they were spoiled. "Things started going downhill after that," Turner remembers. "We couldn't recoup the money we had paid out to the farmers for the cucumbers. White folks would talk to Black folks and try to get them not to patronize us. That started all kinds of rumors that our checks were no good. We was the plague."

In 1972, Turner was asked to direct SWAFCA, and he set out to change its sagging fortunes. SWAFCA's membership was rapidly aging, and many of the younger members were not eager to make a career out of the tedious, low-paying work of cultivating vegetables. Much to his dismay, Turner found himself spending most of his time trying simply to keep SWAFCA solvent. He had to fight a number of lengthy legal battles, leaving him little time to develop the new vegetable canning complex the co-op had planned.

By the summer of 1977, SWAFCA's fortunes had hit rock bottom. A fierce drought parched the state of Alabama, destroying 90 percent of the cooperative's cucumber crop, while a carcinogenic mold called aflatoxin ruined half the corn crop, making it unfit for human or animal consumption. Turner believed that one use for the damaged crops might be alcohol for fuel, and that a distillery could also be an outlet for surpluses the co-op accumulated from time to time. He was able to convince Randy Blackwell, an old friend from the civil rights movement then working in the Department of Commerce's Office of Minority Business Enterprise, to approve an $83,000 federal grant to cover the cost of building an experimental distilling system.

A rudimentary distillery was beginning to take shape on a concrete loading platform outside the cooperative's vegetable packing house when Albert Hubbard, a white ex-moonshiner, first visited SWAFCA. He stepped out of his car and hopped up onto the platform to scrutinize the junkyard collection of trash bins, gas storage tanks, and used pipes that had been joined together. He strongly doubted that the system would ever produce much high-proof alcohol. The next day, Hubbard returned to meet Albert Turner and offer a few suggestions to improve the efficiency of the still's design. Turner listened intently to the soft-spoken man dressed in a leisure suit and wing-tipped shoes as he gave a detailed critique of the system. After a moment's reflection, Turner offered him a job as a project consultant. "I don't have no PhD, but my education sure cost me more than a lot of people who got one in school," Hubbard said with a smile as he agreed to accept the challenge.

For more than twenty years, Albert Hubbard had kept the secrets of his past hidden from all but a few of his closest friends in Selma. He had been born on an 84-acre farm outside of Tuscaloosa, Alabama into a family of accomplished moonshiners who sold enough liquor to keep a steady supply of biscuits and meat on the dinner table during the lean years of the Great Depression.

According to Hubbard, his father was a virtual encyclopedia of information on moonshine lore, a true master of his profession who distilled a quality brew "as clear as mountain spring water and as smooth as the finest Jack Daniel's whiskey." He was a perfectionist who "never settled for second best and would rather make a few gallons less, and make it good, than try to stretch it." Hubbard and his father viewed with disdain the less scrupulous moonshiners who tried to squeeze every last gallon out of a batch of whiskey mash, often producing a rotgut liquor filled with impurities that could cause mammoth hangovers for those who drank it. Some moonshiners used sweat-drenched hats to filter out impurities, while others simply bottled the liquor as it trickled out of the still. A few even encouraged turkeys and chickens to roost in their mash boxes because they believed that the nitrogen in the fowl manure improved the flavor of the final product. Many still sites, according to Treasury agent Morris Stephenson, took on the look of a garbage dump after just a few weeks' operation.

Despite the often slipshod distilling techniques, moonshining was one of the most profitable businesses in the rural South during Hubbard's childhood. Illegal stills were stashed in secret locations all through the pine forests and foothills of the Appalachians. Some were hidden in holes in the ground with only their tops exposed, others were built on wheels for a quick change of location, and a few were concealed in hollows with their entrances carefully camouflaged by branches.

The more successful moonshiners, such as Hubbard's father, were often pillars of respectability in their communities, contributing generously to local churches and charities. Hubbard started out in the family business when he was still a young boy, by helping his father tote gallons of sorghum syrup through the woods to the still site. His first taste of moonshine came when he was only seven years old, after he discovered a bottle of "shine" his father had hidden underneath a hen's nest. "I got the bottle and drank about half a pint and all I remember is how sick I was. . . . I was up all night throwing up. My daddy didn't even whip me when he saw how sick I was, because he figured I'd been punished enough."

Hubbard followed in his father's footsteps, producing moonshine which he marketed through an extensive bootlegging network that stretched across central Alabama and into eastern Mississippi. He sold most of his liquor wholesale and tried to avoid the small street sales that could often lead to arrest. But eventually his exploits drew the attention of a Montgomery, Alabama Treasury agent who was determined to put an end to Hubbard's career. The agent tailed Hubbard's car for days along the desolate back-country roads where the nightly moonshine pickups were made, often stopping the car for quick searches. Eventually, the agent's persistence paid off. In 1959, Hubbard was arrested, convicted of distributing moonshine, and sentenced to prison. When he was finally paroled, he started a new career as a Pepsi-Cola distributor in Selma and later managed the Selma Elks Club, a favorite meeting place of the local white establishment. After purchasing the modest, red-brick building that housed the Elks Club, his life settled into a comfortable routine of bookkeeping, arranging weekly dinners, and tending bar. But Hubbard never lost his love for moonshining and was delighted with the opportunity the SWAFCA project presented to return legally to his first profession.

The months following the November, 1977 meeting between Hubbard and Turner turned into a whirlwind of designing, pipefitting, and welding. By December, the still was turning out 160-proof alcohol, giving SWAFCA a low-grade fuel that, except for the occasional cough and sputter, managed to power co-op trucks and tractors when mixed with gasoline. During 1978, Hubbard improved the system by adding a distillation column made from a piece of used steel pipe. Aside from this new wrinkle, added to bring the proof up to 180, the system was essentially the same as the one Hubbard once used to produce moonshine. He first created a mash consisting of ground corn and water, which was cooked over a wood-stoked fire in an old trash dumpster. Malted corn and barley then were added, since they contain enzymes that help break down corn starches into simple sugars. Then the mash mixture, still containing the corn's protein, was cooled to around 90 degrees and transferred to a 2,400-gallon fermenting vat made from an old natural-gas tank. There it was mixed with a yeast culture which multiplied rapidly and spread throughout the mixture. Yeast takes in simple sugar and creates waste products of alcohol and carbon dioxide that vigorously bubble to the surface of the mash. After about three days of fermentation, the process was usually complete. Turner would periodically thrust his hand into a fermenting batch to gauge the progress of the process. If it felt too sticky, there was still sugar to be converted to alcohol. When the process was complete, the mash was called "beer," and the alcohol concentration would reach about 5 percent, fairly low compared to other farm stills that gave 12 percent.

In April, 1978 Hubbard tested the system's column pipe that separated the alcohol from the water in the beer. He built a fire under the pipe column with slabs of white pine discarded from a nearby sawmill. Since alcohol vaporizes at 173 degrees Fahrenheit, and water vaporizes at 212 degrees, Hubbard could separate most of the alcohol from the beer by keeping the bottom of the column heated to the point of water vaporization and the top at the slightly cooler alcohol-vapor temperature. As alcohol-laden vapors traveled up the column, they cooled and recondensed on a series of perforated metal plates welded to supports at 4-inch intervals inside the pipe. Each time the vapors condensed, a higher proof vapor would travel upwards, boosting the final proof strength of the alcohol.

The vapor finally left the column and traveled through another pipe into an old car radiator submerged in a tank of water. The water would cool the radiator and the vapors to the point where the vapors condensed to liquid form.

The problem with this column was that it was not sophisticated enough to produce more than 75 percent alcohol, or 150 proof. Hubbard then recycled the once-distilled alcohol back through the pipe column to raise the proof. "This is history," an excited Albert Turner exclaimed as he tested the twice-distilled alcohol and found that it registered 180 proof.

A few hours after the first run of 180-proof alcohol was completed, a trickle of curious visitors dropped by SWAFCA headquarters to sample the new fuel. Albert Turner had notified his friends in the community that he would be willing to give out free samples of the brew to anyone brave enough to test it in his or her automobiles.

SWAFCA's first customer was the local minister, who drove his Chevy Impala slowly up to the concrete loading platform, stuck his head out the window, and examined the bizarre looking contraption with a bemused look on his face. Turner filled a gallon jar full of alcohol and carefully poured it into the Chevy's tank. "Now crank her up and see what she'll do," he said. As the minister drove away, Turner, beaming with satisfaction, turned to Hubbard and said, "The mass of people here still think it's a joke, but once the preacher gets hooked on the stuff, the whole congregation will go for it."

Alcohol produced at the SWAFCA distillery was blended with gasoline to form a weak-proof gasohol and was used in its pure form to power an old John Deere tractor which Albert Turner hauled to Washington, D.C. for a gasohol rally on the Capitol steps in 1978.

The rally turned into one of the gasohol movement's first media events as then-Democratic Senator from Indiana, Birch Bayh, uncapped a bottle of vodka and poured it into the tractor's fuel tank. The engine didn't cough or miss a stroke.

After Turner's appearance in Washington, a steady stream of journalists, including correspondents from the *New York Times*, the *Atlanta Constitution*, and *NBC Nightly News*, made their way to Selma to profile the SWAFCA distillery. Turner courted the media with great gusto, and they obligingly portrayed him as the gasohol movement's first folk hero. He was pictured in the press as a down-to-earth man with few pretensions who went ahead and put together a crude but functional distillery at a time when most Washington energy experts didn't believe such a thing could be done on a small scale.

The more retiring Hubbard was somewhat amused by all the hoopla that engulfed the SWAFCA project. Originally he had been afraid of being branded as a crackpot by the Elks Club members for helping SWAFCA build the distillery. But by the fall of 1978, he found he was being treated as something of a local celebrity.

The partnership between Hubbard and Turner proved to be fragile, and after all the media attention faded, the two men went their separate ways. Hubbard was somewhat annoyed by the fact that Turner was spending so much time promoting alcohol fuels and so little time improving on the actual distillery design. In the fall of 1978, he left the SWAFCA project and went to work building his own distillery.

Turner continued to spend a good deal of his time away from SWAFCA headquarters, appearing at gasohol rallys and lobbying in Washington for more funds. He was able to win strong Congressional support for his efforts to turn farmers into energy producers, and in 1979 SWAFCA was awarded a $300,000 federal grant to build another, larger distillery which could produce 260,000 gallons of 190-proof alcohol a year. Turner envisioned alcohol-fuel production as a powerful economic-development tool for black farmers and hoped it would provide SWAFCA members with a new measure of prosperity, which had always eluded them in the past. As the small-scale alcohol fuels industry expanded, Turner planned to introduce new energy crops such as jerusalem artichokes and sweet potatoes to make more intensive use of the SWAFCA members' farmland. Next to the distillery he foresaw the construction of a hog feed lot, which would be partially fed by the rich protein byproducts of the alcohol production process.

But for such an ambitious development plan to succeed, Turner was convinced that the small farmers would have to be able to maintain at least partial control of the fuel production facilities. "It's rare in a movement like this when a guy like myself ever really gets into the production end," he reflected in an interview at his home in Marion. "The big guys use me for all the criticism, they use me to work out all the kinks in the system and to fight with the Treasury agents, and once it really happens, some big twelve-million-dollar-plant goes up, and I go back to the fields. But I intend to have some kind of plant, and I don't intend to shut down the co-op project and forget about the future."

In December, 1980, Albert Turner obtained over one million dollars in additional federal loans and grants to help him realize his dream of constructing an integrated feedlot-distillery complex for the SWAFCA cooperative. It gave him an opportunity to become one of the "big guys" of the farm-based alcohol movement. But some federal officials inside the U.S. Department of Energy had serious reservations about the wisdom of granting SWAFCA a new aid package, for Turner's track record had become increasingly marred by failures. The new distilling system that replaced his first "moonshine" system operated sporadically and with disturbing inefficiencies. And two farm plants Turner had designed in South Carolina also had failed to perform to expectations. One federal energy official who greatly admired Turner's early pioneering efforts commented rather sadly, "Albert is a man of tremendous faith. He thought that the Lord himself could make that mash ferment. But things didn't work out that way. And now he resembles the first runner in a relay race. He's gone that first mile and then passed the torch on for someone else to carry. And if he can't manage to get his act together, the race will soon be all over for him."

The progress of the SWAFCA experiment in small-scale alcohol fuels production was closely monitored by the Treasury Department's Bureau of Alcohol, Tobacco and Firearms (BATF). A metering device was placed on the column to record the quantities of alcohol produced each month, and Turner was required to account for how each gallon of fuel was used. The Treasury Department was somewhat troubled by the notion that farm and even backyard distilleries might become commonplace, fearing that this might trigger a renaissance in the moonshining industry they had fought so hard to stamp out. They recalled that, during the height of Prohibition, enforcement had been a nightmarish task as moonshine stills of every conceivable size and shape had been put into operation. Some of these stills were small enough to fit inside the deep-fat frying well of an old electric stove, while others filled large city warehouses. In one single year during Prohibition, the BATF seized some 51,000 different stills and arrested 152,000 people for violating federal laws against liquor production.

MODERN MOONSHINERS PROLIFERATE

The modest success of the SWAFCA distillery inspired a wave of experimentation in small-scale alcohol production, resulting in thousands of requests for experimental distilling permits to the BATF. The avalanche of applications was unexpected, and the BATF staff slowly sorted through them and considered some kind of system to regulate the new grass-roots industry.

Lengthy delays in issuing permits caused some of the most ardent alcohol fuel enthusiasts to follow in the steps of the moonshiners and flirt with illegal alcohol production. Lance Crombie, a PhD microbiologist turned Minnesota farmer, was one of the first fuel producers to be taken to task by the Treasury Department for illicit moonshining. The story of the BATF's seizure of his solar distilling apparatus made headlines in local papers and created an instant martyr for the gasohol movement's faithful.

Lance Crombie was a man of wide-ranging interests who had served as an associate professor of pharmacy at the University of Minnesota before deciding in 1974 to try farming 600 acres of corn and wheat outside of Webster, Minnesota. In the winter of 1976–77, a spell of frigid weather sent the fuel-oil heating bill for his large brick farm house to over $450 a month, inspiring Crombie to devise a less costly way to keep his family warm. He began developing a solar heating system with an alcohol burning furnace for those days when the sun stayed behind the clouds. He designed a simple solar distilling system which produced a thin trickle of 60- to 100-proof alcohol. Eight to 10 gallons would accumulate on a sunny day.

Crombie's solar still was built out of a shallow plywood box with the inner surface painted black and lined with a heat absorbent black cloth. Liquid beer, produced in much the same way as at the SWAFCA distillery, was dribbled down this black surface. Alcohol would vaporize out of the beer as it was heated by the sun rays, condensing on a plastic sheet which covered the plywood box. By venting the box, Crombie controlled temperatures to separate alcohol and water, and a higher-proof beer dribbled down the cover sheet into a holding compartment.

Shortly after Crombie distilled his first few gallons of alcohol from the system, the local sheriff, accompanied by two Treasury Department agents, confiscated his homemade solar still. Crombie was outraged that the agents would dare to seize his valuable weapon for fighting the OPEC oil cartel. "We had a three-hour discussion, really going back and forth about the energy crisis, the cost of fuel, all sorts of things," Crombie later recalled in an interview published in *Mother Earth News*. "The agents got pretty hot under the collar, too, because I thought the whole situation ridiculous, and I kept needling them. I told the Feds, for instance, that I was going to have them charged with armed robbery. . . . after all, they had guns and were trying to take my property."[2]

The Treasury Department agents eventually decided that Crombie's solar still was too primitive to be classified as a true distillation system and thus could be operated without any type of experimental permit. A month after the system was confiscated, an indignant Crombie walked into the BATF regional headquarters in Minneapolis to retrieve his creation.

After the Crombie case was finally settled, the BATF realized that it was impossible to defuse the growing public sentiment favoring small-scale fuel producers. A new set of regulations, adopted in 1980, made it much easier to obtain permits for alcohol fuel production. Bonds for plants under 10,000-gallons-per-year capacity were removed, and the requirement that BATF inspectors place seals on distillery storage tanks and inlets from distilling columns was revoked. The Treasury Department instructed agents to assist rather than obstruct would-be stillbuilders.

Crombie emerged unscathed from his run-in with the Treasury Department, appearing even to relish the publicity it generated. He embarked on a vigorous national campaign to promote alcohol fuels with a fervor reminiscent of William Hale, a man Crombie greatly admired. Repeatedly, he was called before Congressional subcommittees to testify on the potential of alcohol fuels, and eventually he started a small consulting business. Crombie publicized his neochemurgic philosophy in *Mother Earth News*, which published a lengthy interview:

> Alcohol is the easiest to produce, least dangerous, most practical liquid fuel imaginable. It can be manufactured without a lot of expensive equipment and with very little specialized know-how. About all anyone really needs to learn is how to make mash, and people could just figure that out for themselves. . . . Just imagine. The American farmer has always overproduced everything he or she has tried to grow. However, if we put that farmer in the energy business . . . before you know it there will be so much fuel around that we'll have to have federal price supports for gasoline.[3]

Mother Earth News, a popular back-to-the-land journal of homesteading technologies that was born in the sixties, became a staunch supporter of alcohol fuels after the publication of the Crombie interview in 1978. The journal's founder, John Shuttleworth, asked Emmerson Smyers, the director of *Mother*'s research center in Hendersonville, North Carolina, to test out the feasibility of Crombie's design. Smyers soon found that the system, even with several modifications, did not work efficiently. So he began experimenting with a number of different designs adopted from moonshine stills. Eventually, Smyers developed a distilling column packed with marbles or plastic wadding which could produce from 2 to 6 gallons of 180-proof alcohol per hour with reasonable efficiency. Several of these systems could be built for less than $500 each out of used hot-water tanks, steel pipe, and standard plumbing fixtures, and be put together by a competent welder in less than fifty hours.

Shuttleworth found interest in his publication soaring as plans for these simple backyard distilling systems began appearing in the bimonthly editions of *Mother Earth News*. The plans were accompanied by detailed explanations of the basic steps involved in converting trucks and automobiles to operate on 100 percent alcohol. In the summer of 1979, Shuttleworth sent *Mother Earth News* staff members on a national tour with one of the backyard stills in tow to promote alcohol fuels as a solution to the energy crisis. At each stop on the tour, reprints of the Winter 1978 issue were passed out, complete with a cover manifesto proclaiming that Lance Crombie and his eighteen-dollar solar still could single-handedly save the family farm, convert the United States to renewable sources of energy, substantially cut the U.S. trade deficit, and reduce unemployment.

MORE AN ART THAN A SCIENCE

This kind of gung-ho boosterism, which was not confined to *Mother Earth News*'s promotion department, tended to oversimplify the task of actually producing fuel-grade alcohol from homemade stills. A more down-to-earth account of the trials and tribulations involved in backyard fuel production could be found in a book by Micki Nellis, entitled *Making It on the Farm*.

> Our own learning experience [Nellis worked on the still with her husband] started after we had read all the books, talked to the experts and seen the operating plants. After all that, we thought that we could go out and whip up a batch and have it turn out right the first time. Wrong! The process is still more of an art than a science. You have to develop a feel for it, a judgment, and you have to understand what's happening. The plant we put together is not efficient, not easy to operate and only produces about a gallon an hour. But instead of making a $10,000 mistake, we made a $200 mistake, so to speak. Instead of ruining a 3000 gallon batch, we ruined a 50 gallon batch.
>
> The entire plant is put together out of a 55-gallon drum and things you can find at any hardware store. . . . The first important lesson we learned is to start out in the morning. Don't think you will cook a batch after supper and get to bed before 3 a.m. It never worked that way for us, anyway. . . .
>
> On the first batch, we made a terrible mistake. We added all the water to the corn in the beginning. As a result, it took hours to heat the batch to boiling. . . . On the second batch, we didn't add enough water. The mash was impossible to stir with a 1″ by 2″ piece of yellow pine lumber, and the enzyme did not get good contact with the raw material. The bottom was also scorched. . . .[4]

Other early backyard efforts met with considerably more success than that of Nellis and her husband. Marc Cardoso, founder of the Tennessee Gasohol Commission, began making alcohol in 1978, utilizing waste fruit, corn, and fermentable table scraps. For $1,200 he built a system that produced 4 gallons per hour of 160-proof alcohol. The distilling column was comprised of 30-gallon drums stacked on end and sealed with a silicone substance. The column could be broken apart for cleaning whenever necessary. He went on to build seven more columns which produced over 6,000 gallons of fuel per month.[5]

Some of the early experimenters were not as candid as Nellis about the results of their initial efforts and were less successful than Cardoso. For example, a Minnesota farmer, prominently featured in the pages of *Mother Earth News*, developed a crude farm-scale system he claimed could produce 28 gallons per hour of 160 proof. But the system needed considerable improvement, and before he worked out all the kinks of his early design, he cashed in on his media fame to launch a consulting business, designing some ten farm-scale systems—most of which had serious technical problems. One prospective fuel producer purchased two distillation columns from him but reported that they operated so poorly that he eventually took them to a junk heap. A Nebraskan testified to spending more than $144,000 on a plant designed by the Minnesotan that produced a sum total of only 300 gallons of low-proof alcohol. With his business on its last legs, the consultant and propagator of these unsuccessful systems told a reporter for the *Wall Street Journal*, "I expect we'll go back into farming a lot sooner than a lot of people think."[6]

The failure of the Minnesota farmer to live up to his press reviews was so embarrassingly obvious that *Mother Earth News* editors felt obliged to print a brief notice warning its readers of his mistakes. Unfortunately, he was not the only individual to build defective small-scale stills. For every early success story there are at least two failures of equal magnitude. Such was the nature of this trial-and-error, farmyard engineering movement. The BATF reported in November, 1979 that a spot check of 275 stills in seven southeastern states revealed that they had produced a sum total of only 2,000 gallons of alcohol.[7] Although this report could not be taken entirely at face value, (many of the fuel producers may have purposely underestimated their production figures to the BATF) it was broadly indicative of the ineffectiveness of many of the early systems.

There was also a disturbing element of outright hucksterism surfacing in the small-scale industry as numerous hastily established firms began peddling colorfully named but poorly designed stills to help farmers slay the OPEC dragon. Some of these firms made extravagant claims that their stills could produce fuel for well under 50 cents per gallon, which could not be backed up by any experimental data. Others tried to defy the basic laws of science by declaring that their processing techniques could make a single bushel of corn produce 20 gallons of alcohol instead of the normal 2.5 gallons. As William Holmberg, an official with the Department of Energy, noted to a *Wall Street Journal* reporter, "Alcohol fuels represent a flag at the Alamo, around which farmers and others can rally to regain energy independence for rural communities. That level of enthusiasm and patriotism makes people vulnerable to snake-oil salesmen."[8]

Another problem facing the small-scale movement was how to make clear to new experimenters the potential hazards involved in alcohol fuel production.

For even at its most basic level, alcohol production required a sound grounding in industrial safety regulations. A carelessly tossed cigarette stub, for example, could ignite alcohol vapors seeping out from a poorly built column. And if pressures in the distilling column and boilers were not regulated with proper gauges and relief valves, dangerous explosions could possibly result. Even the seemingly innocent process of fermentation was not without hazard, for the carbon dioxide released by fermenting mash represents a potentially lethal gas. A South Carolina man became one of the first casualties of the small-scale movement when he fell into a fermentation tank and was overcome by carbon dioxide fumes. He died within minutes.

The first attempt to establish a rudimentary training and certification program in small-scale production techniques was established by Dr. Paul Middaugh, a South Dakota State University professor. Middaugh spent the winter of 1978–79 developing a plate-column still and accompanying system suitable for mass production. In May, 1979, while attending a "gasohol day" rally in Colby, Kansas, Middaugh met with Donald Smith, the president of Alternative Energy, Ltd., a small-scale still company, and Dr. James Tangerman, president of the local community college, to discuss training seminars for farmers in the basics of small-scale fuel production. Three months later, Colby Community College, situated on the high plains of western Kansas, became the first academic institution in the nation to offer a formal course in the ancient art of moonshining. Farmers, chemists, engineers, mechanics, federal energy officials, and the brewmaster from a large commercial distillery converged on Colby for the first high-powered seminar. A week of brainstorming produced a course manual and firm dates for a series of seminars in the fall.

In the months that followed, hundreds of students from as far away as California, Florida, and even Italy arrived at the tiny college to attend the seminars. Tangerman did his best to roll out the red carpet for participants, personally welcoming them to Colby and handing out certificates for successful course completion at a brief graduation ceremony. The Colby campus was suddenly filled with middle-aged students who talked excitedly about the miracles of alcohol during their rare breaks in the week-long course.

The weak link in the early Colby seminars proved to be the lack of actual hands-on experience in operating a small-scale distillery. The poorly designed system which the school initially used for demonstration did little to bolster the students' confidence in the state of alcohol technology. It was disturbingly ineffective in early trials. Bacterial contamination of the fermenting mash greatly reduced the alcohol levels of the final beer, and the system was difficult to clean. At one point, the still was extracting only one of the potential 2.5 gallons of alcohol from each bushel of corn that it processed. This problem was common with many early still designs and hardly confined to any single brand.

But the Colby seminars underwent a major overhaul in 1980 after Dr. Middaugh, dismayed by the poor performance of the still, withdrew his support from the training program. By the fall of 1980, the program had been strengthened and expanded to include a two-year program, and a special "hands-on" seminar where students learned how to operate a new 25-gallon-per-hour farm-scale plant donated to the college by a Texas manufacturing firm.

GENE SCHRODER DISCOVERS ALCOHOL FUEL

The Colby Community College program served as a model for forty other community college training programs established in 1980 with help from the Department of Energy. These schools brought academics, industry engineers, and farmers together under the same roof, making possible a much-needed interchange of ideas on improving farm-scale technology. The results were soon apparent as a new generation of more sophisticated farm stills began to appear in the Midwest.

One of the finest examples of these new systems was developed by Gene Schroder, a tobacco-chewing Colorado farmer who drew his initial engineering inspiration from a visit with Paul Middaugh in the winter of 1979. Remarkably, in one year Schroder, his father Derral, and brother Billy produced a highly efficient distilling system which set a new standard of excellence in the emerging small-scale industry.

The commitment to develop farm-scale alcohol systems evolved out of the Schroders' earlier leadership of the bitter wave of farm protests that swept rural America during the fall of 1977. With market prices for wheat less than $2 per bushel, Gene Schroder realized as he prepared to sow his winter wheat crop that the prospects were bleak for making any money from the spring harvest. In meetings that lasted late into the night, the family considered ways to keep the farm solvent. But they concluded that money could not be made with grain prices so low.

Instead of simply griping about the miserable economic outlook, the Schroders decided to take action. "We've got the power, the greatest power of any group in this country—the ability to produce food," Gene told his father, "and we'd best use it. We'll make our demands to the government for higher prices, and if the government doesn't meet these demands, we don't plant our fields."

A few days later, on the highway leading to Springfield, Colorado, the Schroders parked a tractor with a sign on it declaring, "This Farm On Strike." Other strike signs soon appeared on rural highways, and the Schroders' strike call proved to be one of the sparks that set off the most powerful farm protests of the seventies. For two years, tractorcades sporadically blocked streets in state capitals and in Washington, D.C., and at thousands of rallys farm strikers called agriculture officials on the carpet for low farm prices.

Eventually, when all efforts had failed to force the federal government into guaranteeing higher crop prices, many farmers turned to alcohol fuels as one possibility for revitalizing the American agricultural system. The Schroder effort resulted in a showcase model for producing alcohol on the farm.

The Schroder family's quest for energy independence was a natural outgrowth of the tradition of self-reliance typical of pioneer families. Gene Schroder's grandfather drove a wagon to southeastern Colorado at the turn of the century and settled in an isolated stretch of buffalo-grass prairie land a few miles outside the junction town of Campo. The Schroder clan initially prospered on the wide-open prairie, becoming one of the largest landholders in the region. The soil, enriched over the centuries by decomposed buffalo grass, was extremely fertile when first put to the plow and yielded bumper crops of corn whose husks were harvested and woven into brooms. But the Schroders' prime market collapsed in the late sixties when Mexican farmers, using cheap labor, started exporting broom corn to the United States. Derral Schroder traveled to Washington in an unsuccessful effort to obtain protective tariffs to offset the advantage of cheap labor in Mexico, but returned home bitter and disillusioned with federal farm policies.

The Schroders struggled to find new crops to grow on their vast acreage, eventually shifting into sorghum feed grains, winter wheat, and cattle pasture, which proved only marginally viable even in the best years. Despite the small returns on their farm operations, they profited from the increased value of their land holdings, and living costs were minimized by making good use of what the land provided. In a shed behind Gene's grandfather's house, the entire clan would gather to butcher the cattle fattened on buffalo grass, cutting out prime pieces of beef and grinding the rest into hamburger. The products of a long day's work would be divided equally among the clan's family households.

Gene left home after high school to attend the state university in Fort Collins and later veterinary school. He returned to Campo accompanied by his bride Laurie, the daughter of a South Dakota farmer. The couple moved into a small stucco house a few miles down a winding dirt road from Derral's homestead. It was shaded by a few windswept trees and surrounded on all sides by fields of sorghum and winter wheat—a peaceful place for much of the year. But in the spring, when the fields lay bare after the winter wheat harvest, the wind would often whip up the fragile, fine-grained soil into clouds of dust, darkening the sky and obscuring the sun.

Gene Schroder settled back into the familiar rhythm of farm life, keeping part-time hours as a veterinarian and reading political science and economic theory in his few hours of spare time. He viewed much of the nation's dramatic, postwar economic growth as a product of cheap food policy, which had allowed consumers to devote an ever-decreasing percentage of their total income to buying the weekly groceries. He believed this federal policy had resulted in a dangerous underpricing of farm commodities. Schroder was convinced that a strong national economy would not return until farmers received parity prices for their crops. The parity price of any given farm commodity was defined in the Agricultural Adjustment Act of 1938 as "that price which shall give the commodity a purchasing power with respect to articles that farmers buy equivalent to the purchasing power of such commodity in the base period." The base period referred to in the act was between 1910 and 1914, when farm products commanded high prices on the market, and farmers enjoyed a higher standard of living than ever before or since.

Schroder was wary of large corporate interests, which he believed conspired to block farmers from achieving parity prices. He was particularly suspicious of international groups like the Trilateral Commission, made up of prominent international businessmen, politicians, and academics. The 1976 election of Jimmy Carter, a former Georgia governor who was also a member of the Trilateral Commission, only served to heighten Schroder's mistrust of the federal government's farm policy.

Mistrust burst into open hostility a few weeks after Schroder first declared that his farm was on strike. He led a tractorcade of 4,000 farmers to Pueblo, Colorado to confront Agricultural Secretary Robert Bergland with a load of fresh manure and demands for parity pricing. In the months that followed the Pueblo demonstration, Schroder took to the road, speaking at rallies throughout the Midwest and urging farmers not to plant crops in the 1978 season. At each stop along the road, he helped to organize local chapters of the American Agriculture Movement (AAM), and soon tractorcades were being staged around the country. Tractors clogged major interstate highways, blocked the gates of meat-packing and grain-processing plants in the Midwest, and encircled the Capitol building in Washington, D.C.

Although the AAM appointed no official spokesmen, Gene Schroder and a handful of other articulate members took on responsibility for communicating the farmers' concern to the media, Congress, and the White House. A Washington headquarters was established in the Quality Inn motel a few blocks from the Capitol, and Schroder spent much of the winter of 1977–78 trying to push a "flexible parity" bill through Congress. Farmers felt that such legislation would give at least partial relief to their economic headaches. But the Carter administration defeated the flexible parity bill, and the farm strike began to disintegrate early in the spring as southern farmers backed away from earlier pledges to keep 50 percent of their acreage out of production. By planting from "fence row to fence row," they violated an agreement painstakingly negotiated by the AAM's diverse membership, in which each striking farmer promised to withhold 50 percent of his land from production. The cutbacks were aimed at driving up food prices without causing a world food shortage. But by the fall of 1978, it was painfully clear that the farm strike had failed, as only a handful of farmers stuck firmly to their winter pledges.

Instead of abandoning their struggle, Schroder and the other farmers who formed the leadership of the AAM decided to try a number of new tactics to increase farm income. The most important of these was the establishment of a farm-based alcohol fuels industry. The AAM leadership, like the earlier chemurgist movement, hoped that alcohol fuels would provide farmers with a means to remove surplus crops from glutted food and feed markets and stabilize prices at higher levels. The AAM strongly felt that the processing of crops into alcohol should be done by the farmers themselves whenever feasible, and not by oil companies, food processors, or chemical companies.

Don Smith, a Virginia representative of the AAM, spelled out the movement's strategy for alcohol development when he told the U.S. House Agricultural Committee in May, 1979

We believe we have learned enough to justify the conclusion that small and moderately sized plants can be as efficient, and in many cases more efficient, than large-scale plants. For example, by feeding the [grain] mash wet on the farm, the expense and energy necessary for drying [a step large commercial distilleries routinely include in alcohol production] can be avoided. . . . But because our primary motive is to preserve American family farms and a sound agricultural economy within which farming, and not land speculation, is the primary basis of earned income, we are not interested in [federal] programs which simply utilize agricultural commodities without providing farmers a fair return on their efforts. . . . We oppose incentives for an alcohol production program which is feasible only through the utilization of underpriced grain. We are not interested in supporting the development of a new industry for the benefit of those who already have family farmers over the barrel by virtue of their superior market leverage and oligopoly organization. . . . The AAM wants to see the production of alcohol fuel maintained as close to the source of basic feedstock resources as possible, either on the farms or in the small rural communities . . . where participation in the ownership and operations can be broadly shared as a means of restoring economic health and viability to rural America.

In February, 1979, after one more futile AAM tractorcade to Washington, Gene, Billy, and Derral Schroder returned home to Campo to build their first distillery. Farming on the Schroder land had come to an almost complete standstill, since it no longer made economic sense for the family to risk investing the large sums of money required to finance the yearly planting. "We went out and farmed the land, but it don't take much time to farm the way we do now," Gene Schroder dryly remarked as he drove through a sparse field of sorghum after returning from Washington. "We just went out and planted the damn thing, ran a cultivator through it once, and we go ahead and harvest it. If we make 10 bushels an acre [normal yields in that area were around 40 bushels an acre] well, fine. We didn't spend the money so we won't get burned. I am not going to be a slave to this system any more."

Schroder believed alcohol fuels could change this system by enabling farmers to send a product refined on the farm directly to consumers, cutting out the middlemen. Although some of his colleagues in the AAM were favoring the construction of 30- to 50-million-gallon-a-year distilleries that would require multimillion dollar investments, Schroder was convinced that small systems, producing under a million gallons of fuel per year, would prove to be the farmers' best option. "There is no way that farmers could ever finance a large plant," Schroder said cynically. "Even if they managed to get such a plant in operation, everything will be going great and then the price of corn soars to five dollars a bushel and they can't hack it. They won't be able to make the payments and the bank will come in, acquire control of the distillery and replace the board of directors. The new board will wait until the price of corn drops back down and then make a killing."

The Schroder family began their alcohol fuel project in the winter of 1979 with a visit to the farm-scale unit developed by Paul Middaugh in South Dakota.

Then, in the spring of 1979, they selected an abandoned barn and began buying equipment. Despite their lack of formal engineering training, the trio was well qualified for the task at hand. Gene Schroder's training as a veterinarian gave him a working knowledge of the microbiological processes so critical to successful fermentation. His brother Billy was an expert welder, and his father, Derral, had a natural mechanical aptitude, sharpened by decades of working with farm equipment.

The trio started out simply, using a small pot still made from a water heater to learn the basics of fermentation and distillation. After a week's practice with this system, they designed a set of distilling columns similar to Middaugh's. They worked twelve-hour days, seven days a week. In the evenings they would often dine on brown-bag suppers their wives brought to the barn, easing onto tattered old sofas in a corner to review the day's work. Much of the distillery design was developed through trial and error. One experimental column packed with fiberglass material proved to be a total failure, and the Schroders fell back on Middaugh's plate-column design. And an early cooking vat proved to be extremely inefficient and had to be scrapped.

When the Schroders' first complete system began operation in June, 1979, it churned out 300 gallons of 190-proof corn, wheat, and sorghum alcohol in twelve hours, a rate of 25 gallons per hour. By March, 1980, the plant had expanded and could convert 160,000 bushels a year of grain sorghum, southeastern Colorado's most economical feedstock, into 400,000 gallons of alcohol. A bushel of sorghum with a market value of about $2 could be converted into 2.5 gallons of 190-proof alcohol and stillage byproducts with a total value of well over $3. Total capital costs for the distillery construction stood at $400,000, including $89,000 for labor donated by the family. This translated into production costs of $1.20 per gallon of 190-proof ethanol.

The Schroder farm distillery was significantly larger than the SWAFCA unit, and included two 3,500-gallon cookers, three 7,000-gallon fermenting tanks, and three 16-foot-tall distillation columns. Two diesel-powered steam generators supplied the primary process energy for the distillery, which was designed to operate on a twenty-four-hour basis.

To start their alcohol production process, a cooker was half filled with 68-degree (centigrade) water and then 124 bushels of sorghum were added. Steam produced by the diesel boiler was then injected into the cooker, raising the temperature of the mash to 88 degrees, while enzymes were added to begin the starch breakdown process. After an hour and a half of cooking, the mash was pumped into a fermenting tank where additional enzymes and a concentrated yeast solution were added to the mixture. After thirty-six hours of fermentation, the beer was ready to be processed through the distillation columns into 190-proof liquid fuel. The protein-rich byproducts of the process were partially dried to a 65 percent moisture content and sold to local livestock growers as a valuable feed supplement, containing 28 to 30 percent pure protein.

The distillery combined heavily insulated columns with a series of simple but effective heat exchangers which captured low-grade heat and recycled it back through the system. This energy efficient design, which greatly reduced the amount of boiler fuels and electricity required to process grain crops into alcohol, made it possible for the distillery to produce substantially more liquid fuel than it consumed.

An independent investigation of the distillery's net energy balance by Dan Jantzen and Tom McKinnon, two scientists employed by the Solar Energy Research Institute, concluded that the system required only 38,849 Btu of process energy to produce 85,000 Btu of alcohol energy. Their energy study scrutinized the Schroders' unloading, milling, cooking, and distilling operation, as well as the energy required by agitators, pumps, and lights. The Jantzen-McKinnon report concluded that the distillery had a positive energy balance of over 2 Btu of alcohol energy for each Btu of process energy required. The report provided the first detailed, on-site analysis of the net energy balance of a well-built small-scale alcohol system, and helped to counter the repeated claims of oil industry, university, and federal government researchers that more energy was required to produce a gallon of alcohol than the alcohol could possibly deliver as a liquid fuel.[9] The publication of this report in May, 1980, proved to be a milestone in establishing the viability of the emerging small-scale alcohol fuels industry.* See table 3-1.

In the summer of 1980, the Schroders achieved another major breakthrough in farm-scale alcohol production—the distillation of 198-proof "anhydrous" (waterless) alcohol. This highly refined fuel was needed by the commercial gasohol industry to blend with gasoline, since gasohol distributors had been reluctant to buy lower-proof alcohol. In some cases, the higher water content of 190-proof alcohol will cause a separation of gasoline and alcohol blends, known as phase separation, which can cause problems similar to those motorists may have experienced with Agrol blends in the 1930s. The Schroders' success surprised engineers in the commercial gasohol industry who had doubted that farm operators would ever be able to master the more sophisticated distilling technology necessary to produce anhydrous ethanol. Removing the final few drops of water from ethanol had always been a much more difficult and energy-intensive task than producing the lesser-proof ethanol. It required the destruction of a strong azeotropic bond which a small percentage of water molecules form with alcohol molecules. Large commercial distilleries usually accomplished the dissolution of this bond by sending 190-proof alcohol through a final distillation step in a column filled with benzene gas, a dangerous substance that few farmers would wish to use. The Schroders, however, used a safer, less energy-intensive process which replaced the benzene distilling column with a column packed with a series of molecular sieves.

---

*Another report, contracted by the National Alcohol Fuels Commission in the summer of 1980, was to have studied the energy balance question in more depth. However, when consultants hired by the Commission attempted to explain to distillery owners their need to observe farm distilleries in action, they were told, in four out of seven cases, that the farmers were tired of highly paid consultants coming around and stealing their designs. When published, the report omitted energy balance questions. The consultants were not allowed into the distilleries.

TABLE 3-1
Process Energy Balance Calculations (Source: Dan Jantzen, Tom Mckinnon, *Preliminary Energy Balance and Economics of a Farm-Scale Ethanol Plant*, Solar Energy Research Institute, Research Report 624 699R (Golden, Colorado: Solar Energy Research Institute, 1980)

Input		Btu × 10⁶
Grain Input	6,957 lb/cooker × 2 cookers = 13,914 lb or 238.5 bu	
Cooking Energy Input	75 lb diesel × 2 cooks × 18,500 Btu/lb	= 2.78
Distillation Energy	398 lb diesel × 18,500 Btu/lb	= 7.36
Electrical Input	270 kWh × 11,600 Btu/kWh (29% conversion efficiency)	= 3.13
Grinding, Augering, Site Transport Energy	5 gal diesel × 138,700 Btu/gal	= 0.59
	Total Energy Input =	13.96

Output		
190-Proof Ethanol	481 gal × 0.95 × 85,000 Btu/gal	= 38.84
Stillage	11 lb(dry)/bu × 248.5 bu equals 2,733 lb	

Analysis:

Energy Requirements/Gallon of 190-Proof Ethanol	= 13,960,000 Btu / 481 gallons	= 29,020 Btu/gal
Net Energy Ratio (Processing Only)	= 38,840,000 Btu of Ethanol / 13,960,000 Btu of Process Heat	= 2.78

Note: This energy balance test was on a fermentation batch using a new supply of yeast, and the alcohol yield, according to the producer, Gene Schroder, was less than normal.

The installation of the anhydrous distillation column further enhanced the Schroders' reputation for innovative engineering. In slightly over a year's time they had managed to bridge the formidable gap between the primitive moonshine stills of the Prohibition era and those of the modern gasohol industry.

But the Schroders' technological triumphs of the year were tempered by personal tragedy. During the summer of 1980, Billy Schroder was killed in a freak accident. While welding together a new tower to house the anhydrous distillation column, he lost his balance and struck his head against a sharp, pointed piece of metal before hitting the ground. His death was a terrible loss to the close-knit Schroder family, but it did not weaken their commitment to farm alcohol production.

A FARM ALCOHOL INDUSTRY EMERGES

By the fall of 1980, less than three short years after the Treasury Department granted the first experimental permit to Albert Turner to fire up his SWAFCA still, the skeletal outline of a grass-roots alcohol fuels industry was beginning to form in rural America. It involved not just a handful of farmers as it had in the beginning, but thousands involving every state in the union. Over two hundred farm plants were actually under construction, and over 6,000 permits for fuel production had been granted by the Treasury Department.[10] In the Schroders' home of Baca County, Colorado alone, six farm plants were in operation and one community-sized plant with a production capacity of over a million gallons a year was nearing completion. Based on these figures, Baca County proclaimed itself the ethanol capital of America.

In Iowa, another state active in the development of small-scale plants, a corn farmer named Dennis Day developed a 250,000-gallon-a-year distillery that has introduced several new technologies to farm-fuel production. The plant resulted from an eighteen-month research effort begun in March, 1979 when Day and his son Wayne, a graduate of Iowa State University's chemistry program, first distilled alcohol from a tiny glass-tube still.

In the Day plant, corn is processed through a custom-built roller, then placed in a tank full of water where the corn starch is extracted. As the heavy starch particles sink to the bottom of the tank, the lighter bran and germ particles float to the top. This separation reduces the amount of corn to be distilled in the column, thus reducing overall energy requirements.

While Turner's first SWAFCA distillery had utilized an old dumpster trash-can mounted over a wood-fired furnace to convert corn starch to fermentable sugars, Dennis Day experimented with a more sophisticated extruder pressure cooker. Corn is cooked in the extruder at 250 degrees Fahrenheit for ten seconds at a pressure of 895 pounds per square inch and then mixed with water and enzymes. The corn-starch slurry is then pumped into a tank where the cooking process is completed and fermentation carried out. The fermented beer from the tank is then distilled in a towering, 58-foot column which is held at a 26-inch mercury pressure with the aid of a commercial vacuum pump. This keeps the pressure inside the column substantially less than normal atmospheric pressure, allowing both the water and the alcohol in the beer to vaporize at lower temperatures. With this vacuum system, less energy is required to heat the beer during distillation.

Day believes his plant is economic at production levels exceeding 100 gallons of fuel-grade alcohol per day. Below this level, the amount of alcohol distilled simply is not worth the hours of labor involved in its production. Day worked hard to minimize his capital and labor investments and to maximize the fuel production of his plant. "An alcohol plant must be profitable," he concluded. "If it isn't you'd better look to some other enterprise."[11]

The success of Schroder and Day has encouraged the cautious midwestern farming community to believe that one of the lowest priced fuels of the 1980s would not come from fossil fuels but from the conversion of under-priced grains and crop wastes into alcohol. Many analysts believe that the cost of gasoline will rise to above $2.00 per gallon by the mid 1980s. In 1979, the Department of Agriculture sponsored a report by Development, Planning and Research Associates of Manhattan, Kansas which indicated that farm-scale plants could produce 190-proof alcohol at costs of $1.13 to $1.63 per gallon. See table 3-2.

TABLE 3-2

Estimated costs of ethanol production for six highly efficient, well-utilized and well-operated plants. (Source: Manhattan, Kansas report by Development, Planning and Research Associates)

Cost Item	Pot Still	Small on-farm	Large on-farm	Small Community Wet	Small Community DDGS	Large Community DDGS
	—per gallon, 190 proof—			—per gallon, 200 proof—		
Direct						
Feedstock	1.04	1.00	1.00	1.10	1.10	1.10
Labor	.15	.20	.12	.06	.08	.04
Energy	.10	.10	.09	.14	.19	.19
Electricity	.03	.03	.03	.03	.03	.03
Supplies	.09	.09	.09	.09	.09	.09
Subtotal	1.41	1.42	1.35	1.42	1.49	1.45
Indirect						
Repairs, maintenance	.05	.07	.03	.03	.04	.04
Taxes, insurance	.06	.09	.04	.04	.05	.04
General admin.	.00	.00	.04	.03	.04	.03
Subtotal	.11	.16	.11	.10	.13	.11
Byproduct credit	(.25)	(.54)	(.48)	(.46)	(.46)	(.46)
Total operating	1.27	1.04	.96	1.04	1.16	1.09
Capital costs	.36	.30	.17	.17	.22	.19
Total cost	1.63	1.34	1.13	1.21	1.38	1.29

A farmer's actual cost of alcohol production will vary according to a number of factors, the most important being the size and efficiency of the plant, the cost of the basic feedstock, the distance in hauling crops to the plant and byproducts to the market, and the price the farmer can obtain for the stillage byproduct on feed markets.

Given this wide number of variables, a farmer must be careful in assuming that he can turn a profit by building an alcohol plant on his farm. A booklet published by the National Alcohol Fuels Commission cautions farmers, "The financial considerations involved in on-farm fuel ethanol production are fundamentally important. Both tangible and intangible costs must be considered. Assigning a value to tangible costs—capital investment, feedstock, energy, labor—requires a forthright, cold-eyed analysis. Assigning a value to intangible costs and benefits (such as the value of being more energy self-sufficient) is more difficult and is eventually based upon subjective judgments." (See Appendix B on the economics of alcohol.)

The booklet profiled the operations of four farm-scale plants with production capacities of 15,000 to 40,000 gallons per year, concluding that "there is presently no small-scale ethanol production system that is mass-produced, fully warranted, and independently tested to verify the manufacturer's claims. Current small-scale producers are pioneers in the truest sense of the word."[12]

While it was true that by the fall of 1980 the small-scale industry was still in its infancy, an increasing number of efficient backyard, farm-scale systems were being marketed by reputable engineering firms which had undertaken extensive research and development programs. And the promotional claims of these new manufacturers received careful scrutiny from farm organizations. In July, 1980 the Iowa Corn Growers Association (ICGA) sponsored an "Iowa Corn Cookoff" at the Iowa State Fairgrounds, giving midwestern farmers a chance to evaluate eight different small-scale systems. According to *Farm Energy* magazine, "The eight stills represented at the 'Cookoff' varied greatly in cost and capacity—from a small unit producing just a few gallons a day priced at less than $400, to a much larger unit with a 20-gallon-per-hour capacity priced at $80,000." The ICGA provided the corn, water, and electricity for all eight of the still systems and officials from the U.S. Department of Agriculture periodically checked the output of the distilling columns to determine the rate and proof of the alcohol.[13]

In October, 1980 the American Agriculture Foundation, an organization formed by striking farmers to promote research and development of small-scale alcohol fuel systems, sponsored a three-day market in Denver for equipment manufacturers to display their wares. Over thirty different manufacturing firms participated. The $10 admission fee entitled the prospective alcohol fuel producer to view a wide range of products, including process control equipment, stillage dewatering systems, boilers, and different distilling systems set up outside the exhibition hall. The manufacturers who participated in the market were carefully reviewed by a panel of American Agriculture Foundation members which included Gene Schroder. After the market ended, this panel went to work developing the first set of industry-wide certification standards for small-scale manufacturers. The efficiency of farm-scale units is likely to increase as the technology becomes more refined and producers become more fully integrated in the various facets of agro-energy production. Some farmers, including the Schroders, are already working in this direction by researching the use of methane gas produced from cattle or pig manure as the primary source of fuel for their boiler systems. Combined with the advantage of on-site feeding of distiller's grains, some farmers may find the older model of an integrated farm—with various forms of livestock and crops—more economically attractive than one-crop agriculture. Another key to efficient use of capital may be cooperative ventures that can sit idle without soaking up all of a farm's available capital should gasoline be plentiful and market prices for grains improve substantially. Co-op investors could then participate in fuel production or send crops straight to market as prices vary.

Other alternative forms of energy could bring distillery costs down. Methane gas can also be produced from plants cultivated on leftover slops in the distilling process. Solar preheating of cookers and distilling-column steam could lower energy costs, as could waste wood from lumbermills and even coal.

The extent to which the small-scale producers will be able to expand their fuel markets beyond the farm is still in question. To a large extent, it will depend on their ability to compete with—or at least peacefully coexist with—large-scale producers who are likely to dominate the alcohol market. These large-scale distilleries have been built in some of the most productive agricultural regions of the nation. They have the backing of powerful financial institutions and benefit from sophisticated research and development programs which will have a profound impact on the shape of the emerging alcohol fuels industry.

Chapter 4
Rebirth of the Power Alcohol Industry

In February, 1978 the first gasohol pump in the nation was opened by a small, independently owned service station in Lincoln, Nebraska. By the time Gene Schroder produced his first gallon of farm alcohol in June, 1979, some 1,200 midwestern service stations were marketing this modern reincarnation of the chemurgists' power-alcohol blend. And by the fall of 1981, over 10,000 service stations in fifty states were offering gasohol to their customers.[1] A major national gasohol market had been established in less than three years.

Gasohol's rapid rise to prominence was aided by a generous 4-cent-per-gallon exemption from the federal highway tax. This meant that one gallon of anhydrous (pure) alcohol that a distillery sold for $1.60 would, when blended with nine gallons of gasoline, benefit from a 40-cent tax saving, costing the motorist only $1.20. But even with the aid of the tax exemption, gasohol usually sold at prices 4.5 cents to 12.9 cents above unleaded gasoline during the late 1970s.[2] Only in states which included an exemption from their own state highway taxes did gasohol compete head-on with unleaded gasoline.

Gasohol was marketed in much the same way as the Agrol blends of the 1930s. It was advertised as America's homegrown fuel which would both strengthen the farm economy and deliver greater power than gasoline alone. And modern promoters were able to add a powerful new dimension to the chemurgists' traditional sales pitch by stressing the amount of imported OPEC oil that could be saved by using gasohol. By the end of 1979, gasohol sales rose to half a billion gallons a year, a small fraction of the 100-billion-gallon-per-year gasoline market.*

The arrival of gasohol at local service stations was almost invariably greeted with an outburst of publicity. In Wyoming, an Associated Press reporter wrote of a "gasohol craze" sweeping the state and noted that "the ABC Texaco in

---

*This represented about 50 million gallons a year, or 10 percent of the total, in terms of pure anhydrous alcohol.

Torrington has had scores of new customers since it started selling gasohol.'' In Topeka, Kansas, the Highway Oil Company teamed up with a local radio station to offer the first 100 gallons of Kansas gasohol for the bargain basement price of 14.5 cents a gallon. This publicity gimmick set off an early morning traffic jam as motorists converged on the company's lot. The scene, complete with television camera crews, politicians shaking hands, and promoters passing out leaflets and pausing for interviews, was repeated across the Midwest.[3]

A 1978 Iowa Corn Promotion Board survey found that service stations with the new gasohol pumps experienced a 200 percent average increase in sales. This was due partly to motorists' support for farm fuels and partly to the fuel's higher octane and higher performance as compared to unleaded gasoline. Sales in Iowa jumped from 600,000 gallons in November, 1978 to 5.6 million gallons in March, 1979—a 933 percent increase.[4]

Buoyed by strong midwestern sales, gasohol headed east to New York City. A suburban service station opened the first pump in May, 1979, and the *Daily News* editorialized: ''Gasohol is out to make its official entry into the New York market. . . . We couldn't be more pleased. We tested gasohol in a *News* car last year. In 3,000 miles of gasohol driving the car's exhaust was clean and its mileage improved.''[5]

LARGE-SCALE GASOHOL INDUSTRY

Gasohol's overnight success was not so much the result of any market strategy as it was a social phenomenon. And it proved to be a powerful stimulus to the development of a large-scale alcohol fuels industry whose output immediately dwarfed that of backyard and farm distilleries. While a host of farm cooperatives, community development corporations, and newly formed gasohol corporations made plans to build large-scale distilleries, it was the powerful midwestern grain processors that soon emerged as the pacesetters in the field. The grain corporations could produce alcohol from the starchy waste streams of their grain refineries, giving them a substantial edge over would-be competitors who lacked the sophisticated process technology.

The first and most successful of the grain processors to venture into fuel production was Archer Daniels Midland, whose Decatur, Illinois plant distilled more alcohol in its first day of operation than Gene Schroder's early farm plant distilled in a year. ADM's 1979 output of 54 million gallons of fuel alcohol accounted for 80 percent of all the gasohol blending stock used that year.[6]

Company historians trace ADM's birth to a small Decatur flax mill opened by John Daniels in 1902. During the Depression, the Daniels mill was bought by Commander Larabee, then the third-ranking flour milling company in the nation, and the operation rapidly developed into a major power in the grain processing industry. Under the leadership of the Daniels and later the Archer families, ADM steadily expanded its grain processing facilities and set up a chemical division to produce synthetic fertilizers and pesticides. But in the late 1950s, the corporation, still under the tight control of the two families, entered a period of stagnation. In the early sixties, ADM found itself on the verge of financial disaster and was unable to cover stock dividends for three consecutive years.[7]

In February of 1966, the grandson of ADM's founding father turned to Dwayne Andreas, an Iowa born chemurgist, to inject some new life into ADM's aging corporate structure.

Andreas was born in 1918 into a family of Mennonite farmers who lived outside the small town of Lisbon, Iowa. The teachings of the Mennonite religion, which emphasize self-sufficiency and self-denial, resulted in an austere upbringing for the young Andreas; one devoid of such things as colored clothes, jazz music, Sunday-afternoon baseball, and gasoline powered tractors.

All this changed dramatically when Andreas's father was invited by the local Lisbon banker to move into town and take over a bankrupt county grain elevator and feed business. Suddenly, the quiet and solitude of early twentieth-century farm life was replaced by the often frenetic wheeling and dealing which accompanied the task of brokering farm commodities. When a farmer wanted to buy soybean meal for his hogs or sell grain to raise some cash, it was the Andreas family to which he turned. Dwayne Andreas took an immediate liking to his father's new profession; there was never any question in his mind that he wanted to be in the grain business. And the new business prospered as Dwayne and his five brothers mastered the intricacies of farm markets at an early age.

The rapidly expanding business included a new feed plant and a soybean crushing operation by 1938. At the time, soybean was still a minor crop in the United States, but Henry Ford's Dearborn laboratory was beginning to explore the enormous potential of this protein-filled oriental bean, developing soybean livestock feeds and experimenting with soy-based soups, milk, ice cream, meats and plastic automobile shells. This work was noted with great interest by the Andreas family, which was convinced that the soybean market would expand dramatically in the decades ahead.

Dwayne Andreas's first exposure to the power-alcohol movement occurred in 1933, when as a high school student, he read an article written by the renowned British scientist, Julian Huxley, predicting energy plantations would be established in the equatorial latitudes to produce huge quantities of biomass materials that would be converted into alcohol and shipped around the world. In an interview with David Susskind, Andreas recalled being "very much impressed," even "indoctrinated," by Huxley's article, and how he had written a report about it for his high school class.[8] Much later in his life Andreas's interest in power alcohol was revived in a meeting with Joe Pew, one of the founders of Sun Oil Company. Pew was something of a maverick in the oil industry who wholeheartedly supported alcohol fuel development and headed up the Chemurgic Council, an organization formed to carry on the work of the early chemurgists. Pew told Andreas that he was going to retire from the Chemurgic Council and wanted Andreas to serve as its new president. "I am totally distressed with the idea that farmers in the United States . . . are slaughtering their horses and they're going to be dependent on gasoline," Pew told Andreas. "In about twenty or thirty years it isn't going to be here, and it's going to be absolutely necessary to make fuels out of crops. . . . I know because you're a Mennonite . . . you will understand this.''

''He asked me if I would take that chemurgic technology and keep it alive,'' Andreas told Susskind. ''I accepted the challenge . . . and kept it alive and improved it.''[9]

By the time Andreas met with Pew, he was well on his way to establishing a dynamic reputation in the world of agribusiness. He had moved on from his family enterprise to work for Cargill, the powerful Minneapolis-based international grain trading company, and later the Grain Terminal Association, a large midwestern farm cooperative. Andreas put the experience gained from his early involvement in soybean processing to good use and proved to have a genius for determining when to buy and sell soybeans in the volatile commodity markets. When Andreas took on his new job as executive vice-president of ADM, oil prices had yet to begin their sharp escalation and alcohol fuel was not a profitable enterprise. So he focused his attention on finding a method to market texturized soy protein, a versatile meat substitute developed by Paul Atkins, a food processing engineer who had started his career in Henry Ford's Dearborn laboratory. ''It was Dwayne Andreas who had the vision to see that . . . my work had a place,'' the now retired Atkins remembers. ''It was he and only he who promoted it down the road to get the patent.''

Andreas eventually found markets for the textured soy protein in over four hundred different products ranging from artificial bacon bits to high-protein soy cookies. His high-powered management techniques quickly won the respect of ADM's Chairman of the Board John Daniels, who appointed Andreas chief executive of the corporation in 1971 and declared that ''having been involved with Dwayne Andreas had been the most stimulating experience of my life.''

ADM's fortunes soared along with the soybeans in 1973 when a worldwide surge in demand drove soybean prices up from $3.46 to $12.90 a bushel in eight months. When *Fortune* magazine profiled Andreas during that heady period, it found that he ''. . . felt the touch of glamour that, at least in the eyes of outsiders, quickens the lives of conglomerators, movie magnates, and oil wildcatters. For the whole world suddenly seemed to go wild over soybeans, clamoring for the protein-packed little globules with a passion once lavished on the search for pepper or gold.''[10]

Andreas promoted soybeans with near-missionary zeal. A 1973 ADM annual report opened up a vast field of soybeans with a transparent overleaf proclaiming that the corporation was taking upon itself the noble mission of upgrading cereal proteins with soybeans in order to fight world hunger. To fulfill this mission, Andreas kept up a round of business meetings in the U.S. and abroad. During the 1970s, under his guidance, ADM emerged a powerful multinational organization, capable of competing with giants of the grain trade, such as Cargill, for international markets. ADM's corporate empire grew to include a chain of grain elevators stretching from Colorado to Wisconsin, which purchased grains and soybeans that would be processed by a dozen ADM mills strategically situated around the Midwest. A fleet of ADM-owned river barges carried refined food products down the Mississippi to southern ports, where ADM's international export company shipped them out to overseas markets. In Lincoln, Nebraska an ADM subsidiary produced spaghetti, macaroni, corn meal, pet foods, frozen convenience specialties, and packaged prepared dinners for U.S. supermarket chains.

Much of the capital needed for new business ventures could be obtained from the National City Bank of Minneapolis, a bank in which 95 percent of the stock was controlled by ADM. The corporation's overseas holdings included a soybean extraction plant in the Netherlands and a factory in Burnaby, England which produced equipment for European bakeries. In 1978, net sales for ADM topped $1.8 billion and the price of its stock had jumped four times above its 1968 levels. ADM had grown to the point where it was one of the largest diversified grain and soybean processors in the world.

As Andreas's influence in the world of agribusiness expanded, powerful politicians frequently turned to him for advice on establishing a national agricultural policy. Andreas was close to Hubert Humphrey and sometimes accompanied the senator on trips abroad. He was a major contributor to Humphrey's losing 1968 presidential campaign, but in 1972 he hedged his bets. While he contributed $75,000 to Humphrey's early primary efforts, he also contributed $25,000 to Richard Nixon's reelection campaign, in part because he was encouraged by Nixon's dramatic rapprochement with China, which would open vast new markets for ADM products."[11] (He was later embarrassed by Watergate disclosures which indicated that his cash contribution to Nixon had ended up in Watergate burglar Bernard Barker's bank account.)

Andreas's golden boy image in the corporate world was also tarnished in 1975 when ADNAC (an ADM subsidiary owned in partnership with Garnac, a large trading firm) and several other shipping companies were indicted in New Orleans on charges of illegally misgrading and diluting grain shipments to foreign countries. Dan Morgan, a veteran reporter for the *Washington Post*, sketched the dimensions of the scandal in his book, *Merchants of Grain*:

> As the Assistant U.S. Attorney Cornelius Heusel brought witness after witness before the New Orleans grand jury, evidence of the companies' misconduct grew. All kinds of activities came to light that showed how company employees schemed to misgrade or diminish the quantities of grain destined for foreign countries.
>
> These indictments painted a morbid picture of the everyday activities of the grain companies at the grass-roots level. While wealthy merchants and traders in well-tailored suits entertained foreign customers at expense-account restaurants in Geneva, Paris, and New York, employees of the companies down on the Mississippi River were rigging scales and tampering with the inspection samples so they could slip these same customers "junk" or less grain than they had ordered.[12]

Despite the tawdry aspects of the grain business which Andreas was sometimes forced to confront, he continued to take a tremendous amount of pride in ADM's accomplishments which he viewed as carrying on in the chemurgic tradition. However, his interpretation of chemurgy differed markedly from the original chemurgic philosophy articulated by William Hale, who abhorred the international grain trade. With cheaply priced U.S. grains readily available around the world, Hale believed, it would be difficult for many nations to develop self-sufficient domestic farming systems. Andreas, in stark contrast to William Hale, was a vigorous promoter of the grain export trade, and worked constantly to expand foreign markets for refined U.S. agricultural products.

ADM's successful penetration of foreign markets in the 1970s was aided by the grain surplus. The corn market was particularly glutted during much of the seventies, with prices dropping at one point as low as $1.50 a bushel. To take maximum advantage of this abundant corn supply, ADM developed a sophisticated wet milling process which could produce a wide variety of products from a single 56-pound bushel of corn. Each bushel sent through the Decatur, Illinois plant produced enough corn oil to make 1.5 pounds of margarine, 11.9 pounds of high-protein corn gluten, and meal used in bakeries and as an additive to poultry feeds.* The remaining corn starch was either sold for industrial use or experimentally converted into a light fructose sugar which Andreas hoped would someday replace cane and beet sugars in soft drinks and candies.

When sugar prices in 1973 jumped from 12 cents to over 60 cents a pound on world markets, ADM was ready to offer corn fructose syrup to industrial sugar users anxiously searching for cheaper substitutes. Demand for the corn sweetener was strong for a brief period and then slackened after cane prices fell to new lows in 1977. ADM decided to use part of the corn sweetener as a feedstock for producing 190-proof beverage-grade grain alcohol. But the company had little success with the liquor business, as marketers who brand, blend, and distribute liquor shied away from ADM's new venture.[13]

ADM moved quickly into the gasohol market with its surplus beverage alcohol by upgrading its 190-proof level to 200-proof anhydrous alcohol for fuel. By March, 1979 the entire output of the Decatur plant was being soaked up by the mushrooming new market. The plant went into twenty-four-hour-a-day operation, its smoke stacks sending clouds of sweet-smelling vapors billowing into the air. A constant convoy of trucks hauled grain in from some of the most fertile corn-growing land in the nation, occasionally leaving behind a smattering of grain for the small flocks of birds that hovered near the loading docks.

ADM initially priced anhydrous alcohol at $1.30 a gallon, which left a healthy profit on their production costs of less than $1.00 a gallon. But as demand for gasohol soared far beyond the available supply, ADM began escalating the price of its alcohol, charging just under $2.00 as demand peaked, but slipping back to $1.75 in early 1981. One cynical gasohol industry observer remarked, "They are scalping the market. . . . They were losing their shirts in the beverage industry and now they're getting their money back."

---

*A bushel of corn weighs 56 pounds. In processing, it yields 9.2 pounds of gluten feed; 2.7 pounds of gluten meal; 3.5 pounds of germ from which 1.5 pounds of margarine or corn oil can be realized; and 31.5 pounds of starch which is used to produce 2.5 gallons of pure alcohol and 17 pounds of carbon dioxide.

Andreas consistently denied charges of price-gouging, insisting to David Susskind in the fall of 1980: "We lost about thirteen million dollars trying to get into the business before we could get the volume up enough to cover our costs and our overhead. We haven't made that back yet. And right now alcohol is only marginally profitable because corn prices have risen, due to not quite as good a corn crop as we'd like to see. . . . But I do believe, as we get our volume much greater, that we will . . . make a decent return on our investment."[14]

ADM's plunge into alcohol fuels production, and its growing profit, was closely watched by other food processing corporations that had been extremely reluctant to test this unusual new market themselves. But as the size and profitability of the gasohol market grew, a few decided to try it. In early 1980, Staley, ADM's major midwestern competitor, announced the construction of a 60-million-gallon-per-year corn distillery and milling plant in Tennessee. This announcement was followed by another, as the grain trader CPC International disclosed it would build a 50-million-gallon-per-year plant in partnership with Texaco Oil Company in Illinois. And National Distillers, a large Illinois industrial beverage and industrial alcohol firm, announced a joint venture with Standard Oil Company of Indiana (Amoco).

The new partnerships represented a turning point in the gasohol industry as oil companies like Texaco began to compete for a slice of the expanding new fuel market.

THE OLD AGROL PLANT

Although ADM controlled most of the early gasohol market, one competitor supplied fuel alcohol as early as the spring of 1978. This new fuel alcohol producer was neither the offshoot of a rival food processor nor an oil company, but the chemurgists' Agrol plant in Atchison, Kansas. It had come under the management of Midwest Solvents, a beverage-alcohol distilling firm founded by a Detroit investment banker named Cloud Cray. The Chemical Foundation had put the idle plant up for sale just prior to World War II, and Cray planned to tear it down and move it to Trenton Valley, Michigan, where it would produce alcohol for the antifreeze market. But after Japan's surprise attack on Pearl Harbor, the federal government ordered Cray to keep the plant in Kansas and begin alcohol production immediately to aid the war effort. After the war, the federal government withdrew price supports for the wartime alcohol industry and many distilleries went bankrupt. The Cray family managed to keep the Atchison plant solvent by refurbishing it with equipment from other bankrupt plants and shifting into production of beverage alcohol.

In the 1950s the Atchison plant gained a new reputation as a respected provider of neutral spirits (a highly purified beverage alcohol), vodka, and gin. Distillery byproducts included fusel oil (used in paints), distillers dried grains (a yeasty high-protein stillage residue) and carbon dioxide for soft drinks. A wheat mill next to the distillery extracted a protein gluten used in bread baking and an industrial-grade starch used to make charcoal briquets, foundry castings, and wallpaper pastes.[15]

The Cray family gingerly moved the old Atchison plant back into fuel production, upgrading a small portion of the plant's 20-million-gallon-a-year output to anhydrous fuel-grade alcohol. But Cloud Cray, Jr., who had taken over active management of Midwest Solvents in 1975, viewed the gasohol boom with mistrust. He was convinced that alcohol was too valuable a liquid to be used as a fuel, and should be reserved primarily for use as a beverage and for industrial needs. He also realized that the gasohol market, which seemed to be such an attractive outlet for alcohol when grain prices were in a slump, might quickly dry up after a year's drought when grain reserves dwindled and prices soared. A distiller who did not have alternative markets for his alcohol, Cray believed, might soon face financial disaster. Over the long term, Cray was convinced that "gasohol simply does not have a prayer of . . . success until all fossil fuels in this country and around the world are practically depleted." He lashed out at the political pressures brought by gasohol proponents, which ". . . could cause highly subsidized traditional gasohol plants to be built at tremendous loss to our federal and state economy, and, of course, eventually for the farmers as these noneconomic plants fail."[16]

Despite the bleak predictions, Cray was quite willing to take advantage of the business opportunities opened up by the gasohol market. He installed a new distilling column in the Atchison plant to upgrade beverage alcohol into anhydrous, and he directed Midwest Solvents' marketing division to find outlets for the plant's monthly production of 200,000 gallons of fuel alcohol; he even obtained a federal loan guarantee to build a new 10-million-gallon-a-year distillery. Cray's success in obtaining the federal support enraged the increasingly powerful Washington gasohol lobby, which felt that Cray, one of the industry's most outspoken critics, should not have received the coveted loan guarantee. U.S. Department of Agriculture spokesmen said at the time that they felt "more comfortable" with Cray than with other applicants, and Cray defended his record, insisting that he was not so much a critic as a realist. He believed that it made no economic sense to build new distilleries geared solely to fuel production and insisted that the new distillery must be able also to produce beverage alcohol.

THE EAST COAST GASOHOL INDUSTRY

By the winter of 1979 demand for gasohol had far outstripped supply in the Midwest, and jobbers anxious to market the new fuel on the East Coast found shipping costs from the Midwest nearly prohibitive. Few gasohol industry observers were surprised when Publicker, a major eastern distiller, decided to move into fuel-grade alcohol production. Publicker owned a large Philadelphia distillery built during World War II for the synthetic rubber program. After the war, Publicker produced beverage spirits for Old Hickory Bourbon, Inverness Scotch, Amaretto Liqueur, and others until it closed in 1978 after losing $4.5 million. The company hoped to make the Philadelphia distillery prosper once again by converting at least a portion of its enormous total capacity of 60 million gallons a year over to anhydrous alcohol. Company officials found, however, that it would be difficult to purchase enough corn from nearby Pennsylvania farmers to keep the plant running at anywhere near full production. And even when the plant could obtain corn, it lacked ADM's sophisticated refining technology, which played such a large role in the Illinois plant's overall profitability. Publicker's search for an economical feedstock led it to the U.S. Department of Agriculture, which was storing nearly a million tons of raw cane and beet sugar. Under the terms of a complicated domestic price support program, the USDA had been left holding the bag when the bottom fell out of the sugar market. The sugar was stored under tattered tarpaulins and inside leaky warehouses in sites scattered around the country.

In March of 1979, Publicker offered to buy the federal surplus sugar for 8 cents a pound, a little more than half the going market price. It wasn't a particularly generous offer, but there were few other buyers interested in the government's rapidly deteriorating sugar. "We are prepared to put our plant into use for the production of alcohol as part of the present drive to make the U.S. more self-reliant with respect to fuel," Publicker president Robert Leventhal wrote Agriculture Secretary Bob Bergland.[17]

Foreign sugar import quotas legislated by Congress had been slackly enforced by the Carter administration, partly due to pressure from Latin American nations that relied on sugar exports for a major source of income. Cheap imported sugar had been bought up by large industrial sugar consumers such as the Atlanta-based Coca-Cola corporation, which had firmly supported Jimmy Carter's election campaign. This foreign sugar had resulted in a surplus of the higher priced U.S. sugar which the USDA had been forced to buy. Cynical sugar-state senators, headed by the powerful Louisiana Sen. Russell Long, suspected the Carter administration was somehow deliberately tilting its sugar import policy to benefit Coca-Cola, but allegations that briefly surfaced in the press were vigorously denied at the White House.

The Publicker proposal received strong support from the sugar-state congressional block, which hoped that a sugar-based U.S. gasohol industry would provide a new market for their growers. Idaho Sen. Frank Church charged that foreign imports had reduced his state's beet sugar industry to a "shambles," and Florida Sen. Richard Stone declared: "I intend to press this concept until it becomes reality." But Bergland stubbornly refused to give in to the sugar interests and told inquiring reporters, "I think we can do better."[18]

The collapse of the federal sugar deal did not dampen Publicker's enthusiasm to enter the gasohol market. Vice-President Irwin Margiloff, in charge of Publicker's gasohol program, realized that the production of fuel alcohol from U.S. sugar would prove to be marginally successful over the long run; with Brazil converting ever-increasing amounts of its sugar cane into fuel, the volatile world sugar market could once again quickly go from glut to scarcity, causing world sugar prices to repeat their 1973 spiral. With this in mind, Margiloff decided to try a new strategy for bringing the Philadelphia plant back into operation. Abandoning the surplus sugar quest, Publicker bought up large quantities of corn starch from midwestern grain producers, and by the fall of 1979 the company was able to get the Philadelphia plant into at least partial operation. The alcohol it produced supplied some two hundred East Coast service stations, and led to the decision to build a new, more technically sophisticated plant closer to the midwestern corn belt. In May, 1980 Margiloff announced that Publicker would participate in a joint venture with Ashland Oil to build a Southport, Ohio plant that would use corn as a feedstock and coal as an energy source.[19]

ENERGY BALANCE IN LARGE-SCALE PLANTS

From the beginning of the gasohol industry, the energy conversion process which took place when grains were converted to alcohol in a large-scale plant was the subject of constant controversy. Debate raged in scientific journals and the popular press over whether or not these plants actually produced a useful liquid fuel or whether, in the final analysis, they simply consumed energy which could be more efficiently used elsewhere. Oil industry spokesmen, such as Chevron's Robert Lindquist, repeatedly charged that these plants were plagued by what he termed a "negative energy balance." Kansas State University professor Leonard Schroben concluded that since gasohol "is mighty wasteful of a long list of natural resources, . . . a more accurate name would be gas-o-hole."[20]

Gasohol proponents countered that their plants would maintain positive energy balance and dramatically reduce the U.S. dependence on foreign oil. Even the definitions of the terms "net," "positive," and "negative" energy balance were subjects of considerable controversy, further confusing journalists and policy makers sorting through the tangled threads of the gasohol debate. A report released by the National Alcohol Fuels Commission in the fall of 1980 noted that,

> By necessity, any energy conversion process—for example, the generation of electricity from coal or refining gasoline from crude petroleum—reduces the total amount of energy that is eventually available to consumers in a usable form. This phenomenon is commonly accepted in transforming a less desirable form of energy to a more desirable form. Thus, a coal-fired power plant that is only 33 percent efficient is considered acceptable because it translates coal to a more usable form of energy, normally electricity.[21]

Taken at its most basic level, the net-energy-balance equation involved determining how much energy was required to convert crude primary feedstocks, such as coal, oil, and grain, into refined fuels such as electricity, gasoline, and alcohol. Interpreted in the broadest sense, this could involve identifying every conceivable energy input from the moment a feedstock was mined, pumped, or planted until it reached its final destination. Establishing such a global energy balance for gasoline, for example, might require calculating the amount of energy necessary to locate an oil reserve, drill a well, pump, stockpile, ship, refine, and transport the final product to a service station. Establishing the global energy balance for alcohol might require establishing the amount of energy required for every step in the production process: building tractors, producing fertilizer, fuels, and pesticides, trucking the crop to distillery, converting the crop to alcohol, and delivering the alcohol to a service station. Obviously, creating such an all-inclusive net energy balance has proved to be a difficult and imprecise task, resulting in final equations of dubious value.

Many of the studies published on the net energy balance of large-scale alcohol plants have attempted to include the amount of energy required to produce crop feedstocks and the amount of energy contained in the food byproducts of the fermentation distillation process. But the segment of the equation that has received the most serious study concerns the amount of process energy used in the plant itself to convert feedstocks to alcohol. This energy is primarily required to fuel the boiler that produces the steam necessary to cook mash, distill alcohol, and provide electric power to operate pumps and motors. For example, if two units of this anhydrous alcohol energy could be produced by the use of only one unit of process energy, then the plant would have a positive energy balance of 2:1. But if it requires four units of process energy to produce the same two units of anhydrous alcohol energy, the plant would be said to have a negative energy balance of 1:2.

Exhibit 4-1
Primary Fuel Source for Planned 1985 Ethanol Capacity (Source: U.S. National Alcohol Fuels Commission—1981)

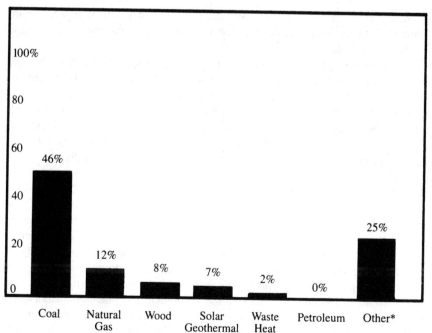

*Includes non-petroleum and non-natural gas energy sources such as bagasse, municipal waste, crop residue, and other biomass wastes.

The ADM plant, which used natural gas for process energy, was certified by the Department of Energy as maintaining a positive energy balance. And new ways are being devised by ADM engineers to reduce process energy inputs. But as critics were quick to point out, some of the early large-scale plants did indeed have negative process-energy balances. The Midwest Solvent's plant required 120,000 Btu of natural gas to produce an 80,000-Btu gallon of ethanol. Another early plant, operated by National Standard Brands in Kentucky, required considerably more Btu of coal in its process energy system than it produced in Btu of alcohol. For Midwest Solvents there was a trade-off of valuable natural gas to produce alcohol, which considerably diminished the attractiveness of the overall process. But plants such as National Standard Brands' unit were examples, gasohol proponents pointed out, of how net energy balance was less of a concern, since relatively low-value coal was traded off to produce a high-value liquid fuel. And the fact that the coal is a domestically produced fuel in plentiful supply further enhanced the attractiveness of that process. An even better trade-off of low-value process energy for high-value liquid fuel is evident at a small farm-scale ethanol plant in Concrete, Washington. This plant, located close to numerous cedar-shake mills, plans to use the waste woods from this local industry to power its small boiler system. See Exhibit 4-1.

DESIGNING A DISTILLATION SYSTEM

John Chambers, a chemist with over three decades of practical experience in distillation technology, has been a pioneer in improving the energy efficiency of alcohol fuel plants.

Chambers, a graduate of the Massachusetts Institute of Technology, helped design some of the first whiskey plants that went up after the repeal of Prohibition in 1933. He also worked with the chemical engineering of coal conversion, designing a 57-column refinery in Brownsville, Texas in 1948 that experimented with synthetic fuel production. After the Brownsville plant was completed, he returned to the beverage alcohol industry, designing columns for National Distillers, Hiram Walker, Midwest Solvents, and Archer Daniels Midland.

The columns Chambers engineered during the postwar era were designed with only one thought in mind—producing a very high yield of beverage alcohol. "I didn't care at all about the energy balance," Chambers remembers. "You had to take the impurities out of the alcohol. To do this took more energy cleaning up the alcohol than it did in the original distilling process. So for twenty years you had a technology geared totally to quality and not at all to efficiency."

As the price of oil began to rise in the early 1970s, John Chambers went to work designing a more energy-efficient beverage-alcohol distilling system which could still manage to maintain the high-quality standards demanded by the industry. But to his frustration, Chambers could find few beverage companies willing to experiment with these new designs. The companies were afraid that by tinkering with the traditional distilling process, the distinctive flavor of their products—flavors that formed the basis for their sales—might be inadvertently altered.

The large-scale alcohol fuel industry launched in the late 1970s created a vast new market for the type of energy-saving equipment Chambers had worked to develop. So he joined forces with his son, Robert Chambers, a University of Illinois physicist with a long-standing interest in energy conservation, to form ACR Process Inc., a firm which specialized in the design of energy-efficient distilling systems. The father-and-son team opened offices in a small suite of rooms above a record shop in Champaign, Illinois and went to work refining a new distilling system geared specifically for fuel alcohol production which could reduce process energy requirements by two-thirds.

The Chamberses found their skills were in great demand. Once the 4-cent federal highway tax exemption for gasohol blends was renewed in 1980, the ACR Process Inc. offices began receiving a flood of phone inquiries. "The phone calls just completely destroyed my day," Robert Chambers recalled. "So I wrote out a little book and then rather than waste a lot of time on the phone I just sent the callers a workbook." It described the basic principles of distillation economics and offered a number of columns for sale, ranging from a relatively small 12-inch-diameter model similar to those used at the Schroder plant, capable of producing 200,000 gallons of 190-proof alcohol per year, to a giant 42-inch-diameter model which could produce over 2 million gallons per year.[22]

Although economists often equate large scale with efficiency, the Chamberses were convinced that intermediate-scale, 1- to 4-million-gallon-a-year plants would eventually be as profitable, if not more so, than the giant 60-million-gallon-a-year plants being built by the oil company and agro-processor joint ventures. These intermediate-scale plants, like their farm-scale counterparts, often would benefit from reduced transportation costs since they generally would be able to haul in their crop feedstocks from much smaller areas than the large-scale plants. They also would be substantial enough to take advantage of certain economies of scale that farm units could not afford.

The Chamberses envisioned the creation of a network of some 40,000 1- to 4-million-gallon-a-year plants in the U.S. by the year 2000. Each of these plants in time would be surrounded by dozens of smaller, farm-scale operations producing 160–190-proof alcohol primarily for farm use. The surplus production of these satellites then would be sold to the intermediate plants for upgrading to the anhydrous, 200-proof level required for the commercial gasohol market. The higher a farmer's alcohol-proof level, the more his product would be worth to the distillery. This system might operate in a fashion similar to a system of price levels pegged at different milk grades already established in the dairy industry.

The team also developed a column to upgrade 190-proof alcohol to 200 proof without using toxic chemicals. While the bond between most of the water in mash can be broken with distilling columns, the last 5 percent has a tougher bond, called an "azeotrope." To break it and obtain a 200-proof alcohol, most industries relied on a process that used benzene to separate the last bit of water. But the Chamberses developed an azeotropic column that used gasoline instead of benzene, which in theory worked well since the fuel was destined to be mixed with gasoline anyway. But the Chamberses found, after experimenting with the column at a Fort Smith, Arkansas plant where they served as consultants, that the amount of gasoline in the final product varied from about 4 percent to 12 percent, reducing the level of quality assurance.

John Chambers's travels took him far and wide, but one of his most interesting jobs was converting a section of the A. Smith Bowman distillery in Reston, Virginia to fuel alcohol production. He met with executives in the spring of 1979 to finalize plans for the new system, inviting Nebraska Gasohol Commission lobbyist Richard Merritt along to explain the prospects of the new gasohol market. The distillery had long produced a highly esteemed bourbon whiskey marketed under the Virginia Gentleman label or, under a special arrangement with the National Press Club, as the Virginia Gentlemen of the Press. Distillery executives planned to produce anhydrous ethanol during a six-month slack period when the plant was usually shut down. They were a bit nervous, however, about committing themselves and wanted to know, most importantly, if the quality of their bourbon being produced in a separate unit would be affected in any way.

Chambers assured the group that the new installation would not alter the flavor of their bourbon and that the $1 million investment to purchase new equipment could be recouped in five years or less. After the meeting, Chambers joined the executives in a nearby restaurant, where, after some hemming and hawing about not normally drinking during the day, they ordered a round of Virginia Gentleman on the rocks and toasted A. Smith Bowman's entry into the gasohol business.

Business boomed for ACR Process Inc., as its concept of intermediate-scale plants with the capability of upgrading lower-proof alcohol from farm stills proved to have a wide-ranging appeal. The Chamberses hoped that this concept could help to ease the tensions between farm and commercial producers by casting what Chambers termed "community" plants in the role of supporting the growth of farm stills rather than competing with them. They found numerous farmer cooperatives and farmer-backed ventures (including one in Gene Schroder's own Baca County, Colorado) that were eager to raise the funds necessary to build 1- to 4-million-gallon-a-year plants. And even Paul Middaugh, one of the original founders of the farm-based alcohol movement, began to move in the direction of intermediate-scale plants in 1980 when he left his professorship at South Dakota State University to supervise construction of a 1-million-gallon-a-year plant in Pasco, Washington. Built from the ground up in only one hundred days, the Pasco plant used distressed grains, potato culls and—on an experimental basis—spoiled fruits and jerusalem artichokes.

ULTRAFILTRATION AND ALCOHOL TECHNOLOGY

While the Chambers and Schroder families concentrated on various mechanical heat-recovery systems to keep costs and energy use lower, a far different approach to the same goal was under development in the laboratories of New York City's Columbia University.

There, engineering professor Harry P. Gregor had discovered that polymer membranes with microscopic pores could separate many different kinds of liquids from solids, often at greatly reduced expense compared to traditional dewatering processes. With pores small enough to filter out a virus some 75,000 times smaller than the width of a human hair, these membranes can dewater sewage sludge, algae, acids, sugary and salty liquids, and alcohol.

Development of membrane technology began in Germany in the 1920s. Chemical engineers experimented with plastic films spread over porous sailcloths that could, theoretically, separate larger molecules from smaller ones. But research was interrupted when two of the scientists working on the technique—Drs. Herbert Freundlich and Karl Sollner—fled to America to avoid Nazi persecution. They continued their work at the University of Minnesota, where Harry Gregor, then a young chemical engineering student, was introduced to membrane research. Gregor was fascinated by the concept and carried the research on into the fifties and sixties at Columbia.

One of the great blocks to the many commercial uses of the membranes was their tendency to foul when large molecules stuck to the surface. Expanding on Freundlich and Sollner's work, Gregor developed a new membrane made of a material similar to heparin, which medical researchers have found prevents blood fouling. He also tried small electric charges, and was successful in extending the useful life of the membranes.

Gregor became interested in the alcohol fuel movement as it developed in the United States and Brazil, and tested membranes on various alcohol production processes. He found that they could remove most of the salt from leftover stillage as well as dry it out; that they could filter solids out of the beer to make distillation easier; and that they could filter water out of high-proof alcohol to bring it up to 200-proof anhydrous grade for blending with gasoline.[23]

Membranes also can be used to bring 160-proof alcohol up to 200-proof via a purification process in which the alcohol flows from the center of a small vat or pipe through layers of membranes. A corn wet milling distillery that installs ultrafiltration membranes can eliminate 96 percent of process energy costs, Gregor estimated, bringing production costs down from about 90 cents a gallon to 15–20 cents a gallon.

Other forms of alcohol production also can benefit from membrane technology. A weak acid hydrolysis breakdown of wood into sugars (which can then be fermented into alcohol) can be made practical, since lignins, sugar, and hydrochloric acid can be economically separated. Single-cell algae, which are highly efficient biomass converters of sunlight, could be concentrated for a fermentation process or for food uses.

BIOTECHNOLOGY AND ALCOHOL FUEL PRODUCTION

Confident predictions of an "emerging age of biology" or a "chemurgic age" were not unusual in the 1930s, and William Hale's vision that such developments would solve political and economic problems was shared by many in the budding new field. "But little could the students of my generation foresee that biology would mature so rapidly," Bernard Davis, a Harvard physiologist on the forefront of biotechnology research wrote in 1980, "while the predicted social utopia would become more distant than ever."[24]

Tremendous strides have taken place in biological research—particularly over the past three decades—but due to the low cost of oil, comparatively little was achieved in the areas of fuel or synthetic materials. However, as oil prices escalated, federal, academic, and corporate laboratories began to sponsor major efforts to apply biotechnology to the alcohol fuels industry.

One of the most important thrusts of this research has centered around the development of new enzymes that can quickly and efficiently convert cellulose (a biomass material found in wood, straw, and paper pulp) into fermentable glucose sugar. Cellulose molecules are made up of long chains of these glucose sugars. Unlike the starch molecules in grain, which are easily broken down, cellulose molecules are linked by strong hydrogen bonds that previously could be broken down only by acids.

The commercialization of the new cellulose technology could rapidly change the face of the alcohol fuels industry, for it would allow it to shift away from costly sugar and starch crops subject to wide price fluctuations to more abundant cellulose.

The scientific research effort that has brought cellulose technology to the brink of commercialization began during World War II. U.S. soldiers fighting the Japanese in the South Pacific found that their shirts, boots, tents, and knapsacks were quickly rotting away in the humid, tropical environment. Some of the jungle molds could eat a hole through a piece of cotton material in an hour, and reduce garments to shreds in days. As a result, precious cargo space needed to bring weapons and ammunition to the battlefront had to be taken up by basic supplies needed to replace those eaten by the voracious "jungle rot."

The Army Quartermaster Corps, determined to find some way to combat jungle rot, set up a research project in the Army laboratory in Natick, Massachusetts to investigate its basic mechanisms and find some way to control it. Over 14,000 different forms of rot-causing microorganisms were collected and identified by Army scientists during the initial phase of the work. One of the most destructive of these was a mold called *trichoderma viride* which was first isolated on a rotted cartridge belt brought back from New Guinea.[25] This mold secreted a powerful enzyme complex called cellulase which quickly reduced the tightly linked cellulose structures in cotton and leather equipment into simple glucose sugars which nourished the mold.

After the war, Army scientists continued a basic research program into *Trichoderma viride*'s complex biochemical activities. Dr. Leo Spano, a stocky Army biochemist with a gregarious wit and a slight accent from his native Portugal, bombarded the mold with radiation to produce new mutant strains that could convert cellulose to glucose more easily. One of the mutants was placed in a water solution with ground-up leaves and left alone for thirty-six hours. When Spano returned to it, he was surprised to find it had been almost totally reduced to glucose. It was an exhilarating moment for Spano, who later wrote:

In the laboratory filled with test tubes and incubators, I felt apart from the world. . . . But it was there . . . that I realized that a tiny enzyme could change the world as we know it. If man could direct an enzyme and improve it, the compounds could eat up our poisonous wastes and convert them to useful substances. Just think of all the waste cellulose produced in this country—sewage, wood pulp, corn cobs . . . it can all be used to better mankind.[26]

The work that followed Spano's breakthrough, although faithfully supported by the Army laboratory, was painstaking and proceeded slowly through the 1950s and sixties. It proved difficult to keep the mold cultures free of contamination and collect enough of the cellulase enzyme complex to carry out the conversion process. Then, in the spring of 1970, George Emert, a biochemistry graduate student at Virginia Polytechnical Institute, visited Spano in Massachusetts to discuss the mold's potential in developing a cellulose-based alcohol production process. Emert, in a new post with the University of Colorado, continued to experiment with the *Trichoderma*'s mutant offspring. In the summer of 1974, he left academia to develop a cellulose-based-alcohol plant for Gulf Chemical, a subsidiary of Gulf Oil Company in Jayhawk, Kansas.

By January, 1976 Emert had produced alcohol in a small, silver fermentor that looked somewhat like a space capsule. Emert found that cotton wastes, corn stalks, paper, and peanut shells all could be converted into alcohol when first treated with the cellulase enzyme. He received the go-ahead from George Huff, director of the Gulf Chemical research program, to proceed with a pilot plant that would use municipal garbage (purchased at $10 per ton) as its feedstock.[27]

A report later released by the National Alcohol Fuels Commission concluded that municipal garbage constituted one of the most abundant and immediately available alcohol feedstocks. By 1990 nearly 10 billion gallons of alcohol, enough fuel to replace almost 10 percent of current gasoline consumption, could be supplied by municipal solid wastes—more than double the amount available from grain resources. Despite this impressive potential, the garbage-to-ethanol project was fought by Gulf Oil's refining division, which was opposed to having any segment of the corporation enter the gasohol business. The Gulf refining executives feared that the gasohol blend would prove to be a technically inferior fuel that would cause problems in their overly burdened refineries.*

The conflict between the refining division and the chemical subsidiary was abruptly resolved by Gulf President Jerry McAfee who decided to abandon the Emert project in late 1978. Explaining his decision to Gulf Oil's Washington lobbyist Spencer Sheldon, McAfee stressed that the primary thrust of Gulf's alternative fuel development program should remain centered around coal and oil shale technologies.

In August, 1979 Gulf donated the project to the University of Arkansas,

---

*Some executives were so wary of alcohol that, when reached by reporters at Gulf headquarters in Houston, Texas in June, 1978, they denied that any research into alcohol was taking place. One week later, Gulf Chemical unveiled its process at a meeting in Washington, D.C.

keeping only a 20 percent share of any royalties from licensing the process.* Although frustrated by his experience at Gulf, Emert was not about to abandon his efforts to commercialize the technology, and in March, 1980 he agreed to work with United Biofuels Industry, a Virginia corporation chartered to build a 50-million-gallon-a-year ethanol distillery that would use agricultural, forest, and paper wastes as its primary feedstocks. Although United Biofuels has been reluctant to discuss costs involved in alcohol production at this stage, Spano provided an estimate in late 1979 of between $.90 and $1.22 per gallon—depending on financing costs—in a plant half the size of the United Biofuels project.[28]

By the time United Biofuels produces its first gallon of alcohol, probably sometime in 1985, the type of enzyme it uses may have been substantially improved by the revolutionary new field of genetic engineering. Wielding the powerful tool of recombinant DNA technologies, scientists have been able to improve dramatically the ability of *Trichoderma viride* and other microorganisms to produce cellulase enzymes.

Understanding the structure of DNA is as important to modern biology as understanding the atomic structure of elements is to modern chemistry, with the first successful gene-splicing techniques being comparable to the first splitting of the atom. Scientists first began to unravel the complicated structures of DNA molecules in genes, but it was not until the late 1970s that scientists could begin to put to practical use the knowledge gained in their laboratories.

The arrival of genetic engineering has generated considerable controversy in the United States. A storm of public opposition greeted General Electric when it applied for a patent on a bacterium created in its laboratories which can degrade crude oil spilled into the ocean by oil tankers. Critics of bioengineering such as Jeremy Rifkin, author of *Who Shall Play God?*, have argued that corporations should not have the power to patent a life form. Rifkin fears that potentially dangerous new organisms might inadvertently be set loose in the environment. Scientists involved in bioengineering have responded that patents are necessary to help recoup the enormous research and development costs that are required to develop useful new life forms and that the potential benefits far outweigh the risks involved in their work.[29] In May, 1980 the Supreme Court of the United States handed down its landmark decision approving the patenting of new life forms, although Chief Justice Warren Burger noted that "a gruesome parade of horribles" had been presented by opponents of bioengineering. Prior to the Supreme Court ruling, the National Science Foundation had conducted a major study of recombinant DNA technology which resulted in stringent federal regulation of genetic engineering laboratories.[30]

Ronald Cape, a Harvard microbiologist, was one of the first of many talented academics who have transferred their work in DNA technology from the university laboratory to private industry. In 1971, he founded Cetus, a laboratory

---

*Emert's attempts to persuade Gulf to continue with the cellulose project were frustrated when, at a top-level board meeting in Houston, one oil executive claimed "No little bug is going to make gasoline!"

which now occupies twelve instrument-jammed buildings in Berkeley, California and employs over 240 people. Standard Oil of California (Chevron), Standard Oil of Indiana (Amoco), and National Distillers provided $30 million of the company's first $35 million in capital funds.[31]

One aspect of Cetus's work that interests Chevron is the development of enzyme colonies that can convert petroleum gases into products such as antifreeze and plastics. With greatly reduced energy and equipment needs, the process is 50 percent more efficient than a comparable petroleum refinery. If successful, the pilot project now under way could change the appearance of a refinery, with all its columns, piping, catalysts, and condensers, into something that resembles a giant laboratory.

A major thrust of Cetus's genetic engineering program—the part that interests Amoco and National Distillers—has been to improve the efficiency of alcohol fuels production. During the late 1970s, Cetus developed a strain of yeast that was 30 percent more efficient than traditional fermentation yeasts. It also explored the potential of new strains of enzyme-producing bacteria such as cullulomonas and the "white rot" Canadian spruce fungus that can effect a more complete conversion of wood products to glucose.

In October, 1980 National Distillers unveiled a $100 million project to build an alcohol plant using Cetus's new yeast strains in a continuous—rather than a batch—fermentation process. Most yeasts expire when the alcohol concentration in a fermenting vat rises above 10 percent and they cannot function properly outside a narrow temperature range. But National Distillers is expected to take advantage of Cetus yeasts, which can survive in alcohol concentrations of over 20 percent and ferment sugars within a much broader temperature range. Richard A. Tilghman, National Distillers' vice-president, refused to discuss the financial benefits which may result from the commercial application of these Cetus technologies, but he admitted to the *Wall Street Journal* that ". . . we wouldn't be going into this first-of-a-kind plant with something like only a 2 percent reduction in costs. We know we are going to get a significant reduction in costs."*[32]

FINANCING AN ALCOHOL FUELS INDUSTRY

The legacy of President Jimmy Carter's "moral equivalent of war" included, by the time he left office in January, 1981, two theoretical goals for replacing imported oil with alternative fuels. According to Carter's synthetic fuels development plan, some 2 million barrels a day of synfuels, mostly from coal, should be in production by 1992. Another goal, approved only months before Carter

---

*Hoechst AG of West Germany, a major chemical corporation, reported ten-fold increases in productivity with their continuous fermentor. Others at work on the same type of research include MIT, Cornell, Rutgers, Dartmouth, Iowa State, Missouri and Connecticut universities; Bechtel, Phillips Petroleum, and other major corporations; and the USDA regional research centers. ADM also has a continuous fermentation process, which, according to an article in *Food Processing* magazine (April, 1981), reduced heat requirements for the distillery from 70 cents per gallon to 12 cents per gallon.

left office, is the replacement of 10 percent of the nation's gasoline supply with alcohol fuels by 1990. Although the goals are not mutually exclusive, they will be competing for scarce investment money in a climate that investment brokers describe at best as "very skittish."

The implications of both goals are staggering. A 2-million-barrel-per-day synfuels industry would require over $60 billion in private capital investment.[33] Based on a $3 billion investment per 50,000 barrels of daily capacity, such synfuels plants involve about $3.86 per gallon of yearly capacity.

On the other hand, replacing 10 percent of the nation's 100-billion-gallon gasoline demand involves production of about one-quarter of the synfuels goal. Capital investment for such plants varies widely. Leo Spano gives $65 million as the total investment needed for a 25-million-gallon-per-year cellulose plant, or about $2.60 per gallon of capacity.[34] Corn-to-ethanol plants have ranged in the neighborhood of $1 to $3 per gallon of annual capacity, although some farmers have achieved much lower rates with scrap materials and their own labor, and some research and demonstration models have spent as much as $22 per gallon of annual capacity.

If the cost of meeting the goals was staggering, the cost of not meeting them was even more staggering. The roughly 7-million-barrel-per-day U.S. imports of 1980 cost over $75 billion, considering prices at $30 per barrel.

The price tag on a few large-scale ethanol distilleries is a relatively minor expenditure for major oil companies that are accustomed to the multibillion-dollar development costs of oil-shale and coal-based synthetic fuels. And as the pattern of joint ventures between grain and oil companies demonstrates, it is no problem for oil companies to acquire the technical and marketing expertise required to operate the sophisticated corn-milling alcohol production units.

BILLIONS REQUIRED FOR METHANOL PLANTS

The price tag on a large-scale methanol distillery constitutes a major expenditure even for an oil company. A coal-to-methanol plant might produce as much as 700 million gallons of fuel a year—more than ten times the capacity of the typical large-scale corn-to-ethanol plant—and cost $2–$4 billion. Preliminary cost estimates indicate that these plants may be able to produce methanol for between 25 cents and 79 cents a gallon.[35] See table 4-1.

During the 1970s the oil industry, as a whole, tended to favor the development of coal-to-synthetic gasoline processes over coal-to-methanol processes. But given the favorable economics of coal-to-methanol production, when compared to both synthetic gasoline and grain-to-ethanol, many oil companies are taking a new look at methanol in the 1980s. Several oil companies, including Arco, Chevron, and Shell, have jointly funded a study carried out by a New York-based consulting firm to determine the best long-term strategy for developing methanol fuels, and Conoco, an oil company with vast coal holdings in the western United States, is considering a coal gasification plant in Louisiana that would feed a 220-million-gallon-a-year methanol plant.[36]

TABLE 4-1

Comparison of Typical Corn-to-Ethanol/Coal-to-Methanol Plants (Source: U.S. National Alcohol Fuels Commision—1981)

	Corn-to-Ethanol Plant	Coal-to-Methanol Plant
Annual (Daily) Output	50 Million Gallons per Year (3,600 Barrels per Day)	639 Million Gallons per Year (50,000 Barrels per Day)
Capital Cost*	$90 to 100 Million	$2 to 4 Billion
Required Land Area for Plant	25 to 50 Acres	300 to 600 Acres
Work Force	75 to 100 Persons	350 to 400 Persons
Process Water Requirements	4 to 5 Gallons per Gallon of Output	6 to 7 Gallons per Gallon of Output**
Output Over Life of Plant	1 Billion Gallons (23 Million Barrels per 20-Year Production Cycle)	13.9 Billion Gallons (330 Million Barrels per 20-Year Production Cycle)
Construction Time	2 to 4 Years	6 to 8 Years
Daily Raw Material Requirement	1,649 Tons of Corn (58,900 Bushels)	23,000+ Tons of Bituminous Coal

*Does not include first-year operating costs.
**Water use for bituminous or sub-bituminous coal. Water requirements for lignite will be significantly less.

Major oil companies and other large energy concerns are also carefully assessing the potential for converting flared natural gas into methanol. Most plans for this kind of methanol production call for the construction of large floating plants next to offshore natural gas deposits or anchored in the coastal waters close to on-shore producing wells. For years, much of the natural gas produced in Saudi Arabian oil fields was simply flared off, but in the 1980s it became a prime prospect for methanol conversion. Another potential site for methanol production from natural gas is in Alaska where plans to bring the state's large natural gas reserves to the lower 48 states via an extensive pipeline may be jeopardized by high interest rates and soaring construction costs estimated in 1981 at over $40 billion. If the natural gas were converted to methanol, it could be shipped to market, thus avoiding much of the tremendous expense associated with the pipeline. The Alaska Energy Center, a research group created by the state legislature, favors this approach for both environmental and financial reasons. Fourteen floating methanol plants, each capable of producing 42,000 tons per day, could be built for a total capital cost of approximately $4.2 billion.[37]

The biggest drawback causing the oil industry to hesitate from a massive push to develop methanol is the lack of a clearly defined market for the fuel. Methanol cannot be blended with gasoline as easily as ethanol to form gasohol-type blends. It also requires more extensive engine modification than ethanol when used as a pure fuel. However, methanol blended in small, low concentrations with other alcohols can be used as an octane booster in unleaded gasoline, providing the fuel for a small but immediately available market during the 1980s. Some 1.1 billion gallons of methanol, synthesized primarily from natural gas, already find a market each year in the chemical industry for use in paints, plastics, and synthetic fibers. Yet, most oil companies remain cautious about making multibillion-dollar investments in coal-to-methanol production when the chance remains that its market might never expand much beyond the limited horizons of octane boosters and chemical feedstocks.

The environmental impacts of methanol fuel production constitute a second drawback to a major production effort, since both the mining and processing of coal feedstocks can have extensive adverse impacts on land, water, and air quality which may prove extremely expensive to combat. Under the lax environmental protection policies of the Reagan administration, coal-to-methanol plants would face little opposition from most federal pollution control authorities. However, by 1990, which is the earliest that most of these plants might be operational, the political winds may have shifted again in favor of more tough-minded enforcement of federal environmental protection laws.

Wood-to-methanol plants face many of the same drawbacks as coal-to-methanol plants: uncertain markets and serious environmental impacts. In addition, the technologies involved in wood-to-methanol production have seldom been attempted on a large scale. However, most research to date indicates that an efficient wood-to-methanol plant could be built for a total investment of $88 million. This plant could produce 40 million gallons of methanol a year at a cost ranging from 67 cents to $1.33 a gallon, but would-be investors in wood-to-methanol projects are wary that their market may eventually be taken away by coal-based methanol if major oil companies decide to move in that direction.[38] See table 4-2.

For the farm-based ethanol movement led by Gene Schroder, the prospects of an industry composed primarily of a few large oil-company-controlled coal-to-methanol plants and oil company-agribusiness joint-venture ethanol plants is frightening. It seems unlikely, though, that these corporations can dominate the entire picture. With such a wide variety of feedstocks available for ethanol production, there should always be a place for the small-scale community-based producers both on and off the farm. To date, a relatively diversified group of medium- to large-scale ethanol ventures have been put together by farm cooperatives, community development organizations, regional governments, and independent entrepreneurs. Many of these ventures appear to stand a relatively good chance of commercial success as they are taking advantage of new engineering technology which can drastically reduce process energy requirements.

TABLE 4-2

1979 Cost of Methanol From Wood Using Oxygen Gasification (Source: U.S. Office of Technology Assessment report, *Energy From Biological Processes Volume II*, page 140)

Fixed investment (field erected)	$80 million
Working capital (10% of fixed investment)	8 million
Total investment	$88 million
	$/bbl
Wood ($15/green ton)	10.35
Labor, water, chemicals	1.10
Electricity (3.8 kWh/gal, $0.04 kWh)	6.40
Capital charges (15-30% of total investment per year	$13.80-$27.60
Total	$31.65-$45.45
	($0.75-$1.08/gal)
Estimated range:	$28-$56/bbl
($10-$30/green ton wood)	($0.67-$1.33/gal)
	($10.50-$20.90/10^6 Btu)

Input:	2,000 green ton/d of wood (50% moisture)
Output:	2,900 bbl/d methanol (40 million gal/yr)
Load:	330 operating days per year

However, most of these grass-roots ventures have faced major funding obstacles. Private lending institutions have proved extremely reluctant to bankroll the young and rapidly changing alcohol industry, particularly groups which lack a solid track record upon which loan officers can base a decision. Even the best engineering firms have had problems raising funds for these ventures.

The respected international engineering firm Davy McKee found it very difficult to deliver to grass-roots ventures the "complete and fully coordinated financing solutions" it advertised in promotional brochures. McKee carried out well-researched feasibility studies for some twenty-five large-scale ventures but found that the studies alone could not shake loose any private investment capital. "We showed the investment firms our feasibility studies," Davy McKee's sales manager Calvin Todd said, "and they said, 'Well, that's fine and that will be an important piece of the total package but that's not all.' Their concern was not about building the plant but about who's going to operate it five or ten years from now when they still have their money in it." Todd has since worked hard to develop "packages" of marketing and managerial expertise for Davy McKee's clients which will bolster the confidence of the banks in these new ventures.

Still, there remains a more fundamental, long-term problem which makes the bankers wary of investing too heavily in the ethanol fuel industry. Investors fear that sudden price rises in the volatile world grain markets in the 1980s might destroy the economic viability of many new plants, since most will use corn as a primary feedstock. Agricultural economists cannot predict with certainty what will happen to the price of corn over the next decade, any more than they could in the past decade. The point was brought out by Raphael Katzen, an engineering consultant, in testimony to the National Alcohol Fuels Commission in 1980. "In 1971 and 1972 we saw corn go up 300 percent in one year and saw it drop back in two years to one-third of what it had been. You can't have that sort of price gyration and have a stable motor-alcohol-fuel industry."[39]

A severe drought in the Midwest during 1980 resulted in a poor harvest which reduced the size of the surplus built up during the late 1970s and sent corn prices from $2.50 to $3.50 a bushel. Another poor harvest, coupled with increased worldwide demand for U.S. corn, could send prices spiraling up to unheard-of levels. Large-scale plants could recoup some of the increased price of their corn feedstock by charging more for their livestock feed byproducts, corn oil, and glutens. But major increases in the prices of corn would invariably result in higher prices at the gasohol pump. This was amply documented in a report by a congressional agency, the Office of Technology Assessment, which noted:

> The largest cost in ethanol production is the net feedstock cost, or the feedstock cost less the byproduct credit. With corn at $2.50 a bushel, the corn grain costs $.96 per gallon of ethanol and the byproduct credit is about $.38 per gallon, resulting in a net feedstock cost of $.58 per gallon. Since farm commodity prices are extremely volatile, the net feedstock and resultant ethanol cost are also variable. A $.50 per bushel increase in corn grain prices (and a proportional increase in the byproduct credit), for example, would raise the ethanol cost by $.12 per gallon.[40]

The congressional report also concluded that "byproduct credits could begin to drop due to saturation of the domestic (livestock) feed market at about 2 billion gallons of ethanol production per year. At significantly higher levels of production, new markets for distillers grain . . . would have to be developed or distillers would lose the byproduct credit, increasing the ethanol costs by $.38 per gallon."

Farmer-owned small-scale plants will have much less vulnerability to the vagaries of world grain markets than the large-scale plants. Small-scale plants with light debt loads may be able to weather a period of high grain prices by simply shutting down or slowing operations. The farmer could take his corn and market it at top prices and the plant could continue to operate on spoiled grains and cull crops. When grain prices dropped, the farmer could once again make his alcohol from corn.

But large-scale plants, with heavy debt loads to repay, must operate on a fixed schedule, and can ill afford to shut down during periods of high grain prices. In most cases alternative crops would not be available in sufficient quantities to maintain production levels. Thus, many of the large-scale plants could be transformed into multimillion-dollar white elephants if grain prices were to soar too high.

To guard against such an eventuality, many of the large-scale ventures are considering installing cellulose conversion equipment so that their plants can operate on paper, wood, and agricultural cellulose wastes when cheap grain is unavailable. But it is still far from clear just where these large-scale plants will obtain the massive amounts of cellulose required to maintain productive capacity. Most of these plants are located in midwestern grain belts far from cheap sources of cellulose, such as urban garbage and forest waste. Some gasohol proponents believe that cellulosic feedstocks can be supplied by corn stalks and cobs, but the economic feasibility of gathering and transporting these materials is still uncertain and they are normally returned to the soil to maintain fertility.

However difficult it may be to finance alcohol fuel plants in the face of fluctuating corn prices, the problem may be insignificant compared to potential oil price fluctuations. According to a 1980 congressional Budget Office report, a shut-down on the Persian Gulf oil supply would lead to pressures on the world oil market that would send prices to at least $100 a barrel, which would make ethanol cost competitive with corn at $8.00 per bushel.[41] "People who put their money in alcohol fuel will be heroes if the Persian Gulf falls," observed one industry insider, "but they will be bankrupt if oil prices fall and corn prices rise."

Along with the potentially unstable price of raw materials, another block to the development of an alcohol industry is the high cost of money itself. With interest rates high very few investors are willing to risk money on even the soundest projects. There are, however, a number of alternative financing avenues that involve varying degrees of government involvement. The United Biofuels cellulose plant, for example, is financed through the sale of Industrial Development Bonds issued by a regional development authority cooperating with several local governments. Interest rates are several percentage points lower than the prime rate, and investors find the bonds attractive since the interest paid to them is tax-free. Unlike a sewer or school bond issue, the city or county working with an industry under such an arrangement is not liable, and thus it risks little if the plant goes bankrupt. But the approval of the city or county hinges on the potential of such plants to create jobs and help the local economy. From that standpoint, alcohol plants look like a good bet. A report released by Employment Research Associates of Lansing, Michigan in 1981 concluded that the establishment of a 6-billion-gallon-a-year renewable alcohol fuels industry, requiring a total investment of $12 billion, would create some 960,000 new jobs in the United States. The report predicted, "Every section of the nation would benefit from the construction and operation of alcohol fuel plants and by the subsequent improvement in our balance of payments when money which would have been spent to purchase foreign liquid fuel remains instead in the United States."[42]

Another financing possibility would involve city, county, or regional governments issuing bonds to cover the cost of alcohol producers.* While such an arrangement would involve some risk to the taxpayer, a secure source of emer-

*In California, a special revolving state fund to help equip fuel distilleries was established. The state provides the alcohol fuel producer with the necessary funding and retains ownership of all equipment purchased. Once the plant becomes commercially viable, the owner buys the equipment from the state. If the plant fails, the state retains the right to foreclose to minimize its risk.

gency fuel for fire, police, and hospital vehicles may be of critical importance during a future oil shortage to protect public health and safety.

Given the numerous uncertainties that lie ahead, the alcohol fuels industry, as well as local and state governments, have looked to the federal government to reassure a skeptical investment community that the future for alternative fuels will remain bright. And after almost five decades of neglect, the federal government has taken the first halting steps forward in helping resurrect Henry Ford's dream of an alcohol-powered America.

Chapter 5
The Politics of Alcohol Fuel

On a cold, damp day in September 1976, Richard Merritt drove his green Porsche up the winding driveway of the Bethesda Meeting House, an historic church on the outskirts of Washington, D.C. He stopped for a moment to reflect on the church where Abraham Lincoln once worshipped then turned his gaze to the faded exterior of the aging three-story parsonage where he had rented a $90-a-month room. In these modest quarters, which he grandly titled the Alcohol Alternate Fuels Institute, Merritt established the first beachhead of the Washington gasohol lobby.

A thin, middle-aged man who sported a salt-and-pepper moustache, Merritt lacked even the most rudimentary tools normally required for a successful Washington lobbying campaign. He had no formal stationery and sent off much of his correspondence in envelopes retrieved from his incoming mailbag. Lacking in the standard collection of well-tailored, three-piece suits, he made his daily rounds in an eclectic array of ill-fitting sports coats and checkered pants. His first-floor office, which doubled as a bedroom in the evenings, had once been used as a practice studio by a rock 'n' roll band, and its walls were still covered with colorful strips of shag carpeting put up for soundproofing. A dozen different bumper stickers touting the virtues of gasohol were pasted on the wall above his bed, and over on one side of the room sat a table covered with stacks of photocopied newspaper clippings describing alcohol fuels which Merritt had quietly run off in the offices of sympathetic congressmen.

There was no doubt about it: Merritt looked like a long shot for success in the high-powered world of Washington's lobbyists. If an official from the American Petroleum Institute's elaborate suite of downtown Washington offices were to have visited Merritt's office, he would have scoffed at any suggestion that this amateurish outfit could ever make much of an impact in the nation's capital. Yet in the spring of 1979—less than three years after Merritt's arrival in Washington—J. P. Smith of the *Washington Post* would write that ". . . the gasohol lobby is becoming one of the most powerful in Congress."[1]

This lobby was a broad-based, at times uneasy, coalition of interests which included a bipartisan block of farmbelt congressmen, the American Automobile Association, the National Gasohol Commission (a newly formed industry and grass-roots association), and a smattering of environmentalists and black civil rights organizations. Although Smith's comments in the *Post* somewhat overstated the influence of the gasohol lobby, the coalition was amazingly successful in obtaining the passage of a wide range of legislation and in arousing unprecedented popular support for the new industry. This was accomplished in the face of a vigorous campaign waged by the Mobil and Chevron oil companies to convince the federal government to withhold support for the fledgling gasohol industry in favor of accelerated development of coal- and shale-oil-based synthetic fuels.

Yet in the fall of 1976, the gasohol lobby was more a creation of Merritt's fertile imagination than a tangible presence on the national political scene. When Merritt arrived in Washington, the city was still reverberating with the political shockwaves left behind by Watergate. The resignation of Richard Nixon had left the nation limping into the 1976 elections with Gerald Ford, an unelected president, in the White House. Ford's Democratic opponent was Jimmy Carter, an obscure ex-governor of Georgia who would win a narrow triumph in the November elections and then suffer a humiliating defeat to Ronald Reagan in his quest for reelection four years later. With the exception of Carl Curtis, Nebraska's senior senator, there were few members of the Washington political establishment who had any memories of the earlier efforts to promote power-alcohol blends by the chemurgists. And in 1976, Albert Turner had not even begun building his first small-scale still in Selma, Alabama, and the first commercial gasohol pump had yet to appear in the Midwest.

Amid the hoopla surrounding the final months of the 1976 campaign, there were few Washington politicians who heard the angry cries from rural America which would later give birth to the farm alcohol movement. Only a handful of dissident scholars predicted the collapse of the Iranian monarchy which would leave two decades of United States policy in the Persian Gulf in shambles. The gas lines that closed in on the capital city in the summer months of 1979 were foreseen only by a handful of harried officials working in the federal energy bureaucracy. But within three years, these events would thrust alcohol fuels into the national political spotlight, helping the gasohol lobby prod a lethargic Congress into action.

When Richard Merritt first launched the Washington gasohol campaign he had none of the inside contacts that help lobbyists get past the secretaries who guard the doors of the capital's power brokers. He was forced to peddle his cause from door to door, much like a traveling salesman. Every day he would leave the parsonage, his battered briefcase packed full of newspaper clippings, and drive to Capitol Hill, where he would drop in on a dozen congressional offices and deliver a ten-minute alcohol fuels sales pitch that combined the fast-talking skills of a former car salesman with the fervor of a true believer. He focused his efforts on congressmen who had already shown an interest in curbing the oil industry's power through a series of divestiture proposals. Backers of the divestiture proposals hoped to inject more competition into the oil industry by breaking up the oil-well-to-gasoline-pump vertical integration of many oil companies, and also by prohibiting a horizontal expansion of the oil industry into coal, nuclear, and solar energy areas.

"I've got something the oil companies hate worse than divestiture," Merritt would declare. "I've got raw-material divestiture. We won't even need the oil companies if we can make alcohol from farm products."[2]

Merritt's exaggerated enthusiasm for alcohol fuels initially drew little response. The handful of congressional aides who took time to listen to what he had to say were somewhat wary of his come on. Fred Wahl, a staff aide to Sen. Frank Church, remembers that the early Merritt had "the right idea but was a bit overzealous. . . . He came through as a guy who believed absolutely in what he was saying. We knew enough about the issue to go along with his concepts and welcomed someone like him spreading the gasohol gospel."

"Walking up and down the halls of Congress and passing out stuff on gasohol was a pretty lonely deal back then," Merritt admitted. "You could tell nobody gave a damn. But I am not complaining. I would damn near have done it for nothing just to be Mr. Gasohol."[3]

Mr. Gasohol was born in 1931 in Webster City, Iowa, a small farming community struggling to weather the lean years of the Great Depression. His first introduction to alcohol fuels occurred at a favorite boyhood haunt—a race-track outside of Ames, Iowa where many of the cars making the rounds of the midwestern racing circuit were powered by ethyl alcohol. When Merritt was a teenager his family left Iowa and moved to Boulder, Colorado where he devoured every auto racing magazine he could lay his hands on. "The experts agreed," Merritt remembers, "if you wanted to go really fast and really race your car, you had to use alcohol."

After a stint as a Jeep mechanic with the Air National Guard in Anchorage, Alaska during the Korean conflict, Merritt finished college and landed a job with the Ford Motor Company, where he was assigned to a management training position in the ill-fated Edsel division. For Merritt, it was a boy's dream come true: "I was sitting in at the top-level decision process where they decided to make the new cars bigger, smaller, shorter, or wider." But the dream did not last long. The Edsel division folded and Merritt was out of a job. For the next fourteen years he drifted in and out of five different auto industry jobs with General Motors, Porsche, Audi, Volkswagen (where he was the top-rated sales-man in the tough Detroit market), and Ferrari.

For most Americans, the Arab oil embargo of 1973 resulted in the trou-blesome but temporary inconvenience of gasoline lines. But Merritt attributes the loss of his job as sales manager of a promising Philadelphia Ferrari dealership directly to the embargo. "The Lord Himself couldn't have sold Ferraris back then," he remembers. "With speed limits and gasoline lines, nobody was going to buy a $25,000 car to drive 55 miles per hour and wait in line for gas."

With few jobs available in the depressed auto industry, Merritt returned to Detroit to write a detailed guide to Ferrari sports cars. He bought a 50 percent interest in a 1952 Formula One Grand Prix Ferrari, powered by a blend of 33 percent alcohol, 33 percent benzene, and 33 percent gasoline, which he often took for workouts at auto club rallys. If the alcohol worked so well in a high-performance car, Merritt wondered what kind of job it could do in powering the rest of the nation's automobiles. His curiosity drove him to the Detroit public library, where, amid the dusty shelves of automotive technical history, he discovered a sheaf of documents dating back to the chemurgist conferences of 1935 and 1936. Among them was Francis Garvan's paper disclosing how Standard Oil of New Jersey promoted the Cleveland Discol alcohol blend in England while attacking U.S. power-alcohol blends as inferior.

From that day, Merritt photocopied every bit of information he could locate on alcohol fuels. He visited obscure libraries and called old acquaintances in the auto companies. By the time he attended a February, 1974 meeting of the Society of Automotive Engineers, he was well versed in the history of alcohol fuels. He hoped to find the major oil and auto company engineers interested in his research but soon found that those who attended the meeting had few kind words for alcohol fuels. The new technical papers which mentioned alcohol were filled with negative comments and devoid of any ideas on getting alcohol out into the marketplace. At the end of the meeting, Merritt recalls, "Two or three guys wearing oil company identification tags started up a conversation in front of me. The first guy said, 'I thought that we got rid of this alcohol thing back in sixty-four.' The second guy said, 'I thought we did too, but it's popped up again.' Then the first man cautioned, 'Maybe times have changed and we ought to have another look at it.' Then they walked off to the bar. That was proof enough for me that the oil companies had done all they could to kill alcohol fuels, and it turns out that there was a big Society of Automotive Engineers meeting in 1964 where several alcohol fuels papers were presented by oil and auto companies telling how terrible the stuff was. I was really hopping mad when I found that out, and furthermore, I said 'I'm going to do something about it.' " And so Merritt's crusade was launched.

Eighteen months after the SAE meeting Merritt arrived in Washington. In the interim period, subsisting largely on royalties from his Ferrari book, he collected more material on alcohol fuels and appeared on a number of radio and television talk shows to discuss his findings. He was even singled out by national commentator Paul Harvey, who noted in August, 1975 that "Richard Merritt, Ferrari expert, is pushing alcohol fuels." He was able to obtain a brief audience with midlevel White House aides in the Ford administration but was unable to arouse much enthusiasm for developing alcohol fuels. If he was ever to make much headway in convincing the cumbersome federal bureaucracy, Merritt realized, he would have to permanently move his base of operations to Washington.

THE MIDWEST GASOHOL LOBBY

For six long months, Merritt lobbied in Washington without pay. When skeptical congressional aides demanded to know for whom he worked, more often than not he would reply "the state of Nebraska."

Ever since 1971, Nebraska's Agricultural Products Industrial Utilization Committee, a state-funded group of farmers, oil industry officials, and Nebraska businessmen, had conducted a series of pioneering studies to determine the viability of alcohol blends in the OPEC-dominated decade of the seventies. The committee was largely the creation of Nebraska state senator Loren Schmidt, an outspoken critic of the oil industry with a long-standing interest in alcohol fuels. A key ally in Schmidt's efforts to persuade the state legislature to establish the committee was Dr. William Scheller, a University of Nebraska chemistry professor who was one of the first academics to argue for a national alcohol fuels industry.

The committee quickly became popularly known as the Nebraska Gasohol Committee, after Scheller coined the term. It was chaired by Holly Hodges, a prosperous state wheat farmer who dressed in the conservative suits of the midwestern businessman. It was Hodges who, in the winter of 1977, heard of Merritt's efforts on behalf of alcohol fuels in Washington, D.C. and invited him to Nebraska for a visit. "I figured that after all the things I had said using their name in Washington, they'd either have to sue me or hire me," Merritt remembers. Hodges cautiously greeted Merritt in Lincoln, Nebraska and, after some lengthy meetings, agreed to hire him as a consultant at $1,000 a month.

Although Merritt's financial base of support would be limited to the Nebraska Gasohol Committee for the next two years, he immediately sought to broaden his political base of support to include other segments of the growing midwestern gasohol lobby for Nebraska's push to become the hub of the new gasohol movement was becoming bogged down in bitter statehouse politics. Nebraska's governor J. J. Exon lent only lukewarm support to the legislature's efforts to construct a prototype large-scale alcohol plant and get gasohol into the marketplace. In August, 1977, after Exon opposed these senate initiatives and called for further study of gasohol's economic feasibility, Loren Schmidt accused him of caving in to hostile oil company pressure. Exon vigorously denied the charge, declaring "I have never known of the oil companies giving to my [election] campaigns in the past and I'm not interested in any contributions in the future."

Little actual progress was made in developing gasohol, as a wave of charges, countercharges, and angry denials was exchanged between the Nebraska Gasohol Committee and Governor Exon. The Beatrice, Nebraska *Daily Sun* editorialized: "No thanks to politics, Nebraska is fumbling away its leadership role in the development and promotion of gasohol. As a result, some other states . . . are picking up the ball and trying to run with it."[4]

Illinois was one of the first states to pick up the gasohol "ball." The Illinois gasohol program was spearheaded by Al Mavis, a burly farmer with a bristling walrus moustache who was employed by the State Department of Agriculture. In November of 1977, Mavis, working out of a small yellow-carpeted office peculiarly located in the carnival section of the Illinois State fairgrounds, arranged for some of the first major sales of gasohol in the Midwest. He persuaded a state industrial chemical distributor to buy ethanol from a Wisconsin distillery, blend it with gasoline, and truck the mixture to a service station in Zion, Illinois. He arranged a festive ceremony to mark the station's first gasohol sales and then met with Archer Daniels Midland officials to urge them to enter the new fuel market.

Mavis's boss at the State Department of Agriculture was John Block, who would later head the U.S. Department of Agriculture under President Ronald Reagan. Block gave his cautious support to the gasohol movement and allowed Mavis a wide range of freedom to promote the concept of a farm-based alcohol fuels industry. Mavis made frequent trips to Washington to testify before congressional committees where he was an outspoken critic of federal policy towards the developing gasohol industry. He also made numerous other public appearances on behalf of gasohol, delivering his speeches in a rapid-fire, emotional fashion which prompted one Illinois newspaper to dub him the "high priest" of gasohol.

Mavis, unlike the more radical Gene Schroder, came from one of the most prosperous farming regions in the nation. His own farm was located in southern Illinois, a region which had been at the center of the agricultural revolution of the post-World War II era. This upheaval had replaced small, diversified livestock and crop farms with much larger farms concentrating on corn and soybeans. Petrochemical refineries had been built outside rural Illinois towns to provide the fertilizers, herbicides, and pesticides which the new energy-intensive cropping systems required. Helicopters sprayed the fields and enormous diesel-powered tractors serviced the farms. Fields that once yielded 50 bushels an acre of corn were made to produce as much as 150 bushels, and those Illinois farmers who survived the consolidation process achieved a measure of prosperity undreamed of in their fathers' day.

During the 1970s, Mavis saw this new farm prosperity threatened by the constantly increasing costs of fossil fuels on which farmers depended so heavily. Crop production, even in the nation's richest farming areas, could not be increased fast enough to compensate for the skyrocketing costs of imported OPEC oil. Mavis became convinced that the farmers' survival would ultimately depend on shifting from oil to farm-based energy sources. But unlike Gene Schroder, Mavis had few qualms about asking agribusiness corporations to help develop the new gasohol industry. He moved easily among the farmers, bankers, food processors, petrochemical dealers, and commercial seed company executives who had importance in the rural Illinois landscape. Mavis saw all of these men and women as part of the "team" that would make the gasohol industry grow.

The more militant wing of the midwestern gasohol lobby was represented by the blue-capped farmers of Gene Schroder's American Agriculture Movement. During the winter of 1978, these angry farmers had concentrated their Washington lobbying efforts primarily on obtaining parity pricing* for their crops—a campaign which made little headway in a Congress dominated by urban representatives whose constituents never failed to vent their anger when food prices increased at the grocery store. Richard Merritt, who frequently encountered the farmers on his daily rounds on Capitol Hill, wanted to persuade them to shift their lobbying efforts from parity pricing to gasohol development. In the long

---

*Parity would mean a return to the relative purchasing power farmers enjoyed in the prosperous year of 1914. In relation to prices of other goods and services, farm produce has been steadily declining ever since, severely straining farm economies and driving millions of farmers off their land.

run, Merritt was convinced, gasohol development would aid the farmers much more, but he was unsure how they would react when he attended an American Agriculture Movement protest rally. The farmers, in a feisty mood, released a herd of goats on the Capitol steps as Merritt struggled to begin a dialogue. "With parity it's you against the American consumer and you're always going to be the loser," he exclaimed. "But with alcohol, it's you and the rest of the nation against the OPEC sheiks. That's got to be a winner and Congress won't vote against you."[5] Merritt's first foray into agriculture territory proved successful, and later helped to influence AAM's leadership to add their support to the midwestern gasohol lobby.*

BEHIND-THE-SCENES HELP FROM 20 MILLION MOTORISTS

One of the most important early allies of the midwestern gasohol movement was the American Automobile Association (AAA), which represents about 20 million dues-paying members as a lobby in Congress. During the 1930s, however, the powerful AAA had been on the cutting edge of the oil industry's antialcohol campaign. In 1933, AAA's executive vice-president Ernest Smith, speaking to a national radio audience on the NBC Grange Hour, denounced power alcohol as "Economically unsound . . . [doing] little or nothing to help farmers while imposing a heavy and unnecessary expense on highway transport."[6]

More than four decades had passed since Smith's radio message, and copies of his speech, along with the negative results of the AAA's 1933 test of power-alcohol blends, now lay buried in a corner of the American Petroleum Institute's library. No one in AAA's new national headquarters, which faces a Washington beltway interchange from a stand of Virginia woods, was aware of these early documents.

Times had changed for the conservative, ever-cautious AAA. The 1973 oil embargo made it clear that the American motorists' cherished personal mobility could no longer be taken for granted. During the oil embargo, panicked motorists who had never asked the AAA for more than an occasional emergency tow and road map swamped the national staff with requests for detailed information on fuel availability. While there might be risks and inconveniences for the nation's motorists in trying to introduce new automotive fuels into the national energy system, there were equally apparent risks in trusting the OPEC nations and the oil industry to keep the traffic moving.

Two men working at the AAA's national headquarters swung the association's executive board behind alcohol fuels. Richard Curry, a former science writer who worked as the AAA's Director of Environment and Energy, did most of the research into the technical feasibility of methanol and ethanol. He authored a series of pathbreaking articles in AAA's widely distributed magazines where

---

*Merritt notes that his courage to deal with the angry farmers "actually came from Bill Krass," a burly carpenter from the U.S. Senate woodworking shop who had become an early enthusiast of alcohol fuels and often helped organize gasohol rallys at the Capitol.

he introduced the gasohol concept to AAA members. The AAA's lobbying and promotional efforts on behalf of gasohol were largely handled by Ray Daley, a Harvard history graduate and ex-Hollywood actor.

When Curry researched fuel alternatives in the wake of the oil embargo, one of the key problems confronting the AAA was its poor public image on the environmental front. Earth Day, 1970 was a particularly uncomfortable memory for AAA executives, who watched in dismay as news clips showed autos being buried at protest rallys. They felt it unjust for the auto to be the only symbol of environmental degradation. "We were being plastered with excess guilt," Curry remembers. Alcohol, he found, was the only alternative that was practical and environmentally benign.

It was no easy task for Curry and Daley to push the staid AAA, which usually lobbied for nothing more controversial than the Highway Trust Fund, into the thick of a heated political fight to establish alcohol fuels as a recognized alternative. They walked a delicate organizational tightrope. The least misstep could have discredited their efforts and left their in-house critics free to steer the AAA back onto a more traditional course of action.

Daley worked closely with the more retiring Curry to launch the AAA's gasohol campaign. As a special assistant to AAA president James Creal, Daley had been involved in a number of consumer issues; he was also chairman of a Federal Energy Administration consumer advisory committee. In the spring of 1977, the committee invited gasohol proponents from Congress, including Rep. Dan Glickman, a Kansas Democrat, and Rep. Virginia Smith, a Republican from Nebraska, to speak to the unofficial advisory panel. Daley had also rounded up a number of experts and spent several days telephoning the Washington press corps about the event. The FEA bureaucrats, who were used to the committee's usually sparsely attended meetings, were surprised to find the government hearing room packed. Daley told the committee: "The idea of going back to grain to help our automobiles has an ironic twist. The guy whose voice was the loudest in shouting for Henry Ford to get a horse was the local grain dealer who was selling the fodder to feed those horses. Well, it may be that the pendulum is swinging back and once again grain will be feeding our horses—the ones that power our engines."[7]

Shortly afterwards, the AAA hosted a breakfast for Senate and House staff members where alcohol fuels experts from the Department of Energy, U.S. Department of Agriculture, and private industry were brought in to discuss various aspects of their work. While Congressman David Emery of Maine introduced such speakers as Gary Shultz from Conoco or Dwight Miller from the USDA's Peoria, Illinois lab, it was Daley who had pulled strings and done the legwork in getting speakers lined up and their travel approved.

An endless series of meetings with government and industry officials evolved, and soon Curry and Daley were working full time on alcohol fuel. In contrast to Richard Merritt, who would publicly accuse the federal energy bureaucracy of conspiring to block alcohol fuels development, Daley met privately with key energy department officials to try to woo their support, usually using material researched by Curry. In January, 1978 Daley told a Senate committee: "It is our [AAA's] opinion that alcohol is the best potential energy source which can be made available in adequate quantity to provide the alternative fuel sources necessary to help power America's motor vehicle fleet in the years ahead."

AAA was never openly attacked by the oil industry for its gasohol lobbying efforts, but there was considerable consternation in the industry ranks over AAA's high-profile support for gasohol. In August, 1978 Exxon vice-president Ed Hess flew in from Houston to try to persuade AAA vice-president John DeLorenzi to tone down the association's efforts. Instead, Hess found himself quickly backtracking as he was called down, point by point, on his uninformed generalizations about alcohol. Hess believed that phase separation was an insurmountable problem and that alcohol would ruin a car's engine—even in a gasohol blend. "In essence," Daley recalls, "Hess knew very little about alcohol fuels. I quoted back to him reports by Amoco, Phillips Petroleum, and Gulf which completely refuted what he had to say. . . . I even used some of Exxon's own statistics." Within the week, Exxon's New Jersey research center had asked for two barrels of ethanol from an Illinois distributor on Hess's orders.[8]

The AAA remained squarely in the gasohol camp and Daley was given permission to organize a gasohol promotional giveaway on the front steps of the Capitol. It took six frustrating months of work to locate an alcohol supply, line up a tank truck, and negotiate with the Washington police, the Capitol police, and the fire department. Daley contacted Publicker for the supply, but the company was not interested. He finally talked ADM into shipping 600 gallons at cost, and called a local gasoline distributor, Dave Fannon, for help with the tank truck. Fannon agreed to help out, but when he contacted his major supplier, Cities Service, he was warned that the tank truck could not bear any company markings, and that the company did not want to be linked in any way to the gasohol controversy.

Finally, on June 28, 1979, the AAA gasohol giveaway was staged. Thirty-five senators and congressmen stopped to fill up their cars from the 8,000-gallon silver tank truck Fannon operated. In the meantime, Daley—flanked by AAA president James Creal and Richard Merritt, who was frantically passing out bumper stickers—preached the virtues of gasohol to the assembled members of congress and the press. Forty-six years after the AAA's power-alcohol-blend test, the association was helping alcohol fuel return to the nation's capital.

THE OIL INDUSTRY VERSUS THE GASOHOL LOBBY

The oil industry's efforts to fight the growing political influence of the gasohol lobby paralleled, in many respects, the earlier campaign waged against the chemurgists. Gasohol was denounced by oil industry spokesmen as a technically inferior, economically unattractive fuel which would help neither farmers seeking higher grain prices nor motorists seeking plentiful supplies of fuel. Many oil company officials conducted strong lobbying campaigns at both the state and national levels to defeat legislation promoting the development of ethanol and methanol fuels. When articles extolling the virtues of these fuels were published in newspapers, they often elicited sharp rebuttals from oil industry public relations experts.

The first major political skirmish between the oil industry and alcohol advocates took place in September, 1977 in California. At that time, a bill was introduced by California state assemblyman Daniel Boatwright which would fund a test fleet of methanol-powered state vehicles. Chevron Oil, a company which had conducted several unsuccessful tests on methanol-gasoline blends, was strongly opposed to Boatwright's bill.

A few hours after Boatwright officially submitted this legislation to the state assembly, he was confronted by an angry Chevron lobbyist, Al Schults, in the hallway outside his office. "Chevron has done all this research before, and our company president ended up with his car stalled," Schults exclaimed. Boatwright, one of the most powerful figures in California state politics, was late for a luncheon speech and in no mood to debate the issue. "If you've got some information," he retorted, "hand it over and leave me alone."[9]

Two weeks after the hallway encounter, Schults visited Boatwright in his office to drop off a stack of Chevron test reports detailing the problems its researchers foresaw with methanol fuels. Shortly after Schults's visit, the Chevron management distributed a leaflet at its Richmond, California refinery (which was located in Boatwright's voting district) urging its employees to write to their state representatives about pending alcohol fuels legislation. "If Rolaids will relieve acid indigestion, what will relieve gasohol?" headlined the appeal.[10] Boatwright began receiving a barrage of letters from Chevron employees and stockholders opposing his methanol bill. "All of the letters were the same," Boatwright recalls. "They said that this methanol project was not something the government should be getting into and should be left to private industry." Despite Chevron's lobbying efforts, Boatwright's bill was passed by the California assembly, and in January, 1978 a team of contract engineers began initial design work on a methanol-powered automobile.

Undaunted by its California setbacks, Chevron maintained an active antialcohol campaign throughout the mid-1970s. This campaign was headed by Robert Lindquist, a Chevron chemist who had conducted much of the company's initial research into methanol-gasoline blended fuels. Lindquist began speaking forcefully against alcohol fuels on radio talk shows, television debates with gasohol proponents such as Al Mavis, and at scientific meetings. He lashed out against both methanol and ethanol with a sense of passionate conviction that shocked some of his more moderate colleagues in the oil industry. He gained considerable attention with his frequent public appearances and was one of the first oil company representatives invited to brief the Ninety-fourth Congress on alcohol fuels. On November 18, 1977 Lindquist handed out a six-page Chevron position paper to a group of twelve U.S. senate aides seated around a large mahogany table in the Senate Agriculture Committee hearing room. The paper charged that alcohol fuels were uneconomical, unsafe, and hampered by severe technical problems. "Alcohol fuels," Lindquist testified, "have been proposed throughout the history of the motor car, but every time they've been rejected on economic and technical grounds. The emotional appeal of running cars on alcohol from agricultural sources has obscured technical difficulties and costs."[11]

Fred Wahl, Sen. Frank Church's staff aide, commented after the November, 1977 meeting that "It seemed that it was Lindquist's mission in life to resist this thing . . . to throw his body in front of a truck if that's what it took to stop this stuff. It was his crusade."[12]

Chevron's campaign to discredit the gasohol movement was joined by Mobil Oil Company in the fall of 1977. Mobil began to promote vigorously an alternative to gasohol—a patented process which could convert either ethanol or methanol into synthetic gasoline. The prototype model conversion unit was built at a cost of $15 million and was housed inside a metal warehouse outside the Mobil research laboratories in Paulsboro, New Jersey. It could convert 4 units of alcohol into 1.5 units of gasoline. The core of this prototype was a slender metal cylinder 4 inches wide and 25 feet tall. Alcohol was funneled through this catalyst's computer-designed chambers at high pressures and temperatures. The simple alcohol hydrocarbon group was broken down and then restructured into the long, complex hydrocarbon chains that make up gasoline.

Before deciding to develop the process, Mobil vice-president John Wise assured reporters, "You can bet we checked into using methanol as a fuel very thoroughly. . . . We weren't about to spend millions of dollars to develop a technology that nobody needs."[13] Other companies in the transportation business, however, had come to quite different conclusions. Volkswagen, for example, concluded in a company report that methanol-gasoline blends could be formulated to operate quite well, and that the pure methanol was a "very attractive, clean-burning alternative fuel for automobiles with relatively minor problems which can be overcome." Although Wise was careful not to say that problems with methanol blends were impossible to overcome, he insisted that they were so severe that "it would be cheaper to simply convert the methanol into gasoline."[14]

With a single technological breakthrough, Mobil hoped to steal some of the gasohol lobby's thunder and slow the gasohol movement. The company hoped to persuade the American public that the best possible use of both ethanol and methanol fuels was not at the service station pump but as a feedstock for factories that would churn out synthetic gasoline. When supplies dropped, Mobil would harness its vast coal reserve holdings to produce methanol which would be converted into "M-Gas."

Much of the press hailed the Mobil breakthrough as what one publication termed "the final link in the coal-to-motor fuel process chain."[15] But the outspoken Richard Merritt demanded, "Why would anyone want to go and take a perfectly good, clean-burning fuel like methanol and transform it, via an extremely energy-intensive process, into a highly polluting fuel like gasoline?" Merritt turned the oil industry's negative-energy-balance critique of gasohol on its ear, likening the Mobil process to "taking four pounds of steak and converting it into a pound and a half of bologna."[16]

Daniel Horowitz, a chemist with the U.S. Patent Office who supervised the patent issued to Mobil for its process, also had doubts about the efficiency of the new technology. When Mobil's top lobbyist, Joe Pennick, wrote a letter to the *Washington Post* in May, 1978, arguing that alcohol fuel development should be abandoned in favor of the Mobil process, Horowitz decided the time had come to speak out. "The overall Mobil process is so inefficient," he wrote Sen. Carl Curtis in an angry rebuttal to the Pennick letter, "that by rough estimates only a fraction of an ounce of gasoline would be produced from a pound of coal." Horowitz said Mobil's claims were "misleading" and insisted, "If Mobil tries to confuse the public by saying the process is an answer to the energy problem they are stretching it to the breaking point."[17]

The Mobil campaign to promote its synthetic gasoline process had also become embroiled in controversy in November, 1977, when the U.S. Senate Republican Conference, a congressional research arm of the Republican party, released an optimistic report on the economics of producing ethanol from grain. The report, commissioned by Nebraska Senator Curtis, recommended that the nation move rapidly ahead to develop a national alcohol fuels industry. Mobil researchers prepared a lengthy critique of the report and delivered it to Curtis with a letter from Pennick stating: "We feel the consumer would not benefit by legislation that requires alcohol to be used as a motor fuel. His car would perform no better, the air would be no cleaner, the costs, whether hidden or not, would be higher because of the higher cost per gallon for alcohol . . . and the expenses of a more elaborate delivery system. We think a better way to expand gasoline supplies in the future would be to make alcohol from coal and then turn it into gasoline using Mobil's conversion process."[18]

Curtis was unconvinced, and in a speech on the Senate floor he derided the Mobil process, which "required $15 million to build a plant with a capacity to produce 100 gallons a day of gasoline. . . . In sharp contrast, a plant producing 54,000 gallons of ethanol a day can be constructed for $26 million."*[19]

Mobil's promotion of its synthetic gasoline process, despite the favorable write-ups in the trade press, fell far short of its mark, in terms of countering the gasohol movement, even though the process is still under serious consideration by coal-developing consortiums. The promotion never seriously weakened the growing political clout of the gasohol lobby and, indeed, only served to strengthen the credibility of Merritt's oft-repeated charge that some sort of shadowy oil company conspiracy existed to suppress gasohol. And the cantankerous Curtis, with the aid of patent official Daniel Horowitz, had succeeded in raising serious questions about the viability of the Mobil synthetic gasoline process.

Although Chevron and Mobil were by far the most vocal of the oil companies participating in the early congressional gasohol debates, they by no means represented an industry-wide position on development of the fuel. While Chevron and Mobil publicly attacked gasohol, several oil company research labs were

*Curtis was, of course, comparing a pilot research plant with a well understood technology. In practice, finished plants producing either M-Gas from coal or ethanol from corn would sell their product at very similar prices.

quietly coming to quite different conclusions about its potential. Foremost among the gasohol movement's allies within the oil industry was Continental Oil Company (Conoco), the same company which had encouraged Gary Schultz to participate in the Nebraska 2-million-mile road test. Conoco's coal subsidiary was one of the largest coal reserve holders in the United States, and its president, Jimmie Bowden, openly advocated using methanol in fleet vehicles, in blends with gasoline and in turbine fuels. The oil company, meanwhile, had been working on a project called Seameth that involved converting natural gas into methanol and then shipping it as a liquid, rather than run the risks associated with transporting liquefied natural gas (LNG).

Conoco's Washington lobbyist, Mike Ware, was pleased that Bowden and Schultz were speaking out. He was convinced that the political winds of the eighties would bear the strong scent of alcohol, and he concluded, "The worst thing that the [oil] industry could do is to take a hard line against alcohol fuels. It would not only be a no-win situation. It would be an absolutely must-lose situation." Ware recalls, "the whole issue has a lot of support with groups outside the [oil] industry that the industry might need somewhere down the road. I felt the best thing we could do was to keep an open mind."[20]

Ware quietly lobbied the American Petroleum Institute, the oil industry's Washington trade association, to adopt a more moderate position towards alcohol fuels development. His efforts were supported by other API member companies such as Occidental Petroleum, whose fiercely independent president, Armand Hammer, had already expressed his support for alcohol fuels to Senate aides. But Ware was still wary of allowing Continental to publicly align itself with the gasohol movement. He strongly opposed allowing Gary Schultz's participation in a volatile panel debate over the merits of gasohol at the National Press Club in the fall of 1978. One of the scheduled participants in the debate was Jack Anderson, who had written several columns sharply criticizing the oil industry's attempts to fight the gasohol movement. One Anderson column, published in August, 1977, charged, "The oil industry, of course, opposes developing rival alcohol fuels. This not only would threaten oil profits but could break the industry [fuel] monopoly."[21] Ware was afraid that anything Schultz said would be taken out of context later and appear in some future Anderson column. "It was a very hard decision for me to make," Ware recalled, "but I felt that Gary was somewhat naive and I didn't want him to get us into trouble."[22]

In the late 1970s, with Conoco interested in methanol, and Occidental, Gulf, and Amoco interested in ethanol, the American Petroleum Institute decided to change its negative stand. Originally, API had based this stand on a task force that involved representatives from six oil companies—Exxon, Mobil, Shell, Amoco, Sun, and Chevron (represented by Lindquist). The 1975 API Alcohol Fuels Task Force report concluded that "alcohol fuels could not be profitably sold" as long as oil was readily available.

This position stood for some years, but was altered in surprising testimony to a DOE Alcohol Fuels Policy Review Task Force hearing in December, 1978. API's Jack Freeman, of Sun Oil, told the DOE group that positive energy balance was possible and that ethanol "has excellent combustion properties." While API did not believe that crop-derived ethanol would make a major contribution in the future, Freeman told reporters, "If this is a change in viewpoint, it's because times have changed."[23]

FEDERAL RESISTANCE TO THE GASOHOL LOBBY

The federal energy bureaucracy which greeted Jimmy Carter upon his inauguration in January, 1977 had done little to develop a U.S. alcohol fuels industry since the 1973 Arab oil embargo. The top echelons of the energy bureaucracy were surprisingly ignorant of the energy potential of alcohol fuels. This was amply demonstrated to Richard Merritt when he first tried to obtain some federal assistance for Nebraska's gasohol program. On April 16, 1978 Merritt arranged a meeting to discuss the Nebraska program with George Hall, one of the Carter administration's senior energy advisors, who was putting the final touches on a new energy plan which Carter would soon unveil to the nation. During that meeting, Hall expressed an almost total lack of knowledge of the technology involved in alcohol fuel production and refused to commit any federal support for the Nebraska program. Thus, Merritt was not surprised when the program announced by Carter four days later made virtually no mention of alcohol fuels. Carter preached that the nation was faced with the "moral equivalent of war" and grimly warned that the United States "will have to rely for at least the next two decades on conventional sources of energy."[24]

As evidenced by Merritt's encounter with Hall, there was a woeful lack of practical information on alcohol fuels technology at the policymaking level of the new Carter energy bureaucracy. But these new energy officials could have quickly educated themselves on the subject by simply reviewing prior federal research reports on alcohol fuels tucked away in the government archives. The first such federal report was issued in 1906, at a time when Congress was debating whether or not to lift the tax on industrial alcohol to encourage its use as a fuel. The Department of the Interior sponsored two studies in 1911 which contained a surprising amount of sophisticated data on engine performance with alcohol fuels. And two detailed Department of Agriculture reports were issued during the 1930s: the first in 1933 during the Great Depression and the second in 1938 when the chemurgists were pushing for legislation to mandate the use of power-alcohol blends. And in 1939, a National Resources and Policy Committee, acting at President Roosevelt's request, did an extensive survey of the potential use of both methanol fuels from coal and ethanol fuels from agricultural products. "Our energy resources," Roosevelt sternly warned in an introduction to the report, "are not inexhaustible, yet we are permitting their waste in their use and production. In some instances to achieve apparent economies today, future generations will be forced to carry the burden of unnecessarily high costs and to substitute inferior fuels for particular purposes. National policies concerning these vital resources must recognize the availability of all of them."[25]

But alternative fuel development would be largely ignored by the federal government until the OPEC oil embargo and the price hikes of the early 1970s. As the OPEC nations managed to steadily boost the world price of oil, the federal energy bureaucracy was finally roused to action. In a special White House report released in the fall of 1973, Richard Nixon outlined a plan to free the United States of its dependence on all foreign oil imports by 1980. This plan made no mention of using ethanol fuels but did include the use of coal-derived methanol. However, at an Interior Department conference staged to review the White House plan, representatives from both the oil and automotive industries voiced sharp opposition to a national energy program emphasizing methanol rather than synthetic gasoline fuels. Only a handful of academics and representatives of European-owned companies objected to their assertions that methanol could not be used as a transportation fuel.

As Nixon's "Project Independence" began to get under way, a major recruitment effort was made to hire talented new individuals to help the energy bureaucracy carry out its alternative fuel program. These new recruits entered federal service with high expectations, convinced that the government was ready to launch a new energy program that would rival the NASA space program in size and scope.

Eugene Ecklund was one of the early recruits to join the Nixon energy bureaucracy. The son of Swedish immigrants, Ecklund was raised in Minneapolis. After receiving a degree in electrical engineering, he joined the Navy and performed technical intelligence work for the Allied forces during World War II. He was appointed to a special team of technical experts authorized to study captured German documents which detailed the technological underpinnings of the Nazis' massive synthetic fuels industry.* When World War II was over, Ecklund turned down an opportunity for a Washington desk job with the Navy to try his hand in the corporate world. Over the better part of three decades, he compiled an impressive record as a capable executive who could take over a slumping corporation and turn it into a healthy, profit-making venture. He was a "can-do" man, an expert at figuring out how to bring innovative new technologies into the commercial marketplace.

In the winter of 1974 Ecklund joined a tiny three-man division set up inside the Environmental Protection Agency's regional office in Ann Arbor, Michigan to assess the potential of alcohols and other promising alternative fuels. He soon

---

*In 1944, Congress approved funds for a pilot plant that would produce synthetic gasoline using these captured Nazi processes, and facilities were set up in Louisiana, Missouri, and Bruceton, Pennsylvania. The Louisiana plant produced gasoline at an astonishing 1.6 cents per gallon. But a 1952 review by the National Petroleum Council concluded the cost of the synthetic gasoline to be 41 cents per gallon. They arrived at this figure by adding all funds spent on the research project—including housing for scientists and staff—and dividing that figure by the amount of fuel produced. The election of a business-oriented Eisenhower administration prompted a complete funding cut in March, 1953, and in June the plants were permanently closed. Democrats in Congress were outraged and charged that the oil companies had influenced government policy. Two decades later, Bureau of Mines chief John O'Leary called it "the most serious error in energy policy in postwar years." The documents gathered up by the allies are now stored at Texas A & M University. Department of Energy funding for the university's Document Retrieval Program was cut in 1979, with only about 20 percent of the documents recovered.

became one of the federal bureaucracy's few experts on the technical performance qualities of alcohol fuels. He enjoyed his work at the small regional office far away from the stifling layers of the Washington bureaucracy.

Then, in January, 1975, Congress set up the Energy Research and Development Administration (ERDA) to speed up the government's program to develop new energy resources. Ecklund's division was transferred to ERDA but soon was becalmed by funding problems and a lack of support from the new agency. At one point funding for Ecklund's division was frozen for a full year, leaving it in "a kind of no-man's-land." Ecklund found that even when ERDA allocated funds for his division, progress was painstakingly slow, frustrated at every turn by bureaucratic apathy and inertia. "If I look at the bottom line, I totally failed at ERDA," Ecklund would sadly conclude. "Throughout my previous career, if I had to get something done, I could always get it done. But at ERDA, I was never able to penetrate the system."[26]

The system Ecklund encountered was dominated by men who were convinced that synthetic gasolines produced from coal and oil shale should be developed long before any effort was made to commercialize alcohol fuels. This view was reinforced by a January, 1976 ERDA report, a follow-up on the 1973 Interior Department conference, that made no mention of ethanol and considered synthetic gasolines to be a far more promising alternative than methanol from wood or coal. The report's summary stated: "The only private institutions likely to undertake synfuel ventures are the oil companies, because they have the most compelling incentives—an existing business that requires a continuing supply. . . . This dominating interest by oil companies will inevitably shape the choices of synfuels to be produced." The summary noted that while "methanol technology is closer to being commercially ready" than the synthetic gasolines (or syncrudes), its adoption by the oil companies would be extremely unlikely since "methanol would not fit readily into the existing marketing system." Therefore, the report concluded in something of a self-fulfilling prophecy, "Syncrude is the most institutionally preferred product and will dominate."[27]

The oil industry's long-standing interest in coal and shale oil syncrudes did not mean that these new fuels would be commercialized without substantial financial assistance from the federal government. ERDA doled out grant awards to oil companies to bring new fuels such as Gulf's Solvent Refined Coal (SRC), Exxon's Donor Solvent (EDS), and Dynalectron's H-Coal to the gas pumps. Numerous experimental plants were built but by the end of the decade in which Richard Nixon hoped to acquire energy independence, there still was not a single commercial-scale syncrude plant under construction in the United States.

ERDA died a quiet death in the winter of 1977 when it was merged into the Department of Energy, the new superbureaucracy created by the Carter administration. During his final days with ERDA, Ecklund worked in isolation from policy experts, such as George Hall, who were preparing Jimmy Carter's energy program. A cautious man by nature, Ecklund felt little compulsion to take a Merritt-style advocacy approach to convince his superiors to give priority to alcohol fuels development. Such an outspoken posture within the narrow confines of the energy bureaucracy would have weakened Ecklund's reputation for dispassionate objectivity, which he prized.

After ERDA's dismantling, Ecklund took a midlevel position in DOE, where he expected to continue to push quietly for an expanded federal test program for alcohol engines. He had little inkling that in less than a year's time after his arrival at DOE, he would find himself thrust squarely into the middle of a stormy political controversy and forced to appear before hostile congressional investigating committees. It would be Ecklund's task to try to defend the energy bureaucracy's muddled alcohol fuels policy. It would be neither Ecklund's nor DOE's finest hour.

THE GASOHOL LOBBY CONQUERS WASHINGTON

In the fall of 1977, one year after his arrival in Washington, Richard Merritt's gasohol campaign began to take on both form and substance. Merritt junked his old sports coats in favor of a closet full of well-tailored suits. He was frequently asked to testify before congressional committees and gave countless interviews to the press. The gasohol lobby was rapidly consolidating under the umbrella of the National Gasohol Commission, which would eventually be headed by Al Mavis. It was beginning to make its political muscle felt in Washington.

First and foremost among gasohol's congressional allies was Nebraska's Sen. Carl Curtis. He strongly believed that the use of surplus grains in the alcohol fuels industry could help stabilize prices on the chaotic grain markets at a level that most farmers could tolerate. This might enable the U.S. Department of Agriculture to begin to cut back on the massive federal subsidies that midwestern farmers required throughout most of the 1970s in order to stay afloat. Curtis waged a solitary battle in the early seventies to develop ethanol fuels, often cutting other appointments and making surprise appearances at hearings if alcohol fuels were on the agenda. This conservative Republican senator could never muster much support from his congressional colleagues until the fall of 1977, when, thanks to Nebraska's road test and European research, a subtle shift in attitudes toward the gasohol issue was taking place. Eventually, a host of other farmbelt senators—including Republicans Robert Dole of Kansas and Charles Percy of Illinois and Democrats Birch Bayh of Indiana and Frank Church of Idaho—began to push for further federal assistance to the gasohol industry.

The congressional gasohol lobby also included, by the fall of 1977, a number of senators representing largely urban constituents. These were led by Republican Jacob Javits, the dean of the Senate Foreign Relations Committee. As the senior senator from New York, Javits represented a large pro-Israel constituency that was concerned about the growing political clout of Arab nations as well as the need to drastically reduce U.S. dependence on imported oil. After studying the available liquid fuel alternatives, Javits decided to make assistance to an alcohol fuels industry a top legislative priority. It was hoped that the creation of this domestic industry would slow the dollar drain to the OPEC nations which a Senate Foreign Relations Committee probe indicated was jeopardizing the stability of the U.S. banking system.

The committee found that U.S. banks had been hard pressed to find sound investment outlets for the flood of petrodollars arriving in their vaults from OPEC nations. Much of this OPEC money was eventually loaned out to financially strapped Third World nations struggling to pay off oil-import bills. The committee warned in a report co-authored by the Joint Economic Committee in the fall of 1978:

> The point may come when one or several countries will find it more in their interest to simply default or repudiate their external debts rather than to have to continue to repay old loans. If this happens, the domino effect could take place in which other debtor countries follow suit; the banks panic and start calling in their international loans; the stock market drops precipitously; and the international capital market collapses."[28]

Committee investigators labeled this a "doomsday scenario" but one that was "taken seriously enough by responsible officials."

Growing tensions in the Middle East and the Persian Gulf were also major factors in the support that Javits and many other senators showed for rapid growth in the gasohol industry. The Saudi Arabian government, for example, warned that any new war between the Arab states and Israel would immediately jeopardize the flow of Arab-OPEC oil to the United States, as it did during the 1973 oil embargo. And the view that a small, well-organized team of terrorists could easily stop the flow of oil through the Persian Gulf was widely held. Privately, Javits predicted this catastrophic event as a certainty. Although the United States obtained only about 11 to 15 percent of its oil from the Gulf during the seventies, international oil-sharing agreements would have put a much larger dent in U.S. supplies if oil to Europe and Japan were to be cut off as well. A shut-down of the Persian Gulf had become almost as unthinkable as a nuclear war, and the response of the Carter administration was to shore up the secret presence of a Saudi-paid, U.S.-equipped army of mercenaries operating in Saudi Arabia, as well as maintaining a strong naval presence in the Indian Ocean.

Meanwhile, many congressmen saw a clear need for alternatives. In the fall of 1977 a flurry of congressional activity began to develop as the Congressional Research Service, at the request of Senator Javits's office, launched an extensive review of Library of Congress documents containing information on alcohol fuels. During the same period the Senate Republican Conference Committee conducted its economic feasibility study of gasohol for Senator Curtis. The first substantive action was taken by the Senate Agriculture Committee, which approved a bill requiring $60 million in federal loan guarantees, to be awarded through the USDA, for experimental large-scale alcohol and biomass energy facilities.

Merritt now found dozens of young congressional aides eager to meet with him and discuss possible legislation to stimulate the new industry. Within a year two key federal tax measures in support of alcohol fuels had been signed into law: a 4-cent-per-gallon federal excise tax exemption on any fuel mixture containing at least 10 percent ethanol or methanol (providing the alcohol was produced from renewable resources) and a 10 percent tax credit for all investments in alcohol production equipment. The combined impact of these two measures provided a powerful federal economic stimulus for the gasohol industry at a critical point in its early development. The 4-cent federal tax exemption, which translated into a 40-cent tax break per gallon of anhydrous ethanol, when combined with numerous state tax exemptions, made the price of gasohol competitive with unleaded premium fuels throughout much of the Midwest. And the 10 percent investment tax credit, when added to an already existing 10 percent industrial tax credit, added up to a whopping 20 percent tax break for entrepreneurs and corporate groups interested in bankrolling the construction of new alcohol plants. In the months which followed the enactment of these two measures, the gasohol industry entered a period of explosive growth.

The early congressional enthusiasm for gasohol was not shared by DOE or the USDA, the two agencies to which was delegated most of the responsibility for implementing alcohol fuels legislation. In an effort to break the bureaucratic logjam inside the two agencies, Merritt helped Sen. Birch Bayh's staff draft a letter that urged DOE Secretary James Schlesinger and Secretary Bob Bergland to "undertake an immediate and comprehensive effort to tap the energy potential of our nation's renewable resources." Signed by nineteen senators, it went on to note that:

> In recent weeks we have witnessed a burst of interest in a very old energy source. Alcohol fuels derived from the produce of our nation's farms and forests, and by reclamation of wastes, are the most apparent and exploitable renewable energy sources available today. Ethanol and methanol have been used successfully in internal-combustion engines for decades.
>
> Whatever may be the results, we cannot afford to give less than a major national commitment to develop fuels from the productive capacity of our farm and forest lands. This is especially true at a time when our government is asking farmers to leave part of their farm land idle due to excess supplies of farm products.[29]

James Schlesinger, the pipe-puffing Secretary of Energy, initially viewed the sudden burst of enthusiasm for alcohol fuels with an air of somewhat bemused detachment. In gentlemanly discussions behind closed doors, his top aides pegged gasohol as a crackpot issue, taken seriously only by a handful of farmbelt fanatics. Deputy Secretary of Energy John O'Leary bluntly told the AAA's Ray Daley that alcohol fuels were "a myth" with little practical importance in solving the national energy crisis, and later told reporters that gasohol was "the new laetril."*[30] Dale Myers, Assistant Secretary for Energy Research and Development, sent out a memo requesting that alcohol fuels be excluded from a DOE study

---

*In a National Press Club breakfast for reporters, O'Leary responded to questions on alcohol by saying, "It's not strategic to use food for fuel," and he refused to answer follow-up questions. Cornered by a group of reporters at the elevator, O'Leary claimed his source for such a statement was DOE research. But DOE researchers contacted by these reporters exclaimed, "That's a disasterous way of approaching it."

targeting new technologies ready for commercialization. But Myers was forced to back down from this directive after Sen. Frank Church obtained a copy of the memo and threatened to release it to the press, along with a bitter blast at DOE. Church, who had introduced a Senate bill mandating 10 percent gasohol blends in all fuel by 1990—a goal that may have been politically unrealistic—was so frustrated by DOE's intransigence that he called Assistant Energy Secretary for Policy Al Alm into his office and delivered a bitter hour-long personal attack at the DOE leadership.

Senator Church, like most elected representatives from strong agricultural states, was an outspoken advocate of alcohol fuels for reasons of political survival as much as energy development. His home state of Idaho grows potatoes and sugar beets, proven feedstocks for alcohol, and he knew there was sorely needed political capital to be gained in supporting alcohol fuels development. He expected a difficult reelection campaign in 1980. Indeed it was; he lost his seat in the Senate, as did Birch Bayh of Indiana and George McGovern of South Dakota, other vocal advocates of alcohol fuels.

The intense bureaucratic arm-twisting attempted by Church and others in Congress, such as the young Democratic Congressman from Kansas, Daniel Glickman, did yield one tangible, albeit modest result: the creation of a special DOE Task Force to study alcohol fuels. DOE promised Congress in July, 1978 that the final report of this new task force would be ready by January of 1979.

Marilyn Herman, a young veteran of the Washington political scene, was chosen for a key leadership position on the task force. She had worked as a political aide to Sen. Abraham Ribicoff, a Connecticut Democrat, before moving on to a job as a congressional liaison officer for the Federal Energy Administration. Herman was partial to, and liked by, the gasohol lobby, and her close ties to Capitol Hill made her highly suspect in the byzantine world of the DOE bureaucracy. Her meetings with Capitol Hill "gasoholics" were usually held in a small delicatessen north of the Capitol.

When Herman joined the DOE task force, veteran bureaucrats within DOE refused to take the new group seriously, viewing its formation as little more than an overblown political gesture to the powerful congressional gasohol lobby. And they resented the intense politicization of what they viewed as a technical issue that should be left to the experts. As the task force progressed in its research, it became apparent that Herman would push for a favorable final report on alcohol fuels, and several attempts were made by DOE officials to remove her from her job. After several bitter internal agency battles, they failed, primarily due to Herman's support on Capitol Hill.

Meanwhile, the White House domestic policy staff began to take an active interest in alcohol fuels, viewing it as an outstanding example of the type of alternative energy development that a proposed windfall profits tax on the oil industry could help finance. Moreover, it was clear that their popular appeal in the Midwest might translate into farmbelt votes in the 1980 presidential election. The White House staff began to work with the DOE task force to bring out the final version of the alcohol fuels policy document as soon as possible. With each day that passed, the need for a clearly stated Carter administration alcohol fuels policy became more urgent. The Shah of Iran was overthrown and world oil supplies tightened up, while a confusing DOE gasoline allocation threatened to exacerbate the impacts of the temporary shortage. During this tense period, Herman and her boss, Al Alm, were directed to bypass the DOE bureaucracy and work directly with the White House to develop a coherent national alcohol fuels policy.

By the time the gasoline crunch hit in the summer of 1979, public interest in alcohol fuels had skyrocketed. National news networks broadcast reports on the opening of midwestern gasohol service stations and invited oil industry representatives to their studios to debate with midwestern gasohol advocates. Finally, on July 12, 1979, the DOE task force belatedly released its report. A press conference was called in a seventh-floor conference room of DOE's For- restal Building where Al Alm fielded a barrage of questions from reporters. Alm said DOE was ready to launch an intensive effort to encourage production and use of renewable alcohol fuels. As table 5-1 illustrates, the report indicated that new production technologies would make it possible to produce over 100 percent of the nation's automotive fuels by the year 2000. It should be noted that this projection used the most optimistic assumptions about the country's ability to divert its natural resources into alcohol. One analyst who had been a high official at DOE and involved in the preparation of the report called the projected max- imum figures "absurd."

In its introduction, the report noted that "a genuine grass roots movement" had developed to give alcohol fuels "an unprecedented broad base of support."[31] The strength of this movement was reflected in the blizzard of legislation pro- posed for alcohol fuels in the Ninety-sixth Congress (1979–80). Eighty-two separate bills were introduced in the House and Senate to establish a broad range of federal incentives, ranging from greatly relaxed regulatory restrictions of the Treasury Department, making licenses to produce alcohol easier to obtain, to massive loan programs to help finance the construction of new large-scale plants. Over 10 percent of the bills passed—an excellent record by Capital Hill standards. (Typically, only about 2 percent of bills introduced actually pass and become law.)

One piece of congressional legislation created the National Alcohol Fuels Commission (NAFC), which conducted a series of hearings around the nation and released its findings to the President in the winter of 1981. The NAFC, chaired by Senator Bayh, was composed of congressmen representing farmers, auto makers, and organized labor. Its final report strongly supported stepped-up federal support for both large- and small-scale ethanol producers and also urged that pure methanol- and ethanol-powered vehicles be developed by the U.S. automotive industry. While the report pleased the gasohol lobby, it got the cold- shoulder from the incoming Reagan administration.

As support for alcohol fuels built towards a crescendo during the gasoline- starved summer of 1979, the Carter White House attempted to make good on its new commitments toaid the industry. Charles Duncan, a former Coca-Cola

TABLE 5-1
Projected Maximum Alcohol Production from U.S. Biomass Resources¹ [Billion gallons per year] (Source: The Report of The Alcohol Fuels Policy Review, June 1979, U.S. Department of Energy)

	1980		1985		1990		2000	
	Ethanol	Methanol	Ethanol	Methanol	Ethanol	Methanol	Ethanol	Methanol
Wood	23.5	86.3	21.8	80.2	20.2	74.2	25.8	95.0
Agricultural residues	9.1	33.4	10.3	38.1	11.3	41.5	13.1	48.1
Grains:								
Corn	2.3		2.1		0.9		2.0	
Wheat	1.2		1.4		1.6			
Grain sorghum	0.4		0.3		0.3		0.3	
Total Grains	3.9		3.8		2.8		2.3	
Sugars:								
Cane			0.2		0.7		0.7	
Sweet sorghum			0.2		3.0		8.3	
Total Sugars			0.4		3.7		9.0	
*MSW	2.2	8.6	2.3	9.2	2.5	9.9	2.9	11.6
Food processing wastes:								
Citrus	0.2		0.2		0.3		0.4	
Cheese	0.1		0.1		0.1		0.2	
All Other	0.2		0.3		0.3		0.3	
Total processing wastes	0.5		0.6		0.7		0.9	
Total	39.2	128.3	39.2	127.5	41.2	125.6	54.0	154.7

¹Based on the following biomass—alcohol conversion factors: Wood and agricultural residues—173 gal. methanol per dry ton, 47 gal. ethanol per dry ton. Corn—2.6 gal. ethanol per bushel. Wheat—2.7 gal. ethanol per bushel. Grain sorghum—2.6 gal. ethanol per bushel. Sugars—136 gal.

executive who replaced James Schlesinger at DOE in October of 1979, immediately granted greater access to the agency's few in-house alcohol advocates, and he traveled to the Midwest to inspect both large- and small-scale plants in his first official trip outside Washington, D.C. While visiting with instructors from community-college training programs, Duncan enthusiastically declared: "We have to make gasohol one of our primary alternative fuels. And we have to make available to people the know-how to produce gasohol on a small scale as well as on a large scale."[32] Duncan also ordered an inventory to be made of shut-down beverage distilleries and alcohol plants under construction which could be used in case of emergency, reasoning that alcohol would be the swiftest replacement for gasoline in a time of crisis.

The change in Carter-administration policy was also reflected in the attitude of Agriculture Secretary Bob Bergland, who had once stated that he was morally opposed to converting grain into fuel.[33] Bergland began to sing the praises of gasohol at public appearances in the Midwest, and even made some tentative efforts to convince his largely hostile federal bureaucracy to take a more positive approach.

The Soviet invasion of Afghanistan in December, 1979 proved to be an indirect but nevertheless powerful stimulus to the new Carter administration alcohol fuels policy. The centerpiece of the U.S. response to the Soviet invasion was an embargo of all U.S. grain shipments, especially corn, which the Soviets needed for livestock to keep a modicum of meat on Soviet citizens' tables. To soften the impact on farmers in an election year, the Carter White House hastily put together a number of federal measures designed to help the gasohol industry absorb the surplus grains.

On January 11, 1980 the White House announced a new alcohol fuels program targeted at an increase in ethanol production of 400 percent by 1981 (that is, to 300 million gallons per year) and 600 percent by 1982. The White House hoped to see these dramatic production jumps achieved by a variety of federal measures, most of which were patterned after congressional legislation and included federal loans and loan guarantees to speed up the construction of new plants and stepped-up research. Most of the proposals were grafted onto the massive energy package that the administration was trying to push through a recalcitrant Democratic Congress. The Carter White House was advised by congressional leaders that its only hope of getting the various proposals through Congress was to lump them together in one omnibus bill. The primary thrust of this "syn fuels bill," which Congress approved in June, 1980, was the creation of a quasipublic Synthetic Fuels Corporation, which would award some $17.5 billion in federal loans and loan guarantees to accelerate coal and oil-shale syncrudes development. But a special biomass section provided $1.27 billion in federal aid to renewable alcohol fuels and other biomass energy projects, along with over $15 million for new research in this area.

These funding initiatives, combined with the 4-cent-per-gallon tax break on gasohol and the 10 percent investment tax credit which included the purchases of alcohol production equipment, constituted a strong show of federal support for the alcohol fuels industry. This support began to draw fire not only from the oil industry but also from liberal academics who saw the Carter administration, desperate to at least make a show of action on the energy front, bowing to pressure tactics from the gasohol lobby. "It is becoming increasingly clear that this politically motivated program is certain to lead us down a costly blind alley in search of alternative energy sources," complained Fred Sanderson, a guest scholar at the Brookings Institution, in a column in the *New York Times*. "The legislation went through Congress without much thought to costs as against benefits," he wrote. "The exemption of gasohol from the Federal tax of 4 cents per gallon sounds modest enough but it really amounts to a 40-cent-per-gallon subsidy on the 10 percent ethanol contained in gasohol. Other federal incentives will add another 10 cents or more. If we assume that ethanol production will in fact rise to 10 billion gallons, the total cost of subsidies to the Treasury over the next ten years would exceed $30 billion."[34]

Sanderson and others have noted that the loss of these revenues would come at a time when a major repair effort is needed to maintain the interstate highway system. The American Automobile Association, usually firmly in the gasohol camp, also agreed that the tax break should be dropped to prevent erosion of the Highway Trust Fund. The AAA's support for the tax break was initially seen as a temporary expedient to get the gasohol movement on its feet, AAA's Curry said. Supporters of the tax break note that even using the $30 billion figure, which they say is high, the subsidy is less than that given the nuclear industry since the 1950s.

THE OFFICE OF ALCOHOL FUELS VERSUS THE ENERGY RESEARCH ADVISORY BOARD

Shortly after the White House policy pronouncements of January, 1980, an Office of Alcohol Fuels (OAF) was created inside DOE to serve as the lead group in implementing the new federal alcohol fuels programs. Veteran alcohol fuels specialists such as Eugene Ecklund found themselves shunted aside as OAF recruited a new staff which included a vocal advocate of small-scale technologies named Bill Holmberg. A tall, thin, ex-Marine colonel, Holmberg had long been conducting his own private campaign to aid the farm alcohol producer from a position within DOE's office of consumer affairs. In the spring of 1979, Holmberg helped fund a week-long exhibition of appropriate technology systems called ACT-'79 (for "appropriate community technology"). The exhibits included a laboratory-scale demonstration of Harry Gregor's membrane ultrafiltration system, a full-scale version of Paul Middaugh's plate-column farm still, and projects by other small-scale alcohol pioneers, such as Albert Hubbard, the Alabama moonshiner. Holmberg affectionately dubbed these alcohol fuels pioneers "the crazies."

Holmberg was an anomaly in the straitlaced federal energy bureaucracy. As a former Marine colonel and intelligence officer, he had the background—and the bearing—to command respect. Yet he held views that, for DOE, were unorthodox. He was an outspoken critic of the oil industry who passionately believed that, given half a chance, the forgotten "little guys"—America's countless backyard tinkerers and small entrepreneurs—could usher in a new prosperous era of renewable energy technologies. And, he argued, this wave of good old-fashioned Yankee ingenuity would be necessary to build decentralized energy production systems that would, sooner or later, prove essential to national security by reducing the impacts of sabotage.

Holmberg found his star rising in the energy bureaucracy when Charles Duncan assumed control of DOE. Holmberg organized Duncan's midwestern alcohol plant tour in November, 1979 and in the winter was appointed to a key position in the Office of Alcohol Fuels. But his unconventional administrative style made him a highly controversial figure even among those DOE officials who sympathized with his goals. His willingness to promise federal support to small-scale producers who were often lacking in technical competence infuriated old-line Schlesinger appointees such as Under Secretary John Deutch.

Deutch hoped to undercut the growing influence of the alcohol advocates in DOE with the publication of a report by the agency's Energy Research Advisory Board (ERAB) which was sharply critical of alcohol fuels. Prepared by a group of industry and academic researchers, the ERAB report basically restated all of the oil industry's arguments against grain-based alcohol fuels and concluded that the only promising avenues of development involved construction of large-scale plants to process forest products and coal into methanol. Dissenting remarks by some academics on the panel, such as Nebraska's William Sheller, were not included in the final report. After reviewing the report in December, 1979, Deutch prepared an 11-page memo which concluded that federal support for gasohol was "not justified on the basis of cost-effective contributions to our near- or long-term energy problems."[35]

Holmberg responded to the Deutch memo with his own 23-page blistering rebuttal attacking the ERAB report and charging that Dr. Paul Weisz, one of the key ERAB advisors, was guilty of conflict of interest since he was then employed by the Mobil Research and Development Corporation. "The position outlined in your memorandum," Holmberg wrote Deutch, "could . . . give rise to the argument that the Department [of Energy] is really only interested in energy systems that will be controlled by a very limited number of major corporations. I know you would agree that under all circumstances, we must avoid the impression that scientists and scientific data are being used to further long-established institutional or corporate interests rather than the best interests of the general public."[36]

Holmberg carried on his fight against Deutch and the ERAB gasohol report when he assumed his new position with the OAF. The acting director, Steven Potts, supported Holmberg's campaign by lobbying Secretary Duncan to disavow the ERAB report. The DOE infighting over the ERAB report became public in a nationally syndicated Jack Anderson column of May 24, 1980 that accused influential DOE advisors who had ties to Mobil Oil of trying to sabotage the federal alcohol fuels program. In June, Sen. George McGovern of South Dakota called for hearings to investigate whether ERAB task force members "with ties to Mobil Oil . . . would rob hundreds of thousands of American farmers of the opportunity to benefit from gasohol development."[37]

These allegations brought an indignant response from Mobil in a widely published advertisement entitled "Science and Politics Don't Mix." The Mobil rebuttal decried Holmberg's attacks on Weisz, a man described as "an honored scientist with impeccable credentials, who is neither a proponent nor an opponent of gasohol." Lashing out at Anderson and McGovern, Mobil publicists wrote that attacks on the integrity of private-sector scientists offer politicians readymade dragons to slay and provide publicity-seeking columnists with "exposés" with which to build circulation. But they "also produce a chilling effect on the reporting of scientific findings."[38]

The cautious Duncan, disturbed by the unfavorable publicity that the ERAB controversy was drawing to DOE, was unwilling to repudiate the report. Instead, he issued a statement expressing "the highest regard for the conduct of the ERAB and . . . full confidence in the technical expertise, objectivity and integrity of the members."

Neither Holmberg nor Potts emerged unscathed from their battles with Deutch. Potts was pressured into leaving his job with the OAF, and Holmberg found his power base within the agency eroded.

ALCOHOL FUELS AND REAGAN ECONOMICS

As the Carter Administration entered its final months in the fall of 1980, the first of the $1.27 billion in federal loans and loan guarantees provided for biomass fuels in the Energy Security Act (also known as the synfuels bill, since it contained another $17.5 billion for coal- and shale-based synthetic fuels) were hastily approved. DOE, which had the task of awarding loan guarantees for large-scale alcohol plants, chose seven from a field of fifty-seven applicants for a total of $392 million on the eve of the November 4 election. The USDA, meanwhile, had pushed $342 million for fifteen slightly smaller projects through the bureaucratic maze at the close of the 1980 budget year.

Large-scale producers were particularly anxious for the guarantees, since they would ensure wary private bankers that, in case of financial disaster, the U.S. Treasury would make good on most payments to creditors. This would, in turn, allow borrowers to obtain lower interest rates and increase the overall profitability of their projects.* A wave of successful first-generation alcohol fuels plants would help convince skeptical investors that the industry could make a profit on their investment, and would help move the industry's emphasis into the private sector and away from dependence on the government.

Oil company joint ventures fared well in the first round of competition for government support—too well, critics said. Kentucky Agricultural Energy Com-

---

*One large producer, Midwest Solvents, turned down a loan guarantee it had been awarded by the USDA in 1979, commenting that their bankers felt so uncomfortable with the strings attached by the government that they had insisted on a higher, not a lower, interest rate.

pany, a joint venture with Chevron, received a $30 million loan guarantee and a $9.8 million low-interest direct loan.* Ashland Oil, in its joint venture with Publicker, was awarded a $25 million loan guarantee from DOE and another $32 million loan guarantee from the USDA. Other oil company applicants included Texaco, Diamond Shamrock, United Refining, and Derby Refiners, as well as an alcohol fuels company formed by the Hunt brothers, Texas oil millionaires. Although the USDA was to handle on-farm and farm cooperative stills, none of the initial round of grants was awarded to small companies.

Meanwhile, the actual 1980 ethanol production figure of 160 million gallons fell short of the 300-million mark President Carter had hoped to achieve in the wake of the Soviet grain embargo, although it did represent a two-fold increase over 1979 levels.[39]

The fact that DOE and the USDA generally favored the large-scale plants in their loan guarantees caused mixed feelings among gasohol proponents, and these mixed feelings were even further compounded by the decision of the incoming Reagan administration to freeze the DOE and USDA scheduled loans pending an investigation by the new appointees to the Energy and Agriculture bureaucracies. Republicans were philisophically opposed to government interference in the private sector, and a transition-team energy study, led by petroleum geologist Michel Halbouty, recommended, "Let the producers compete in a free and open market, and let the consumers choose the winners." The Halbouty report, insisting that grain ethanol takes more petroleum to produce than it delivers, recommended a shift to a coal-based methanol industry.[40]

David Stockman, the youthful Michigan congressman chosen to head the Office of Management and Budget at the Reagan White House, soon emerged as the sharpest critic of the ambitious federal alcohol fuels programs developed during the Carter administration. These Carter-era programs formed part of a broader energy policy which called for a gradual decontrol of oil prices and generous federal aid to stimulate the development of both alcohol fuels and synthetic gasoline fuels.

When Stockman arrived at the White House, he immediately announced that he was opposed to offering federal loan guarantees to either the gasohol industry or the synthetic fuels industry. In this new era of Republican-Reagan economics, he proclaimed there was no room for such massive federal intervention in the private sector. He proposed an energy development strategy which would rely on the stimulus of the marketplace rather than on government aid. The Reagan administration then moved to abolish immediately all price controls on domestic crude-oil production and seek a "congressional recession" or cancellation of the billion-dollar loan guarantee program to build new ethanol plants. Given the enormous political power of the new administration, few veteran Washington observers felt that Stockman would have much problem pushing the necessary measure through Congress.

---

*The *Washington Star* hinted that the approval may have been influenced by a donation of $100,000 to the Democratic National Committee in July, 1980, made by James R. Wade, son of the company's president.

In the spring of 1981, a broad-based coalition of gasohol interests formed under the umbrella of the newly created Renewable Fuels Association to try to save the embattled federal loan guarantees. The association was headed by David Hallberg, a former legislative aide to Congressman Berkely Bedell, who had played a key role in obtaining the passage of alcohol fuels legislation during the Carter administration. Hallberg teamed up with Richard Merritt to wage a furious lobbying effort to protect the federal loan guarantees. By rallying conservative Republican senators to their side, and by meeting quietly behind closed doors with the new Secretary of Energy, James Edwards, Hallberg and Merritt were able to win a second lease on life for the gasohol aid program at a time when dozens of other once-sacred federal assistance programs were withering under the budget-cutting scalpel of Stockman. On August 17, 1981 the Energy Department gave preliminary approval for $706 million in loan guarantees for eleven large-scale ethanol plants with a total future production capacity of about 365 million gallons a year. With the aid of these federal loan guarantees, the sponsors of these projects were expected to be able to raise as much as $2.1 billion in private construction loans. This was an encouraging sign for the hard-pressed gasohol industry, which suffered from declining sales during the 1981 oil glut.[41]

In the meantime, another federal liquid-fuel development program was emerging, somewhat scarred but still basically intact, from the Stockman purge. In the spring of 1981 the Reagan administration announced its intention to provide $20 billion in federal assistance, primarily in the form of federal loan guarantees and price purchase agreements, to promote the rapid commercialization of synthetic fuels derived from coal, oil shale, and tar sands. Although this represented something of a retreat from the full scope of the Carter administration commitment to $88 billion in federal assistance through 1990, it still represented twenty times the level of federal financing assistance available for alcohol fuels.[42]

By saving some of the provisions called for by the Carter administration, the gasohol lobby had demonstrated that even in the new Washington world of Ronald Reagan, it could still wield some influence. Haulberg and Merritt still had powerful friends in Congress. Despite the lofty free-market proclamations of budget director David Stockman, it was clear that the federal policies developed by the Reagan administration would favor the development of synthetic fuels over renewable ones. In the budget-cutting frenzy of 1981, however, not even synthetic fuel programs were certain of being funded.

Chapter 6
Agriculture: The Limits of the Land

There once was a time when America's agricultural productivity seemed as vast as the oceans which lap at the nation's shores. The pioneers who ventured west were treated to seemingly endless vistas of fertile prairies stretching out towards the horizon. As the lands were put to plough they yielded an agricultural abundance unparalleled in world history. Surplus—not shortage—has been the curse of the American farmer as his products have glutted first national, then international markets.

The earlier vision of boundless agricultural potential was something of an illusion, for the nation's farmers have always depended upon the limited resources of water, energy, and top soil to produce their crops. And it is this rapidly shrinking agricultural resource base that now forms the rationale of the alcohol fuels industry. Almost all of the large-scale plants scheduled for construction during the 1980s are geared to produce ethanol and will be located in the nation's prime farming regions, with the majority in the breakbasket states of the Midwest. If current projections prove accurate, corn will be the feedstock for 82 percent of the 1.4 billion gallons of ethanol expected to be produced in 1985, with other agricultural crops and related byproducts accounting for the remaining 18 percent.[1]

The prospect of the United States turning to agriculture to help satisfy its voracious appetite for energy has come to be the single most controversial question surrounding the development of the alcohol fuels industry. Journalists have conjured up scenarios in which greedy farmers form a new cartel with the powers of OPEC, diverting grains from food to alcohol as fuel for automobiles takes precedence over feeding the world's poor. This "food versus fuel" controversy has become part of a broader debate over the impact of ethanol production on agriculture in the decades ahead.

Many farmers, fed up with low grain prices that fail to compensate them for their labor, welcome the dependance that large-scale ethanol plants are developing for corn and other grains. They hope that by offering an alternative market for their grain surpluses, the industry will help to ensure a fair return at harvest time and end the need for costly federal price supports. They argue that both consumers and farmers would benefit from a new era of farm prosperity that could result from a growing ethanol industry.

Others do not share this optimistic outlook; they question the wisdom of our turning to agriculture to provide energy at a time when agriculture's ability to meet increased worldwide grain demand over the next decade—or even to sustain current levels of productivity in decades ahead—is in doubt.

THE LAND BASE—A THREATENED RESOURCE

A popular bumper sticker plastered onto tractor cabs in the Midwest proclaims alcohol as the "solar fuel," but it could also be dubbed the "soil" fuel, since it uses that resource as well.

Unlike solar energy, the soil is quite finite and easily destroyed. This was amply documented by W. C. Lowdermilk, an official with the U.S. Soil Conservation Service, who studied the struggle of ancient civilizations against soil erosion. His report, entitled "Conquest of the Land Through 7,000 Years," was the result of a worldwide journey to find out what U.S. farmers might learn from the ruins of past civilizations that could be of help to them during the dust-bowl years.[2]

One of Lowdermilk's first stops on his two-year trip was the Jordan Valley, a region the Bible describes as "a promised land of brooks . . . of fountains that spring out of valleys and hills, a land of wheat and barley . . . where thou shalt eat bread without scarceness." But Lowdermilk found only devastation in the modern Jordan Valley. He reported "soils of red earth washed off the sloped to bedrock over more than half of the upland areas"; "only dregs of the land were left behind in narrow valley floors." In the deforested highlands above the valley, Lowdermilk trudged through ruins of village sites abandoned because "the soil, the source of the food supply, has been wasted away by erosion."

In what had been the ancient land of Phoenicia, once one of the most prosperous areas in the world, Lowdermilk found the mountainsides stripped of timber and barren of soil. In Mesopotamia, a region that may have sheltered the Biblical Garden of Eden, Lowdermilk discovered that civilizations crumbled as intricate life-sustaining canals became clogged with eroded soil and ceased to function.

Lowdermilk returned from his journey convinced that the United States would have to redouble its efforts to preserve the nation's farmlands, which he viewed as "an integral part of the nation, even as its people are." He was alarmed to find that soil in three-quarters of the 400 million acres then under cultivation was being worn away faster than new soil was being formed. He worked to strengthen the federal soil-conservation program that established shelterbelts of trees to form windbreaks in the ravaged dust-bowl areas. He promoted new farming techniques, such as contour plowing and terracing on hilly lands, which reduced soil erosion from water run-off. Millions of acres of marginal farmland particularly susceptible to erosion were taken out of row-crop production and planted in more protective pasture.

Although the erosive forces of nature were temporarily slowed by these efforts, in the decade of the eighties erosion once again poses a serious threat to farm productivity. One USDA report found that cropland was eroding "at nearly twice the rate considered acceptable" during the 1970s, and that the U.S. had already lost one-third of its native topsoil by 1980.[3] James Risser, a Pulitzer-prize-winning farm journalist for the *Des Moines Register*, explains:

> Topsoil varies in depth from inches to dozens of feet. Depending somewhat on the soil type, the scientists say that an acre of land can lose five tons of soil per year to erosion without permanent harm. The five ton "tolerance value" sounds enormous, but spread out over an acre it amounts to a layer only a fraction of an inch deep. That amount of soil is matched by formation of new topsoil and the processes of nature, so that productivity is maintained. . . . What scares the experts today is that in many agricultural regions of the U.S., the erosion rate is far above tolerance.[4]

The threat of a new dust bowl has surfaced in some areas of western Nebraska where the windbreak trees have been removed in order to install large center-pivot irrigation systems. Much of this dry prairie land was first put to the plow during the early 1970s, in response to high grain prices. Benny Martin, a soil expert with the USDA, estimates that over 200,000 of these acres are now filled with huge green circles of corn. But the land in fact is too fragile to support row-crop production.[5]

Much of this irrigated center-pivot system was developed by out-of-state investors who have, at times, tried to defy even the most basic laws of soil conservation in their efforts to extract short-term profits. Seventeen irrigated circles of corn planted in a pocket of volatile "blow sands" near Tilden, Nebraska were destroyed in 1978 when a prairie wind whipped up the sandy soil and cut down seedlings. A corporate farm crew replanted the crop and, again, the seedlings were cut down. When the third crop held, Stan Grubb, a local farmer, remarked, "The company's leasing the land, taking the good out, and when they give it back to the farmer, it'll take years to go back into production." Knowledgeable family farmers in that region have tried to cultivate these sandy soils with more care, planting their crops in narrow strips between long stands of trees which shelter them from the wind.

In other areas of the Midwest, water erosion is the primary force of destruction. In some areas of Iowa, 5 to 6 bushels of topsoil are washed away for every bushel of grain harvested. Much of this earth eventually ends up in the Mississippi River; indeed, soil scientists say the river receives the equivalent of a 100-acre farm every day.

In western Tennessee, less than 20 percent of the croplands are adequately protected against erosion, and some fields suffer soil losses of 150 tons per acre—thirty times the rate at which new soil can be formed. In the Palouse region of eastern Washington, a prime producer of soft white wheat, annual erosion losses of up to 200 tons an acre have been recorded as spring thaws pry huge chunks of soil loose from the hillsides. The soil invariably carries with it the chemical fertilizers and pesticides which pollute local watersheds, threatening wildlife and the quality of drinking water.[6]

The increasing rate of soil erosion on U.S. farms is closely linked to the turbulent economic events of the 1970s that prompted Gene Schroder and other disgruntled farmers to organize their tractorcade protests. The boom years of 1973–74 encouraged farmers to cultivate millions of erosion-prone marginal acres which had been left in pasture. The bust years that followed forced farmers to crop their lands in ever more intensive fashion if they wished to survive. Rarely could a farmer justify in a short-term economic sense the decision to invest time and money in conservation practices. The soybean, sorghum, corn, and cotton crops that caused the most severe erosion problems also happened to be some of the most profitable crops to grow.

The relationship between economic stresses and soil erosion has been summarized by Earl Heady, an agricultural economist at Iowa State University, who told a *New York Times* reporter, "The rate of soil loss was much increased in the 1970s. And that is related to the changing structure of agriculture. Since the mid-1970s we have had a tremendous increase in the price of land and the cost of farm equipment. So we now have farmers who look upon farming as a real estate game. They buy the land and farm the hell out of it to meet their heavy payments, not worrying about preserving it because they believe it will keep going up in price."[7]

The dilemma facing many farmers was graphically spelled out by Otis Chapman, an Arkansas farmer who told the U.S. House Agriculture Committee, "We are, by necessity, today raping our farmland in order that we might survive. We are raping our water supplies, our natural resources and our wildlife."

One method that has been used to push land beyond normal limits is irrigation. Over 83 percent of the water consumed in the United States each year is used in agriculture, and most of it goes to watering crops. An acre of corn under arid growing conditions can require hundreds of thousands of gallons of water during the growing season. This water is often pumped out of underground aquifers which are now rapidly approaching depletion. The most important of these aquifers is the Ogalla Sioux, whose waters irrigate fields from Texas north to the Dakotas. If water use continues at its present rate, the Ogalla Sioux will be exhausted within forty years.[8] Already, the high price of pumping water to the surface has forced some farmers to revert to less productive dry-land farming. The high cost has forced others farmers to forgo the option of growing less erosion-prone cover crops of alfalfa in favor of the row crops which yield a higher return on investment.

The row crops grown on many irrigated lands are of dubious value in helping solve the nation's energy problems. When the costs of fertilizer and of electricity needed to operate irrigation pumps are factored in, the crops prove to have a negative energy balance when converted into alcohol. On an estimated one-third of these irrigated lands, producing some 300 million bushels of corn, sorghum, and wheat, the energy equivalent of 3 gallons of crude oil are required to produce each bushel of grain. This has led some, like professor of land economics Folke Dovring, to conclude that these energy-intensive irrigated crops should definitely not be considered as feedstocks for the alcohol fuels industry.[9]

Another problem with irrigated crops is the environmental impact they can have on drinking water. In Nebraska, water is being polluted by nitrogen fertilizers: nitrates from the fertilizers seep through the sandy soils into the aquifer pools which also serve as underground reservoirs. A U.S. Geological Survey study of the drinking water in Holt County, Nebraska indicated that 40 percent of the tap water contained nitrate levels exceeding federal safety standards. These nitrates have been labeled potential carcinogens by the Environmental Protection Agency.[10]

In several western states, a major source of irrigation water is the Colorado River, which will be tapped in the decades ahead to aid coal and oil shale development. Many farmers in the region fear that since they cannot afford the high prices the energy corporations will pay for the use of the river water, they will be forced to revert to dry-land farming. In Arizona, farmers in Pinal County have already lost out in the competition for scarce surface water supplies with local industries and suburban communities. With ground-water levels dropping an average of 10 feet a year, one-third of the county's croplands have been taken out of production.[11]

Another major threat to the land base comes from the conversion of prime farm lands to nonagricultural uses. During the decade of the 1970s, some 30 million acres of farmland were withdrawn from production. This land represents an area the size of Vermont, New Hampshire, Massachusetts, Rhode Island, Connecticut, New Jersey, and Delaware combined. A report by the National Association of Counties Research Foundation concluded that "this loss has touched every corner of America; New England has witnessed the disappearance of half her native farmland, the Mid-Atlantic states have lost 22 percent of theirs, and even the vast Midwest has suffered the loss of 9 percent of its cropland." This land was taken from a total cropland base of 431 million acres, of which 230 million could be classified as "prime." During the past decade, the sight of shopping malls, housing developments, and industrial parks springing up in fields that once produced grain or provided pasture became all too common.[12]

The conversion of these farmlands has had a broad-ranging impact on the overall viability of many rural communities located close to urban fringes. A report by the National Agricultural Lands study group pointed out that "the basic infrastructure of the local farm economy may be disrupted, since farmers typically rely on local businesses to provide credit for production expenses, veterinary services, machinery repairs, and other needs. These services provide an essential economic undergirding of the smaller-sized agricultural operations. In turn, this relationship between farm and community determines to a significant degree the style of life offered (non-farm) residents of the rural community. This mutual interaction requires some 'critical mass' of farmer demand for these agricultural support services below which the services cannot be maintained. As agricultural land is converted or idled in anticipation of future development, the aggregate demand for these 'infrastructure' services may fall below this critical level."[13]

The increasing urban demand for farm land has also played a role in the phenomenal increase in the value of farm land, which often pushes its market value up far beyond its productive crop value. This inflation of land prices has emerged as the major barrier to new farmers who are trying to get a start. The land cannot produce enough crops each year to repay the loans needed to finance its purchase. And the inflated value of farm lands has increased property value taxes in many counties to the point where even those farmers who already own their land find it more profitable to sell out to real estate interests than to continue operating marginal farms.

A variety of local and state programs have been established to deal with the difficult task of farmland preservation in a free market system. Some counties have enforced rigorous zoning statutes which prohibit the conversion of agricultural lands to other uses without the approval of planning commissions. But these efforts have often been undercut by townships eager to attract new industries which annex county lands and then rezone them for urban use. A few localities, such as King County, Washington, have attempted to purchase development rights from local farmers as a last-gasp effort to preserve remaining farms. This type of program, used in predominantly urban counties, has proven to be an extremely costly measure that would be difficult to use on a broad scale. To date, the federal government has been reluctant to do anything more than study the problem. The USDA's position is that "The relatively small proportion of the cropland base being converted may not affect the geographic distribution of the production of major crops or the structure of American agriculture in the aggregate. But, the state and local effects are very important, particularly in areas where non-farm influences may seriously affect the viability of farming."[14]

FOSSIL FUELS IN AGRICULTURE—AN INCREASINGLY EXPENSIVE OPTION

Between 5 and 6 percent of all the energy consumed in the U.S. each year is used on the nation's farms. While farming has always been an energy-intensive process, it is only in recent decades that it has come to rely so heavily on nonrenewable fuels. Many of the tasks once accomplished with energy supplied by horses, oxen, and men now rely on fossil fuels and their petrochemical derivatives. This transition was encouraged during the postwar era by agribusiness, land-grant universities, and the federal government as farms became larger and the rural labor force left for urban areas. It is the massive injection of fossil-fuel energy into farming, along with the introduction of high-yielding grain varieties, that has helped to make possible the dramatic increases in yields per acre.

One third of the fossil fuels consumed by the nation's farmers goes into the production of chemically synthesized nitrogen fertilizers; it takes 30 cubic feet of natural gas to produce a pound of fertilizer, which can replace animal manure and nitrogen-fixing legume crops in enriching the soil.[15] Basic changes in the structure of agriculture were propelled by the widespread introduction of these fertilizers, since it was no longer necessary for a farmer to have ready access to a local supply of animal manure or to rotate crops often. Farmers who tended livestock and engaged in a diversified production of feed grains and pasture crops for grazing switched to row cropping, which proved to be more profitable and less time-consuming. Small livestock farms were replaced by larger, more centralized enterprises which were often unable to return all their manure wastes to the fields. Dairy farms which once flourished throughout the Midwest became more concentrated in a few states with particularly lush pasture land, such as Wisconsin and Minnesota.

Today, there is mounting evidence that the limit to grain (particularly corn) production through the use of chemically synthesized nitrogen fertilizers has been reached. In many areas of the Midwest additional increases in application rates of these fertilizers are showing only marginal increases in productivity. Dovring notes that "In Illinois, . . . use of nitrogen fertilizer increased from around 600,000 tons in the early 1970s to nearly 900,000 tons in the late 1970s. Nearly all of the increases went to corn. Up to 1971, nitrogen used on corn was close to one pound per bushel, which is not excessive since the uptake of nitrogen is close to 0.8 pounds per bushel." All told, an extra 548 million pounds of fertilizer were spread on the fields between 1970 and 1978, for an increased yield of a total of 250 million bushels of corn.[16]

Pesticides, another petrochemical input, have also triggered major changes in farming techniques. One billion pounds of pesticides, the Btu equivalent of a billion gallons of crude oil, were sprayed on the nation's fields in 1979. In recent years many scientists have become disillusioned with the usefulness and safety of pesticides. For years, USDA extension agents recommended stronger and stronger doses of various poisons to deal with pest outbreaks. But in some instances farmers found that insects and weeds developed genetic resistances which allowed their offspring to survive intensive chemical spraying. Sometimes pests eliminated by one chemical were replaced in the field by a new, even more damaging pest that was able to flourish once its predator was gone.[17]

Thus, many farmers found themselves thrust aboard a "chemical treadmill" in which they were forced to use ever increasing amounts of pesticides to combat ever increasing problems. In the meantime, there has been growing concern about the potential long-term health hazards that these chemicals may pose to humans. Traces of pesticides frequently turn up in food, and it is extremely difficult to determine how much of a safety threat these residues may pose. Many pesticides once thought safe have now been banned and still others on the market have been linked in laboratory tests to genetic mutation, birth defects, cancer, and nervous disorders.

As well as being a heavy financial burden for farmers, pesticide use is believed by some scientists to play a role in declining soil fertility. Organic matter is often slow to decompose in soils heavily treated with anhydrous ammonia nitrogen fertilizers, pesticides, and fungicides that inhibit the activity of bacterial organisms and earthworms. This decomposition process is vital in maintaining soil texture, or tilth, which in turn helps the land resist erosive forces and compacting caused by heavy machinery. As petrochemicals have been increasingly substituted for animal manure, the level of organic materials has been dramatically reduced, with soil compacting and erosion tending to increase as organic materials decline. William Tucker, in an article published in *Atlantic* magazine, found that "some areas of Kansas and Nebraska have reported hardpan almost two feet deep. The agricultural machinery companies have responded by building bigger tractors to pull the plows, but the heavy tractors are causing even greater compaction. Where hardpan occurs, rainwater no longer soaks into the ground but runs directly off the land, taking large quantities of soil with it.[18]

FOOD, FUEL, AND THE INTERNATIONAL GRAIN TRADE

Soil erosion, water shortages, and petrochemical dependence are all factors which affect agriculture's ability to respond to the demands of the ethanol industry for grains. But perhaps the most important short-term factor affecting their availability is the exploding world demand for U.S. grains, which threatens to send grain prices to unheard-of levels, casting a shadow on the economic outlook for grain-based ethanol plants.

When looking at the international grain market, the past cannot be considered prologue. The 1980s are likely to differ radically from the previous four decades, in which surplus U.S. grains frequently glutted world markets. These surpluses were a constant headache to the federal government, which paid farmers to set aside lands and paid the bill for storing grain reserves. In an effort to raise worldwide demand for U.S. grains to a point that more nearly equaled the farmers' productive capacity, the USDA embarked on what one author characterized as a "ruthless and aggressive" international marketing campaign. Exports through the multinational grain-trading companies were subsidized by the USDA to make them available at bargain-basement prices. The subsidies helped crack tough European markets that had high protective tariffs on imported agricultural products.

Countries with less developed tariff systems, such as Iran, became particularly dependent on U.S. wheat, corn, and soybeans during the late sixties and early seventies, through a foreign aid program labeled Public Law 480. In his book, *Merchants of Grain*, Dan Morgan found that this program succeeded in transforming a nation that had once been largely self-sufficient into a nation that "was buying so much grain that parts of the Persian Gulf were clogged with grain ships that were obliged to wait for weeks before unloading their cargos" in Iran.[19]

The USDA also succeeded in marketing U.S. grains to nations whose traditional diets made little use of wheat or corn. In Japan, this marketing coup was accomplished with the aid of a high-profile public relations campaign in which the U.S. agricultural attaché promoted western-style breads and grain-fattened beef.

Although the USDA marketing efforts did help to greatly increase exports, they fell far short of their goal of eliminating the chronic surpluses. Between 1950 and 1970 U.S. corn exports increased from 3.5 million to 12 million tons. But during that period, as much as 62 million acres of U.S. farmland, about one-fifth of the total productive capacity, had to lay fallow to reduce surpluses. The fact that grain supplies exceeded demand, however, did not mean world hunger was being held at bay, but merely, as one USDA report coldly noted, "This hunger could not be translated into [economic] demand."[20]

This point was underscored by Dan Morgan in *Merchants of Grain*. Analyzing the desperate plight of hungry people in Haiti and Honduras, just 800 miles from the United States, he writes:

People are not starving in those nearby countries because the world is running out of food. They are starving because they are poor—poor and beyond the reach of the vast commercial system that produces food and transfers it from one country to another.

The U.S. Department of Agriculture said in a report issued in 1974 that if the estimated 460 million malnourished people in the world each received only 500 additional calories a day, their hunger would be alleviated. Cereal grains could provide these calories. An extra .15 kilograms per person a day of either wheat, rice, corn, sorghum, or millet would give this additional allotment of 500 calories. This, in turn would require 21.9 million tons of cereals a year.

This amount of food is well within the production ability of the world's farmers. The 21.9 million tons amount to only 2 percent of average annual world cereal production in the last decade.[21]

During the first half of the 1970s, the volume of U.S. grain exports increased dramatically as the devaluation of the dollar made U.S. products more competitive on world markets. A major turning point in the decade occurred in 1972 when the Soviet Union bought $750-million-worth of American grain to blunt the impact of its crop failures that year. Carried out in an atmosphere of high secrecy, the Soviet grain sales caught much of the world by surprise and ushered in a new era of volatility in world grain markets. Virtually all idle U.S. cropland was rushed back into production after the Soviet grain deal as U.S. reserves were reduced to their lowest levels in decades. The U.S. consumer felt the pinch resulting from tight supplies as food prices rose; but the price hikes were only partially passed on to farmers, and the bulk of the increases lined the pockets of shrewd grain traders.

The federal government took advantage of the high grain prices triggered by the Soviet sales to reduce the size of the farm subsidy program. Demand for U.S. grains remained strong for the remainder of the decade, although vastly increased production resulted once again in depressed prices. By the end of the decade, U.S. exports had grown to the point where they made up almost half the total volume of grain traded in the international market, with the U.S. supplying over 80 percent of the world's corn exports.

Given the United States' dominant role in the international grain trade, it was perhaps only natural that the news media began to characterize the mid-western farmbelt as the "breadbasket of the world." The public began to perceive the U.S. farmer as the last line of defense against the vicissitudes of world hunger. But, as Frances Moore Lappé and Joseph Collins point out in their book, *Food First*, this popular perception was more myth than reality. "There are three gaping holes in this national self-image," Lappé and Collins wrote. "First, what food we do export on an aid basis (with long-term, low-interest financing) is only a tiny fraction of our commercial exports. . . . Secondly, over 56 percent of our agricultural exports go to industrial countries, not to the underdeveloped. Third, although it is true that we are the world's leading food exporter, we are also the world's top food importer."[22]

U.S. food imports include tropical crops, such as bananas and sugar cane, and winter vegetables which are often grown on land that would otherwise be used for staple crops in Third World nations. Mexico, for example, now supplies the U.S. with over half its supply of winter and early spring vegetables; meanwhile its infant mortality rate associated with poor nutrition remains high. Over-all, fully half of Central America's agricultural land produces food for export, primarily in large plantation-style systems which drain self-reliance and impose a feudalistic structure over the quest for an improved life.

It was against this complex international backdrop of tightening world grain markets and rising hunger that the alcohol fuels industry was born in the 1970s. The abundant grain that was available in that decade will not be so available in the future, many experts have predicted. Thus, the industry expansion appears to pose the direct threat of a "food versus fuel" conflict which could have disastrous effects on the consumer and—for those who believe that America is the breadbasket of the world—on the poverty-stricken masses of the Third World. These fears were capsulized in a satire by Ellen Goodman, a widely read *Boston Globe* columnist, which previewed life in 1995 when every car would run on ethanol.[23]

All through the 1970s and early 1980s editorial writers had mournfully pre-dicted that sooner or later the American people would have to choose between putting food on the table or food in the car. By 1983, the American people had chosen.

Indeed, it was a source of national pride to most Americans to see the waves of grain above the fruited plain waiting to be turned into mileage. In November of every year people gave thanks because their tanks runneth over.

Not that there hadn't been a brief and heated ethical debate—there had been. But not many people were shocked at the ease with which Americans fed food to their cars. . . .

Each individual family, according to its income level and number of cars, had maintained the freedom to choose between dining and driving. This led to a few minor inevitable inequities. The rich, always worried about the proper levels of nutrition for their Mercedes, spoon-fed them distilled beef wellington. The middle class tanked up on tacos, pizzas and Rolaids.

The major change in the social structure had come from people who used to be called farmers but were now food-fuel engineers. It came to pass that every child of the Food-Fuel Basket of America had an international conglomerate to call his or her own. The son of a magnate from Kansas had recently bought Harvard.

Where once these same families had come to Washington aboard tractors, tying up traffic and looking for federal aid, now they came in private jets to go shopping. There was a story in the paper in the summer of '95 about the farmer who had Bloomingdales sent back to Iowa in the biggest brown bag anyone had ever seen.

As for the hungry of the world, people tended not to think of them as they filled up and drove about. Besides, as the President had said so eloquently in 1992, "let them eat gasoline."

Newspaper articles such as Goodman's, even when written in a humorous vein, irked the corporate public relations men at Archer Daniels Midland, who took great pride in their company's role in providing both food and fuel for a hungry world, and who could point to the company's pioneering efforts to create a soybean market. It was obvious that Goodman and other journalists had made little effort to understand ADM processing techniques, which extracted corn protein, oils, and some starches before converting waste starch into ethanol. "The bloated bellies of starving children come from having too much starch and not enough protein," ADM's Richard Burkett exclaimed in an interview. "The hungry countries—and I've been to them—have enough manioc, cassava and other starch crops. What they need is more protein. Shipping a cargo of unprocessed corn that is only 8 percent protein doesn't really help the starving people."

Instead of sending so much unprocessed, starch-laden grain overseas, ADM officials have proposed that the U.S. ship more high-protein (27 percent) corn products refined in "wet milling" plants like ADM's. Rather than aggravate the world hunger situation, as suggested in Goodman's satire, Burkett hoped that a major shift into ethanol fuel production by food processors would increase shipments of protein-rich products to Third World nations. But in order for distillers byproduct grains to begin to relieve world hunger they would first have to be refined into some type of food product that could be readily integrated into the diets of poor people. To date, several interesting recipes have been developed, but no marketing on a commercial scale has occurred. Even if these products were put on the international markets, it is doubtful that those in greatest need could afford to buy them. Most of ADM's distillers grains are exported to Europe to fatten livestock, since they are not taxed like whole grains.

Another way in which ADM sought to both turn a tidy profit and at least indirectly combat world hunger was by developing large greenhouses which could produce winter vegetables for U.S. consumers. These greenhouse vegetables can supplement some of the winter vegetables imported from Mexico—thus helping to free up more of that nation's prime farmland for the production of basic foodstuffs rather than export crops.

Two byproducts of ethanol production—carbon dioxide and waste heat—can make it economically feasible to grow these vegetables in U.S. greenhouses even during the harshest of midwestern winters. In the past such ventures have rarely been profitable due to high energy costs and the limited productivity of vegetable plants in the indoor environment. However, ADM has pioneered a new hydroponic technology in which greenhouse vegetables are grown without soil in a nutrient-filled water medium. The first experimental greenhouse, one acre in size, was built adjacent to the corporation's Decatur fuel distillery and was a resounding success. ADM researchers found that it was possible to reduce energy costs by 75 percent through the harnessing of waste heat from the cooking and distillation processes. Pumping in carbon dioxide released during fermentation helped improve plant photosynthesis activity and resulted in increased growth rates of as much as 20 percent. In this supportive environment, a single acre of greenhouse can supply up to 300,000 pounds of tomatoes a year, more than ten times the amount that an acre of field-grown plants could produce in the same length of time. Thus, every acre of greenhouse production can free over ten acres of Mexico's prime farm land for domestic food production. In March of 1981, ADM harvested its first commercial crop of greenhouse lettuce and trucked it to nearby grocery stores for sale. The success of this prototype has encouraged the corporation to invest in construction of a much larger 20-acre greenhouse at another distillery site.[24]

The basic thrust of Burkett's response to food-versus-fuel critics—that the world is shorter on proteins than on starches—was generally well accepted by the national media until it was challenged in a report released in March 1980 by the Worldwatch Institute entitled "Food or Fuel: New Competition for the World's Cropland." Lester Brown, author of the report, argued that the world hunger situation stemmed from a shortage of starches as well as proteins. Starches are needed to provide malnourished people the minimum amount of calories they need to be able to function. In the absence of starch calories, proteins are metabolized and burned as if they were starch. Since the conversion process from grain to alcohol requires starch—which makes up over half the total volume of a bushel of corn—the process has the potential to divert calories which could be used to aid malnourished people into use as fuel.[25]

"The stage is set," Brown concluded, "for direct competition between the affluent minority, who own the world's 315 million automobiles, and the poorest segment of humanity, for whom getting enough food to stay alive is largely a struggle. As the price of gasoline rises, so too will the profitability of energy crops. Over time, an expanding agricultural fuel market will mean that more and more farmers will have the choice of producing food for people or fuel for automobiles. They are likely to produce whichever is more profitable."

Brown was not totally opposed to the development of a U.S. alcohol fuels industry, but argued rather for a more modest growth that would not be so heavily dependent on grain crops. "The question is not whether there should be an alcohol fuels industry," he wrote. "Clearly there are many possibilities for converting agricultural wastes and other sources of plant materials into automobile fuels that need to be urgently pursued. . . . At issue is whether government can encourage the production of alcohol fuel without inadvertently launching an industry that competes directly with food production."

The Worldwatch Institute report generated a storm of controversy which

Brown's previous articles and books on world hunger had never encountered. A soft-spoken, scholarly man who preferred bow ties to neckties, Brown suddenly found himself on the receiving end of a series of verbal attacks from the powerful congressional gasohol lobby. Congressman Berkely Bedell, an Iowa democrat, labeled Brown's report "unsustainable doomsdaying" and said that in some instances Brown's report "seriously erred in [its] ready assumption that the diversion of cereal crops to alcohol production will mean their total disappearance from the food/feed stream."[26]

As the former director of the USDA's International Agricultural Development Service, Brown was no stranger to world hunger issues, and felt that the gasohol lobby's assumption that there would be *no* loss to the food/feed stream was misleading. In most wet-milling ethanol production systems, only 16 pounds of the original 56 pounds of a bushel of corn are recovered, and despite their higher protein content, their calorie content is quite reduced.

Sen. Birch Bayh, as Chairman of the National Alcohol Fuels Commission when the Worldwatch report was released, also took Brown to task for his warning. Bayh's principal disagreement with the report centered around its assumption of direct competition between U.S. fuel needs and Third World food needs. The great majority of U.S. food exports, Bayh argued, did not wind up in the bellies of starving children but rather in the bellies of increasingly well-fed livestock in Europe, Japan, and other industrial nations. "If we are talking about the narrow line of survival that [we] are concerned about," Bayh lectured Brown at a Washington, D.C. hearing, "we are not concerned about how we in the United States can grow more corn. If you look at the production of corn, we exported 32 percent of our corn and most of it went not to the hungry nations but to the well-to-do nations that ran corn through their livestock. . . . Most of what the United States is contributing to alleviate world hunger is not to keep the hungry nations from starving . . . but [to keep] the fat nations fat."[27] (Exhibit 6-1 shows the major uses of U.S. corn in 1980.)

The rancorous food-versus-fuel debate is likely to continue well into the 1980s, and any forecast of the impact of an alcohol fuels industry on food supplies faces the unknowns of future U.S. crop-production levels, movements in the world grain trade, and the growth of the alcohol industry itself. The National Alcohol Fuels Commission's report forecast a 1990 ethanol production of 9 to 10 billion gallons, perhaps half coming from corn. This would require about 2 billion bushels of corn, a little less than half of the 4.35 billion bushels now fed to livestock. Stillage byproducts from this corn would be returned to feed markets and the overall impact of the industry on world hunger would be negligible. While Agriculture Secretary Bob Bergland fears that this stillage will disrupt feed markets that rely primarily on soybeans, possible decreases in soy meal prices could make more food available internationally. Overall, the forecast 1990 level for grain alcohol would imply a negligible effect on world hunger in terms of U.S. grain. A much greater impact on world hunger will clearly be the demand for imports of fresh produce into the U.S. that prevents some Third World croplands from being put into production of basic food crops.

Exhibit 6-1
Major Uses of U.S. Corn in 1980 (in billions of bushels) (Source: U.S. Department of Agriculture)

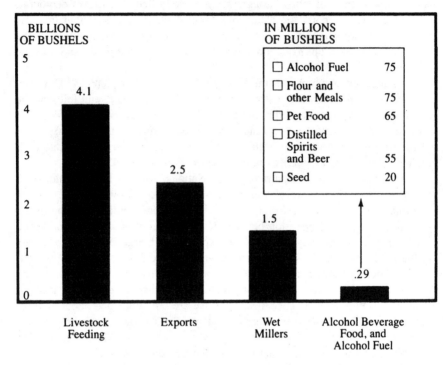

If there is a major crop failure in the U.S. during the 1980s, or another massive Soviet grain purchase, it is likely that the grain ethanol industry will grow at a slower rate than is now predicted. In such a period of scarcity, the world demand for grains vital to food markets would push the price of corn up past what many distilleries could afford to pay. But if political instability in the Middle East causes oil prices to soar once again, and if bumper harvests in the Midwest increase the U.S. grain surplus, the corn-based ethanol industry may enter a period of explosive growth.

CELLULOSE CROP RESIDUES

Given the many uncertainties surrounding the grain trade in the 1980s, many proponents of alcohol fuels in industry, government, and academia have urged a rapid transition away from grains to nonfood cellulose feedstocks, many of which currently have little economic value. Major research and development expenditures have been made to perfect the difficult process of breaking down cellulose fibers into fermentable sugars. Thanks to these efforts, it may be possible to begin this transition to cellulose feedstocks well before the end of the decade.

The list of suitable feedstocks for cellulose conversion is vast, ranging from cotton gin "trash" to garbage to forestry wastes. The total volume of feedstocks is impressive. The Department of Energy's Alcohol Fuels Task Force estimated that over 55 billion gallons of annual alcohol production could be achieved from cellulose by the year 2000.

Perhaps the most immediately available cellulose feedstock is the wheat straw, corn stalks, and other plant materials left over in fields after the harvest. These crop residues will be the closest cellulose feedstocks available to the new generation of grain alcohol plants, which can simply install new front-end processing equipment to handle the cellulose.

While this approach to broadening the feedstock base of the ethanol industry shows great promise, it is not without dangers. Crop residues help maintain fertility and fight erosion when returned to the soil each year. Robert Rodale, a leading critic of current petrochemical agricultural techniques and publisher of *Organic Gardening* and *New Farm* magazines, is staunchly opposed to any use of crop residues for fuel. "The crop wastes that would be put into stills . . . are a main bulwark against erosion," Rodale wrote. "They add humus to the soil so hard rains can soak in instead of washing the soil away. And these so-called wastes recycle minerals and nitrogen back to the soil, elements which otherwise would have to be replaced at a high cost in fossil fuels."[28]

The importance of crop residues in fighting soil erosion was well documented by Lowdermilk's early research and later confirmed in a major study by seven teams of soil scientists working with the USDA and universities around the country. They concluded that, "When returned to the soil, crop residues retain plant nutrients and help maintain soil porosity and tilth for easy tillage and good plant growth. Residue removal also inhibits water infiltration which affects soil water storage and plant use." The scientists stressed in their final March, 1979 report that only a small percentage of the crop residues could be safely removed from the land for conversion into ethanol.

In the corn-belt states, the scientists calculated that some 36 percent of the residues could be harvested without damaging long-term soil fertility and only 21 percent could be harvested in the Great Plains states. In Virginia, the Carolinas, and Georgia, some 40 percent of the crop residues might be used for alcohol production, while in Alabama and Mississippi, only a minuscule 10 percent could be safely taken off the soil. Nationwide, an estimated 80 million tons of crop residues, representing a potential of 4.4 billion gallons of fuel, might safely be made available for alcohol production.[29]

The collection of these scattered, bulky residues may prove to be more costly and energy intensive than simply harvesting grain. While a ton of corn can be processed into 90 gallons of ethanol, a ton of crop residues contain only about 55 gallons. And Cornell economist David Pimental estimates that the amount of energy required to collect the residues, combined with their energy value as a fertilizer when left in the soil, represents fully 38 percent of the ethanol energy they might yield.[30]

A headlong stampede into cellulosic alcohol production from crop residues could result in a major overharvesting. In an ideal world, farmers would surely heed the warnings of soil scientists and remove only safe amounts of crop residue from their land. However, when financially strapped farmers are suddenly offered cash payments for their entire output of crop residues, they might well decide to trade off long-term soil fertility for more immediate profits. A soil conservation official would criticize the farmer for such a move, but the farmer might well reply: "It was either sell the entire field of crop residues to the distillery this year or go bankrupt next winter when I can't afford to pay off my bank loan."

RENEWABLE AGRICULTURE: AN URGENT NEED FOR THE 1980S

It is clear that the long-term expansion of the alcohol fuels industry on an agricultural base consisting primarily of corn and cellulosic crop residue feedstocks is not without major risks. The risks will be born not only by the ethanol industry but also by society at large, which may be forced to deal with the disruptions caused by soil erosion and tight supplies of grain. The costs may be somewhat hidden to the consumer, who is unlikely to see them fully reflected in the pump price of gasohol.

A more promising agricultural path to an expanded alcohol fuels industry requires the use of a much broader range of feedstocks—including jerusalem artichokes, sweet sorghum, and fodder beets—grown in a manner that minimizes the use of fossil fuels and maintains long-term soil fertility. Such a path would make less use of the intensive row-cropping approaches to farming and increase alternative approaches that have been neglected by the U.S. agricultural establishment. Some of the most promising approaches to soil, water, and energy conservation have come from the ranks of farmers who have turned away from chemical fertilizers and pesticides. These farmers, sometimes labeled "organic" farmers, have long been treated with suspicion by USDA officials who actually knew little about their methods. The basic orientation of agricultural research policy, in contrast, involves "the promotion of large-scale energy-capital chemical-intensive agriculture which seeks to maximize return per dollar invested," according to an unreleased 1978 U.S. Senate Judiciary Committee report. The problem, said the report, was that "this emphasis has precluded more active efforts to promote research on alternative farming methods." The bias was most evident in a remark made by former Secretary of Agriculture Earl Butz, who asserted: "Before we go back to an organic agriculture in this country, somebody must decide which 50 million Americans we are going to let starve or go hungry."[31]

No one is starving on the Roman Wortmann farm, a dairy operation that spreads across 230 acres of gently rolling land in northeastern Nebraska. Wortmann, a devout Catholic of German ancestry, was once a confirmed believer in traditional farming practices. But he became disillusioned with petrochemicals in the spring of 1972 when he noticed that the pesticide he sprayed on his fields apparently had killed several of the robins that lived in the trees near his house.

After this incident Wortmann began to question what effects the chemicals he used—largely unknown in his father's day—might be having on his family, his livestock, and the microbial life in the soil. Since that time, Wortmann has worked to develop farming systems that free him of his petrochemical dependence. Soil fertility is maintained through the use of animal manures and a carefully planned crop rotation which alternates crops which demand large amounts of nitrogen, such as corn, with legume crops, such as soybeans, that naturally fix nitrogen. Nutrients in crop residues are carefully returned to the top 4 to 6 inches of soil, a practice which also helps to maintain soil tilth. Insects and weeds are kept under control primarily through a varied crop rotation which avoids monoculture susceptibility to pests.

Wortmann's dairy cattle are an important ingredient in the success of his farm. They give him a ready outlet for alfalfa, a grass crop which reduces soil erosion caused by row crops and adds nitrogen to the soil. The cows convert the alfalfa hay and grains into milk, Wortmann's major source of income, and manure, which he hauls back out onto the fields.

Although Wortmann's five-year rotation plan forces him to grow less corn and soybeans over the long run than a traditional farmer, his yields of these crops in any one year equal and sometimes surpass those of his neighbors. And while his gross cash income is lower than many of his neighbors', his production costs are also much lower, allowing a greater profit margin. His local banker, Vince Rossiter of Hartington, Nebraska reports that Wortmann used to be considered a poor credit risk and that lending guidelines had to be bent in order to continue lending him money each year. But Rossiter indicates that in the years since Wortmann switched over to his new farming system he has developed a sound credit rating at the bank.

The soil on Wortmann's farm is a rich, black loam, filled with earthworms which he delights in digging up from the middle of a deep-green corn field and showing to visitors. Wortmann reports that the higher organic content of the soil has made it possible to farm a sloping hillside field with minimal erosion. "Knowing what a man can do working with nature rather than against her," Wortmann declares, "I'd rather quit than go back to farming the way I used to."

The Wortmann success story has been repeated by thousands of farmers scattered across the country who share similar concerns about the effects on the soil of intensive petrochemical farming. One survey by *New Farm Magazine* indicates that at least 24,000 farmers now practice some aspects of organic technology, with 11,200 farmers identifying themselves as "purely organic." The once aloof members of the agricultural establishment have begun to investigate seriously the claims of these unconventional farmers. In July of 1980, Secretary of Agriculture Bob Bergland wrote that "We in USDA are receiving increasing numbers of requests for information and advice on organic farming practices. . . . Many large-scale producers as well as small farmers and gardeners are showing interest in alternative farming systems. Some of these producers have developed unique systems for soil and crop management, organic recycling, energy conservation, and pest control. We need to gain a better understanding of these organic farming systems—the extent to which they are practiced in the United States, why they are being used, the technology behind them, and the economic and ecological impacts from their use.''[32]

University studies conducted in the U.S. and Europe are beginning to substantiate the claims of Wortmann and other alternative farmers. One five-year study conducted at Washington University in Saint Louis by Professor William Lockeretz and a team of researchers concluded that the organic farmers they surveyed had slightly lower average crop yields than their petrochemical counterparts, but that they earned a roughly equivalent net income per acre. The soil on these farms contained significantly higher contents of organic materials and suffered, on the average, one-third less erosion than their counterparts in the study.[33]

Considerable research is also under way to find efficient biological pest-control methods. One promising route involves the use of ladybugs and other predator species which feed on pests that threaten crops. Other promising biological controls are based on the use of viruses that devastate pest populations but leave plants unharmed, as well as use of sterile male insect pests that are introduced into the environment to cut back on egg-laying by female insects. The life cycles of major pests are carefully studied by researchers in this new field of integrated pest management (IPM) so that they can draw up planting schedules which avoid having crops mature when insects are at their strongest. One Texas cotton and grain farmer was able to reduce his pesticide use by 75 percent with an IPM program without sacrificing yields, and many others have reported similar results.[34]

During the Carter administration, efforts were made to bring IPM and organic farming techniques out from the fringes and into the mainstream of U.S. agriculture. One hundred USDA extension agents were trained in IPM techniques, and research dollars for the development of biological pest controls were significantly increased. Bob Bergland commissioned a major USDA study on organic farming which surprised many of the Department's critics with its optimistic conclusions. The report reviewed case studies which pointed out that ''organic farms are strongly committed to soil and water conservation and used the latest and best technology available to control run-off and erosion. Terraces, grassed waterways, strip cropping and contour farming were commonly used and we saw little evidence of erosion on these farms. Critical areas such as steep slopes or shallow soils were usually maintained in sod. Most of the farmers said that since they had converted to organic methods infiltration was noticeably improved, and there was more water available for crops.''[35]

On most of the midwestern organic farms studied in the report, intensive row cropping of soybeans and corn was replaced by less demanding rotations which included pasture grasses, oats, and alfalfa. The net income returns for these types of rotations on a typical 300-acre farm ranged from $34,432 to $40,432 while net returns from a conventional corn and soybean rotation averaged $49,433. Thus, a commitment to practice organic farming techniques often requires a willingness to accept short-term reductions in cash returns in return for long-term soil fertility.[36]

In the short run it is clear that a major shift towards the diversified rotations practiced by organic farmers would do little to aid the growth of the farm-based ethanol industry. If 30 percent of the total acreage now harvested in corn and soybean rotations were shifted to a seven-year organic farming rotation of alfalfa, corn, soybeans, and oats, total U.S. corn production would be reduced by 900 million bushels, with the potential of producing 2.25 billion gallons of ethanol. Such a shift would cause corn prices to rise by as much as 23 percent due to reduced supplies while the price of oats (another potential alcohol feedstock) would fall by as much as 80 percent due to increased supplies. While a sudden shift towards alternative farming practices is unlikely to occur in this decade, a gradual shift would ultimately benefit the ethanol industry by helping to ensure the continued productivity of U.S. farmlands.[37]

RENEWABLE ENERGY IN AGRICULTURE

Both the Lockeretz and USDA studies found that organic farmers require less fossil fuel to produce their crops than do farmers using traditional methods. Lockeretz concluded that organic farmers used 60 percent less fossil fuels to produce each dollar's worth of crops, and a study cited in the USDA report found that organic farmers in New York and Pennsylvania used 15 percent less energy than conventional farmers to produce a bushel of wheat. A major factor in this energy savings was the substitution of animal manures and nitrogen-fixing crops for chemically synthesized fertilizers. Ironically, the energy used to create the chemical fertilizers is derived from domestically produced natural gas, while the energy used to spread animal manures presently comes from gasoline and diesel tractor fuels. Thus, a farmer may actually increase his dependence on imported oil when he shifts from chemical fertilizers to natural fertilizers.

This dilemma illustrates the need for increased use of renewable liquid fuels on the nation's farms. Ethanol can readily be substituted for gasoline tractor fuels, while farm-pressed vegetable oils and another type of high-carbon alcohol—butanol—may prove useful in replacing diesel tractor fuels.

The fuel potential of vegetable oils has been studied since the 1930s in South Africa, Brazil, and the United States, but the economic picture has only recently begun to look attractive.

Considerably less equipment and energy is needed to press oil from seeds, beans, and nuts than is needed to ferment and distill ethanol from starch and sugar crops, as one might expect. A crude pressing operation can be put together with a small investment in a press (some sizes retail for $5,000) and a filter to remove gummy substances that could clog fuel lines. Researchers are finding that vegetable oils, or combinations of oils and diesel, can be a more efficient fuel than diesel alone. A 1980 Southwest Research Institute report, for example, found that a 60 percent blend of peanut oil in diesel used 11 percent less energy than straight diesel to accomplish the same task. Most other vegetable oils and vegetable-diesel blends were of equal or greater efficiency than straight diesel fuel.[38]

South African researchers found that blends of 15 percent ethanol and sunflower oil achieved a greater thermal efficiency than diesel fuel and about 73 percent less smoke. Ethanol and other additives, such as esters, were used to cut the viscosity of the sunflower oil.*[39]

Sunflowers thrive throughout the Midwest and do not require irrigation in most cases. A harvest of some 1,600 pounds per acre of seeds yields some 67 gallons of oil. With production costs at $135 per acre, and the leftover meal worth $90 to $95 as a livestock feed, a farmer could well afford to produce all of his fuel requirements on about 10 percent of his land. South African researchers call this "tithing," (an expression usually connected with donations to a church). Researchers in South Dakota are experimenting with new varieties of sunflower with higher oil yields which could bring costs down, and yields of 2,500 pounds per acre or more may become commonplace.[40]

Butanol is another possible replacement for diesel fuel, since it has a high cetane (or explosivity) rating. It is a four-carbon alcohol, as compared to methanol with one and ethanol with two carbon atoms per molecule, and it blends easily with diesel, even in the presence of water.

Butanol fermentation from corn or other agricultural materials is a more complicated process than ethanol fermentation, since strict anaerobic (airless) conditions and great care to avoid contamination are needed. However, once the mixture is fermented, butanol can be separated from the "beer" simply by adding salt. This simple separation step changes the polarity of the liquid and decreases butanol's ability to absorb water, causing phase separation. The butanol floats to the top and is recovered, without distillation.

Despite the apparent simplicity of the butanol production process, it has several technical drawbacks. Microorganisms that produce butanol are not very tolerant of their waste product, and more than 2 percent butanol in solution will stop fermentation. This is far lower than an ethanol yeast's 10–12 percent tolerance, and less fuel can be recovered from each batch. Butanol also has an unpleasant odor, which nearby homes or businesses might find objectionable. Whether butanol processing can be attractive enough for farmers, especially in comparison with vegetable oils, remains to be seen. One experiment, funded by the Community Services Administration, is under way at Floyd County, Virginia's Agricultural Energy Co-op. Meanwhile, I. S. Maddox of Massey University in New Zealand suggests that cheese whey is an ideal feedstock for butanol, since it is so dilute that it must be concentrated to make ethanol. By using an already dilute feedstock in a butanol process one turns to advantage the apparent disadvantage in butanol fermenting microorganisms.

---

*Esters are, like alcohols, part of a chemical family. Several Brazilian researchers are focusing on the use of esters as cetane improvers for pure ethanol in diesel engines. Esters can be made from alcohols, and are usually used for flavorings and perfumes; ethyl butyrate smells like pineapple, for example. Some observers have wryly suggested that esters be used with vegetable oils to make metropolitan bus exhaust tolerable or even pleasant, but fuel scientists strongly dislike the idea of making fuel seem palatable.

During the 1980s, as oil costs continue to climb, ethanol, vegetable oils, and butanol fuels are likely to gain increasing acceptance as alternative fuels. The widespread use of these fuels in agriculture—combined with conservation-oriented farming practices—can go a long way to reducing the fossil fuel needs of farmers.

AGRICULTURAL FEEDSTOCKS OF THE FUTURE

The fundamental strength of the farm-based ethanol industry lies in its ability to convert a broad variety of different crops into fuel. This type of flexibility allows ethanol fuel production to be economically feasible in a wide variety of different situations, and reduces dependence on crops which already have a high value in grain markets.

A number of new crops are being explored as possible ethanol feedstocks with a view towards their usefulness in organic farming situations. One of the most promising is the jerusalem artichoke, which thrives in a variety of climates and soils. The plant is native to North America and is considered something of a nuisance in the Midwest.* Even so, the jerusalem artichoke may prove to be the major feedstock for ethanol plants. Certain varieties will produce up to 500 gallons of fermentable sugars. If it is planted in the spring, a jerusalem artichoke's sugar can be harvested in the stalks during the summer or in the roots throughout the fall and winter. This flexibility in harvesting seasons is a major asset for farmers in wet regions such as the Pacific Northwest, who often have a difficult time getting into their soggy fields in the fall. By spreading out the harvest time, the need for above-ground storage is reduced, and a constant harvesting and processing operation is possible.

A major strength of the jerusalem artichoke is its ability to thrive with a minimum of fertilizer, which could be easily available from manure or legume crops. Another strength is its ability to get along without pesticides. Tom Lukens, director of Fuel Crops of Seattle, said, "The jerusalem artichoke has few known insect pests." Even the corn borer, the scourge of midwestern farmers, has "no negative effects on yield," Lukens said. Finally, the artichoke competes very well with weeds, and stands of it rapidly establish themselves. This eliminates the need for fuel-intensive cultivation with tractors and the use of herbicides.

One of the most intriguing possibilities for the artichoke is its ability to help control erosion. Unlike corn, it's a hardy perennial whose tubers send out a myriad of shoots each year. After the first-year planting, it is unnecessary to carry out annual spring plowing, which exposes soil to wind and rain. If jerusalem artichokes were planted on erosion-prone sloping hillsides, their tops might be harvested in the summer, leaving tubers and root systems in the ground to hold soil in place.

---

*Some chemurgists such as Leo Christensen pioneered early research into the potential of jerusalem artichokes. The plant was an experimental feedstock for the Atchison, Kansas Agrol plant in 1938 and the subject of later experiments by the University of Nebraska's Chemurgy Department.

The jerusalem artichoke's ability to produce a reasonable yield on marginal soils also increases its economic attractiveness. Most vegetable farmers find, after careful analysis, that they usually do far better to plant their prime lands in food crops and to utilize only cull crops for ethanol production. A grower may be able to earn $3,000 per acre by raising organic carrots in his prime land—a net return per acre that no fuel crop can come close to matching. However, this same grower may have some land with poor soil that is unsuited to intensive carrot production and which he might want to put to less demanding jerusalem artichoke production.[41]

To date, remarkably little modern research has been carried out in the United States to develop jerusalem artichokes as an ethanol feedstock. The technology for crop harvesting is undeveloped and methods for fermenting are still experimental. There has also been little work done by agronomists on higher-yielding varieties. It appears that a modest amount of research in these areas could greatly aid both farmers and the ethanol industry.

Fodder beets are another potential energy crop which has received little attention until recently. They were developed in the 1920s by European agronomists who wanted to give farmers a crop for livestock feed that would grow well in cold, wet Northern European climates and help reduce the farmers' dependence on imported grains. Several varieties were developed by crossing sugar beets with a large, turnip-like root crop called a mangel. Fodder beets produced more usable livestock feed per acre than mangels and more sugar per acre than sugar beets. Some sixty different varieties were eventually developed and fed with great success to pigs and cattle in Denmark, England, Holland, Sweden, and New Zealand.[42]

When the OPEC price hikes of the 1970s aroused new interest in ethanol, New Zealand began experimenting with fodder beets as an alternative ethanol feedstock. Research trials indicated that certain fodder beet varieties had the potential to yield 950 gallons of ethanol per acre, far exceeding the 633 gallons per acre obtained from sugar beets and the 250 gallons per acre average yield obtained from corn.[43]

The New Zealand experiments caught the eye of U.S. researchers who imported some of the seeds and conducted planting trials in eastern Washington, Idaho, and Oregon during 1980. Many of the farmers in these states had been hard hit by the collapse of the U.S. market for sugar beets and were eagerly searching for new crops to fill their vast expanses of irrigated acreage. One fodder beet variety, the mono rosa, produced enough sugar to yield 900 gallons to the acre of Idaho land in addition to 40 tons of green feed (worth approximately $7 a ton) from the beet leaves. A final byproduct from this fodder beet harvest was some two tons of dried beet pulp stillage which had a market value of about $280 as a livestock feed.[44]

But, as researchers working under the direction of John Gallian of the University of Idaho soon discovered, it was far easier to determine theoretical alcohol yields from the sugar contained in fodder beets than to actually convert the sugar to alcohol. In their first year of research, it proved difficult to find a way to economically extract all the sugar contained in the beet. Some helpful technical information was obtained by studying the sophisticated processing employed in large sugar-refining plants. But most of this equipment proved to be too energy intensive and too costly to be used in an alcohol plant of modest size. Mastering the sugar extraction process will prove to be a key to the overall economics of fodder beet ethanol production. If only 50 percent of the sugar in the beet can be extracted readily by a farm-scale plant, an acre of beets with a potential 900 gallons would in reality yield only 450.

From the farmer's point of view, the fodder beet apparently has some major drawbacks. The plants are highly susceptible to nematode worms which attack roots and also to a leaf disease called curly top which has devastated many a western crop. Sugar beets have been bred to develop a resistance to curly top but this type of breeding work is only beginning with fodder beets. Another problem is their heavy dependence on nitrogen fertilizers to obtain a high yield. This places a heavy strain on the soil and requires a farmer to replace large amounts of nutrients drained out of the soil. Farmers are cautioned to grow the fodder beets on a four-year rotation basis, balancing their demands on the soil with nitrogen-restoring legumes.

Unlike jerusalem artichokes, fodder beets mature slowly in the field and are not competitive with weeds, leaving farmers the choice of either cultivating them or spraying them with herbicides. Their slow growth pattern, combined with their shallow root system, provides little protection against soil erosion, and they would be a poor choice for marginal, sloping farmland.

Despite their numerous drawbacks, however, their extraordinary yields may eventually prove attractive enough for fodder beets to come into widespread use.

Along with fodder beets and jerusalem artichokes, the humble potato has also come to the attention of western farmers interested in energy crops. When potatoes are grown as a food crop they must be carefully protected from cosmetic damage by insects and prevented from growing to unmarketably large sizes. But no such constraints are placed on farmers who would grow potatoes for energy. Mark Martin, a USDA researcher in Prosser, Washington, reported, "The crop which shows the greatest promise for ethanol potential, at least in the Northwest, is potatoes." Martin found that certain potato varieties which he cultivated in small test plots produced enough starch to yield 1,200 gallons of ethanol per acre. When the results of his tests were released, many of the Northwest news media mistakenly reported that he had developed some sort of giant "super" potato which would revolutionize the potato industry. In reality, Martin had simply taken several standard potato varieties and cultivated them with the aim of producing maximum yields regardless of appearance.[45]

He concluded, "The cost of producing a crop of potatoes for ethanol should be several hundred dollars less than producing them for culinary purposes since only practices which maximize production need be used." The technology for cultivating and harvesting potatoes, unlike that for jerusalem artichokes and fodder beets, is well established. And the technology for converting them into ethanol also is well known. But potatoes, like fodder beets, place a heavy strain on soil nutrients, are susceptible to nematodes and other diseases, and require extensive weed control.

Other crops with great potential include sweet sorghum and sudan grasses, both of which produce readily fermentable sugars. Sweet sorghum grows well in dry farming regions and produces up to 400 gallons per acre per year. Its stalk can be pressed to extract a sugar juice similar to that produced from sugarcane. But reliable harvesting and processing equipment for sweet sorghum is not yet available, and agronomists have only begun the lengthy breeding trials which could produce new varieties with much higher yields of sugar per acre. A major question mark for both sudan grasses and sweet sorghum, as well as for fodder beets, is what percentage of sugars can be economically removed from the stalks.

Many agricultural byproducts, including largely unmarketable diseased or damaged "cull" crops, discarded vegetables, fruits, candy, and breadwaters from supermarkets and food processing centers, are also potential sources of ethanol. Nationally, their ethanol production potential totals less than 10 percent of current gasoline consumption.* Yet in certain regions, the impact of these agricultural byproducts could be significant. In the fertile Skagit Valley, situated in northwestern Washington, two farm-scale ethanol plants plan to operate exclusively on carrot, sweet corn, and pea wastes, along with cull potatoes, as feedstocks.

There are major economic incentives for the two Skagit Valley plants to use cull crops and food-processing wastes rather than contracting with farmers to grow crops directly for the distillery. The marketplace value of cull crops and cannery wastes as livestock feed usually ranges from $20 to $30 a ton, with the price largely determined by their relative moisture content and nutrient value. In many instances these culls and cannery wastes yield from twenty to twenty-five gallons of anhydrous ethanol per ton of feedstock—a raw feedstock cost of well under $1.00 a gallon when byproduct stillage is sold back to the farmer for livestock feed.

As long as cull crops and cannery wastes are readily available in sufficient quantities, ethanol plants in the Skagit Valley have little incentive to seek out substantially higher priced food-grade feedstocks. A ton of commercial grade potatoes would cost a distillery at least $1.22 per gallon—while a ton of organically grown potatoes could sell for over $400 a ton, translating into an astronomical $16 per gallon of feedstock cost.

For organic farmers in the Skagit Valley, the premium price they receive for potatoes makes it ridiculous even to consider converting food-grade tubers into fuel. And since they receive as much as $10 a bushel for organically grown wheat, rye, and barley, these grains are also unlikely candidates for the distillery. Gene Kahn, an organic farmer from Rockport, Washington, plans to send only cull potatoes, which in a good year constitute less than 10 percent of his crop,

---

*This 10-percent figure excludes the potential ethanol output from corn processing plants such as Archer Daniel Midland's Decatur facility.

to a local distillery operated by the farmers' cooperative to which he belongs. But in bad years, Kahn and other farmers have experienced cull rates as high as 20 to 40 percent of the entire crop for certain varieties of potatoes. Kahn hopes the culls will provide enough fuel to reduce substantially the amounts of diesel and gasoline fuels he must now use to farm his land, but he is not seeking to produce large surpluses to sell to motorists. In the underutilized acres of prime farm land surrounding Rockport, Kahn would like to see more food crops produced—with fuel stocks increased only indirectly by the greater cull crops available as crop acreage is expanded. Only in marginal acres of rocky, weedy, and nutrient-poor soil does Kahn foresee the possibility of growing fuel crops like jerusalem artichokes for the distillery. And even the jerusalem artichokes would go to the distillery only after the demand for these tubers as food crops in regional produce markets had been exhausted.

In the Midwest, where underpriced grains—not high priced potatoes—are the dominant crops, shifting the agricultural base of the ethanol industry to a broader mix of crops will not be an easy task. Short-term market economics in these breadbasket states still favor the use of grain crops, whose cultivation and fermentation systems are well understood, rather than experimenting with some of the newer fuel crops. However, should grain prices rise on international markets, the farm-based ethanol industry may find corn feedstocks priced out of reach and might be forced to find new feedstocks to survive. A rapid transition away from a restrictive reliance on corn could be assisted by federal research efforts to improve cultivation and processing systems for cellulosic wood crops such as poplar trees in addition to jerusalem artichokes, sweet sorghum, and fodder beets.*

Linking the growth of the alcohol fuel industry to the development of a renewable system of agriculture will be a difficult goal requiring a broad public awareness of the hazards posed by many current farming practices and their potential for being compounded by the ethanol industry. Innovative farming techniques will not gain widespread acceptance if farmers cannot afford to look beyond immediate financial survival to the broader issues of long-range productivity. This dilemma was spelled out in a report issued by outgoing USDA Secretary Bob Bergland in January of 1980. This report, titled "A Time To Choose," concluded:

> . . . one of the most important tasks before us is maintaining the productive capability of our resource base over the long term. It is also clear that the market may fail to adequately reflect the full costs of the resource use over the long run. Intensive production in response to temporary market signals may cause irreparable damage by severely reducing the resource base's productive capability at some future time.
>
> The intensiveness with which resources are used is inextricably linked to the quality of the environment. Farming practices that seriously erode land reduce water quality; pesticides and chemical fertilizers are moved into streams; wildlife

*The alcohol potential of poplar and other wood energy crops is explored in Chapter 10.

and their habitats are adversely affected, and the ecological balance is seriously altered. So, it is not only the present and future productive capacity of our resources that concerns us, it is the quality of the environment, the quality of life, for future generations as well.[46]

As exhibit 6-2 illustrates, the most rapid short-term growth of the ethanol industry can probably best be achieved by continuing its present reliance on grains. Embarking on an alternative path stressing ecologically sound farming practices, substitution of renewable for nonrenewable energy inputs, and the use of new feedstocks will undoubtedly result in a slower development of the industry, yet will ultimately provide a more stable foundation for future growth.

Exhibit 6-2
Raw Materials for Estimated 1985 Ethanol Capacity (Source: U.S. National Alcohol Fuels Commission—1981)

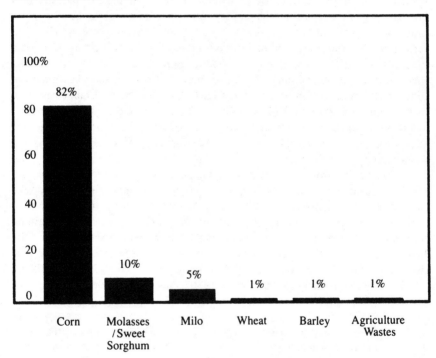

Chapter 7
Brazil: A Quest for Self-Reliance

It was dinnertime inside the elegant restaurant on the forty-first floor of the spindly skyscraper towering over the smog-enshrouded city of São Paulo, Brazil. A wine-flushed delegation of Louisiana sugarcane growers, businessmen, professors, and state officials sat down to a meticulously served meal of paté, shrimp, and steak. After the last dish of ice cream was polished off, Octavio Prado, a silver-haired Brazilian with a patrician profile, stood up to propose a toast. "Here's to alcohol," he said simply as glasses were raised.

That brief toast Prado proposed in March, 1979 set the tone for the Louisiana sugar delegation's whirlwind tour of the Brazilian alcohol fuels industry. The tour was sponsored by the Dedini Corporation—a powerful Brazilian sugarcane equipment manufacturing conglomerate. It included a visit to a federal engineering research institute, inspection of a sugarcane distillery, and a high-level briefing by government energy officials.

The Louisiana delegation's visit was one of many tours Brazilian industry has hosted for alcohol fuels enthusiasts from the United States eager to get a first-hand glimpse of the most ambitious alcohol fuels development program in the world. This program might be called Brazil's version of the Apollo moonshot. It is a multibillion dollar commitment of industrial and agricultural resources aimed at ending this Latin American nation's reliance on imported oil; specifically, it aims at providing fuel for Brazil's "manifest destiny" of western expansion.* It involves the use of both ethanol from starch and sugar-based crops and methanol from wood. The program was also expanded to include other renewable liquid fuels such as vegetable oils from sunflowers, peanuts, soybeans, and rape seed. Alcohol and vegetable-oil fuels are being used both in blends with gasoline and diesel and, in their pure forms, to power a new generation of vehicles developed by U.S. and European auto manufacturers.

The Brazilian sugarcane industry has a long history of involvement with alcohol fuels development. A vigorous turn-of-the-century alcohol fuels movement, similar to the one led by Henry Ford in the United States, took place as

---

*Brazil produces only 20 percent of its 1.2-million-barrel-a-day oil consumption. Most of the imports—about 80 percent—come from Middle Eastern OPEC nations. Over one-third of the imports were cut off by the Iran-Iraq war of 1980.

local officials and plantation owners promoted alcohol fuel use. Cross-country tours of pure-alcohol-fueled cars were staged. Those few individuals who could afford to own private automobiles were, especially in rural areas, more likely to stop by their local sugar mill than their local service stations to find fuel. Power-alcohol blends were commonly available in different gasoline brands, and the government gave its official sanction to a blending program. As early as 1919 the governor of the northeastern state of Pernambuco ordered all official 'vehicles to operate on alcohol. By 1931, the federal government had ordered all gasoline importers to mix a minimum of 5 percent alcohol into their fuel.

The number of Brazilian distilleries producing fuel-grade ethanol increased from 1 in 1933 to 31 in 1939 and to 54 by 1945. Fuel alcohol production increased from 100,000 liters in 1933 to 77 million liters at the height of World War II, out of a total beverage and industrial alcohol production of 819,000 liters. Mandatory blending levels were as high as 50 percent in 1943, when German submarine attacks ravaged world oil-tanker fleets. When the war ended, cheap imported oil was once more readily available and alcohol blends were used only sporadically.[1]

In the wake of the 1973 OPEC oil price hikes, Brazil has revived and greatly expanded its earlier programs. The alcohol production agenda for the 1980s now constitutes the most ambitious national effort ever made to break the oil industry's hammerlock hold on liquid fuel markets. Brazil's annual gasoline consumption of some 15 billion liters represents less than 5 percent of the annual U.S. consumption. And while it would require almost the entire annual U.S. corn harvest to achieve a nationwide 10 percent gasohol blend, Brazil's entire annual gasoline consumption could be produced from crops grown on about 12 percent of Brazil's 45 million hectares of crop land. And alcohol could also be produced from cellulosic feedstocks harvested from Brazil's vast tropical forests. Although rooted in fragile soils, these forests are located in the humid equatorial latitudes where high rates of photosynthesis and carbon fixation take place, resulting in rapid plant growth.

In November of 1975, two years after OPEC's first round of oil price hikes, Brazilian President Ernesto Geisel announced the broad outlines for the National Alcohol Program (PNA). The decision to move ahead with PNA was not based on any short-term economic benefits—at the time, alcohol distilled from sugarcane cost more than twice as much as a gallon of Brazilian gasoline refined from imported oil. But with the national debt hovering at record highs of $30 billion dollars, there was little foreign exchange available to pay for the imported oil. And the world price of sugar, which had averaged 50 cents a pound in 1973, had plunged dramatically to less than 8 cents a pound.[2] Suddenly, the country's major exporting industry found itself stuck with massive surpluses that could find no outlets on glutted world markets. President Geisel hoped that the expanded Brazilian alcohol fuels program would both soak up this sugar surplus and create a decentralized fuel industry which could stimulate new jobs in depressed rural areas.

The rapid expansion of this industry would, Geisel hoped, reduce the sharp imbalances between Brazil's industrial south and other less developed states. It was also hoped that the program would reduce equally sharp imbalances within the agricultural sector between large sugarcane growers who earned millions of dollars in profits each year, and the small, subsistence farmers who barely produced enough food to feed their own families.[3]

Brazil's rural poor had received few of the benefits of the nation's vaunted "economic miracle," which saw the gross national product jump at a 9 percent annual rate between 1969 and 1974. This dramatic economic growth was based largely on a policy which was enunciated by then finance minister Antonio Delfim Neto as "export without pangs of conscience." Miguel Arrares, a former Brazilian governor, found that the workers' real salaries in 1970 represented barely 69 percent of what they had been ten years earlier. The consumption of meat in some cities had dropped to half 1963 levels. More than 60 million men, women, and children—over half Brazil's total population—received less than one-tenth of the national income. Most of the land remained in the hands of a few, with 66 percent of the private lands controlled by 2.8 percent of the population.[4]

"The economy is doing fine, but the people aren't," former Brazilian president and general Emilio Medici remarked.[5] Explosive tensions were building up among the nation's rural poor, and the military government's push for industrialization had ignored the demands of urban labor unions. A student revolt in 1968 was brutally put down and in the 1970s, left-leaning students were forced underground into a resistance force whose leaders were systematically tracked down and executed. By the millions, rural people packed up their meager belongings and migrated to the great coastal cities. Shantytown *favelas* made of huts built from plywood, discarded boards, and corrugated tin sprang up on the outskirts of São Paulo and straggled up the hillsides above Rio de Janeiro. The tide of rural poor overwhelmed city officials who were powerless to satisfy the basic needs of their new citizens for food, housing, and jobs.

Developing a national alcohol program to somehow slow this rural exodus would require that the benefits of PNA trickle down into the pockets of the poor, helping them to improve the standard of living in their rural communities. This would be no easy task to accomplish in Brazil's rigidly structured rural society, which contrasts sharply with the more egalitarian rural societies that characterize much of North America. The midwestern United States was settled by an independent-minded mix of European immigrants who staked out small land holdings, held rancorous town meetings, and gave birth to a fiery brand of prairie populism. But the sugarcane plantation system that dominates Brazil's rural economy was developed in the once fertile soil of Brazil's northeastern coast by an old sixteenth-century Portuguese aristocracy.

A GENEROUS LAND

While searching for a shortcut to the riches of the Orient, the Portuguese explorer Pedro Alvares Cabral inadvertently stumbled onto Brazil. His ship was forced off course by unfavorable winds and currents, and Cabral landed somewhere on the northeast coastline to replenish supplies. On his return to Portugal, he reported to the king, "The land is so generous that, if someone wants to take advantage of it, it will produce anything." The land was covered with forests of the valuable brazilwood trees which flourished in the rich black *massape* soil of the northeast. The Portuguese initially used Brazil as a supply point for sailing ships which explored the Latin American coast. A brisk trade in brazilwood developed after the Portuguese, with the aid of various trinkets, were able to persuade the local Indian tribes to fill their ships with wood.

The rapidly shrinking brazilwood forests could not form a stable base for Portugal's early colonial efforts. As the brazilwood trade intensified, the coastal forests were decimated, leaving large tracts of land cleared for sugarcane cultivation. In the middle of the sixteenth century, the Portuguese began setting up coastal sugarcane plantations and constructing mills to refine the cane into sugar. The new industry was controlled by Portugal's landed aristocracy whose power back home was being challenged by a new class of urban entrepreneurs. But when the King of Portugal used his royal powers to deed out land to his feudal lords, their power in the new world was unquestioned.

In his book *Death in the Northeast*, Jose de Castro notes that both "property and states were set up overseas [in Brazil] according to the homeland's feudal ground rules. Vast reaches of territory were handed over to masters endowed by the crown with absolute authority over everything in sight, including people." De Castro, a harsh critic of the sugar plantation system, charges that "the landless masses were viewed like oxen at the mills, as two-legged animals to be yoked to the heavy feudal cart."[6]

The plantations flourished as the demand for sugar in Europe exploded. Slaves were brought in from Africa to work in the fields and every available parcel of land was planted in cane. The monoculture system of sugarcane cultivation, based on an export market, differed radically from the more self-sufficient agricultural economy which the chemurgists, such as William Hale, had fought to establish in the United States.

Efforts to broaden Brazil's agricultural economy to include subsistence crops such as breadfruit, manioc, mango, peanuts, pineapple, and corn were greatly reduced once sugarcane was introduced. A few small fruit orchards were preserved next to the plantation mansions, and cattle ranches in the arid northeastern interior provided a steady supply of meat, but basic foodstuffs had to be imported. At times, the African slaves who worked in the cane fields were actually forbidden to grow their own small gardens and had to find out-of-the-way spots to secretly plant food crops.

The early sugarcane cultivation in Brazil's northeast took a heavy toll on the environment. Essential organic materials and nutrients were drained from the soil by the intensive sugarcane cropping system, forcing the growers to abandon once fertile lands. The northeastern coastal region degenerated into one of the most severely eroded regions of the Western Hemisphere. The wildlife disappeared as the forests were cleared, and soil-clogged rivers went on periodic rampages, overflowing their banks and flooding the surrounding lands.

The sugar industry, despite the destruction it eventually wrought, initially flourished, and by the end of the sixteenth century it produced 65 million pounds of sugar a year—helping to establish Brazil as a major world producer. According to Brazilian economist Celso Furtado, some 90 percent of the profits generated from the sugar industry stayed in the hands of the plantation and mill owners. This sugar elite, however, was slow to reinvest its profits to build up a more diversified local economy.[7]

Today, Brazil's northeast is an economic disaster area. Its once powerful sugar industry has entered a period of decline and stagnation. Many of its impoverished citizens have left—some for the interior to try their luck on the rubber plantations of the Amazon basin, and others to the more industrialized south. In rural areas, peasant leagues, founded in 1955 to try to assure the sugar plantation workers the right to a small plot of land for their graves, push for land reform. In the mangrove swamps outside of Recife, the northeast's largest city, the main source of food for some 100,000 people is a species of tiny crab that lives in the river mud. *New York Times* reporter Tad Szulc visited Recife in 1960 and found:

> When the tide recedes in the Jordao River, one of three rivers criss-crossing this city and flowing into the bay of the harbor, the dirty brown waters of the tidelands become suddenly alive with thousands of men, women and children submerged up to their waists. They comb the bottom for cranjuegos, the tiny crabs that are their main source of nourishment.[8]

Despite the economic collapse of the northeast, sugarcane cultivation still maintains an important place in the Brazilian economy. The temperate south, settled by a diverse group of European immigrants unshackled by a feudal heritage, became the home of a dynamic new sugar industry. In the late nineteenth and early twentieth century there was plenty of room in southern Brazil for the type of freewheeling entrepreneurship that spawned the industrial empires of Henry Ford and John D. Rockefeller in the United States.

THE SUGAR BARONS OF PIRACICABA

Armando and Mario Dedini left their native Italy for southern Brazil in 1920, settling in Piracicaba, a small town in the state of São Paulo with narrow, hilly streets and red-tiled roofs reminiscent of their home. Sugarcane grew well in the red soil surrounding Piracicaba, and the Dedini brothers opened up a small mechanic and carpentry shop repairing sugar-mill equipment. Armando Dedini died in 1931, but his brother Mario saved up enough money to launch a small factory which produced sugar-mill parts. With a flair for innovation and extraordinary business acumen, Mario Dedini built his single factory into an industrial empire encompassing some forty-three diversified subsidiaries with holdings in every facet of the São Paulo sugar industry. Mario Dedini died in 1974, but the Dedini corporation, one of the twenty-nine largest private firms in Brazil, still remains under tight family control. Dedini companies now market the fertilizers needed to nurture sugarcane, control large tracts of prime farm land, and own many of the sugar refineries in the state. Dedini is also the world's largest manufacturer of industrial distilleries.[9]

Before he died, Mario Dedini built for his family a hilltop estate overlooking the cane fields. The entrance to the estate is forbidding, fringed with barbed wire and guarded by security police, but the inside of the courtyard is pleasant. The family houses cluster around a central courtyard which contains a swimming pool, basketball court, and radio tower. Dedini's corporate guests were entertained on an open-air veranda where elaborate buffets are set by white-jacketed waiters. Sometimes, small mementos such as a Dedini key chain are passed out, to visitors to this inner sanctum of power.

The bulk of the family's corporate power still rests with its industrial enterprises. In enormous foundries outside Piracicaba, Dedini workers fabricate 80 percent of Brazil's sugar mills and export the rest to other sugar-producing nations. Dedini's pacesetting industrial activities, which employ some 8,700 people, have helped preserve Brazil's dominance in the international sugar markets in spite of the collapse of the northeastern sugar industry.

But the south's sugar industry, while absorbing some of the out migration from the northeast, did little to ease the plight of Brazil's rural poor. About the best a skilled worker could hope for was a $4.50 a day job in a Dedini factory. The unskilled plantation laborers were paid even less, and income remained concentrated in the hands of the new rural elite which Dedini helped to create. As the south's sugar industry grew, the sugar mills of São Paulo state added land to their holdings at a 7 percent annual clip. Meanwhile, Brazil has had to rely on foreign producers for such basic foods as beans. The rural poor, with little money in their pockets to express an economic demand for food crops, could do little to slow the trend towards increased sugar production.

The PNA was launched in 1975 at a time when the price of sugar was beginning to recover from its plummet to record lows. The decision to move ahead rapidly with an alcohol fuels industry was greeted with great enthusiasm by the Dedini family, which viewed the program as a means for restoring the prosperity they had enjoyed in previous years. They would not be disappointed.

AN ENGINEER'S GREATEST PERFORMANCE

The PNA's staunchest advocate in Brazil's scientific community was an intensely philosophical, bald-domed engineer named Urbano Stumpf. He was a researcher at a laboratory located at the Centro Tecnico Aerospacial (CTA), located outside São Paulo. A longtime advocate of alcohol fuels, Stumpf was something of a showman. To demonstrate the cleanliness of 100 percent-ethanol fuel, he would bend down next to the exhaust pipe of an idling ethanol engine and inhale deep draughts of the sweet-smelling fumes.

In 1974, Stumpf pulled off what will probably prove to be the greatest performance of his career, when then-President Ernesto Geisel paid a short ceremonial visit to the CTA. Geisel, who was also a former president of Petrobras, Brazil's national oil company, had a solid technical background in the energy field. He was also deeply worried about the failing sugar industry and eager to find new outlets for sugar surpluses that were costing the government millions to store. A scheduled five-minute meeting between Geisel and Stumpf turned into a lengthy, freewheeling comparison of ethanol and petroleum-based fuels. Geisel took the time to tour Stumpf's laboratories, where he learned of CTA test findings indicating that the blending of ethanol into gasoline in 20 percent ratios boosted the fuel's octane rating, thus eliminating the need for environmentally hazardous lead additives. This 20 percent blend, with a proper carburetor tune-up, would, according to Stumpf, give an automobile engine the same mileage as regular gasoline. But it was 100 percent-ethanol's potential as a motor fuel that intrigued Geisel the most. Stumpf test-drove a car that had been converted to run entirely on 196-proof ethanol and gave enthusiastic reports of its technical performance. The ethanol engines, Stumpf's research indicated, could deliver 18 percent more power per liter than gasoline, burn cleaner, and give almost the same mileage when properly modified. Ethanol, Stumpf would later confidently declare, "is the fuel needed to reduce the burden of petroleum."

Within months after his encounter with Stumpf, Geisel had upgraded a small research group inside the Brazilian Commerce Department into a National Alcohol Commission which would work under his personal direction. Despite the president's encouraging actions, Stumpf remembers that "doubts still remained in the minds of the general public, of investors and even in certain government circles, about the practicability of substitution of petrol by alcohol." So Stumpf organized a national "integration tour" which took three ethanol-fueled cars on a 5,000-mile journey through the Brazilian interior. The cars performed well on the rugged road trip and sparked a good deal of national interest in the new government project.[10]

There was nothing timid about the way in which Geisel pushed Brazil into alcohol energy production. There were few of the study reports, congressional debates, or public hearings which characterized the lethargic start-up of the U.S. federal program to develop alcohol fuels. Indeed, Geisel's critics complained that the government decrees and documents which accompanied his November, 1975 announcements of the PNA were the jumbled product of rushed research.

Geisel's PNA plan called for a vastly expanded use of both ethanol-gasoline blends and pure ethanol fuels. Although never directly stated in the official proclamations, it was widely assumed that the PNA called for blends of 20 percent ethanol and 80 percent gasoline by 1980 and the rapid phasing in of 196-proof ethanol vehicles. Eventually, the PNA would have renewable liquid fuels replace the entire annual gasoline consumption of 15 billion liters and 80 percent of the annual 20-billion-liter diesel fuel market. It would require 31 billion liters of ethanol.[11]

These staggering production figures could be met only by a massive program to build new distilleries. Prior to the 1975 PNA announcement, alcohol production in Brazil hovered around 550 to 700 million liters per year, with about half of this going to gasoline blends. From 1 to 3.5 percent alcohol was likely to show up in any given gallon of gasoline.

The federal government, signaling the seriousness of its commitment to PNA goals, earmarked $15 billion through 1986 to fund the program. By November, 1978 the National Alcohol Commission, composed of key cabinet ministers, had approved construction of 208 distilleries with a combined production of 3.9 billion liters of alcohol per year. Each would have a capacity of about 30,000 gallons per day and cost between $10 and $20 million to construct.[12]

The alcohol commission awarded low-interest, long-term loans for distillery construction, even extending this credit to other Latin American nations that would buy Brazilian equipment. Nearly every sugar mill in the state of São Paulo scrambled for a piece of the PNA pie, and Dedini's distillery manufacturing subsidiary, Coddistil, soon found itself swamped with orders. Traditionally, the fuel alcohol had been distilled from the molasses syrup byproduct of sugar production. But this feedstock would quickly be exhausted as the program expanded. A large proportion of the new alcohol supply would have to come from raw cane.

BUILDING A NEW INDUSTRY

In a giant factory on the outskirts of Piracicaba, Dedini workers cut and welded together distilling columns for the new industry. Unlike the simple perforated plates that Al Hubbard set in an old pipe in his Alabama still, the Dedini columns were of an advanced industrial design. The perforations in the plates were covered by ''bubble caps'' that distributed the frothing alcohol-water mixture as it became increasingly pure at successive stages.

The columns were loaded on trucks and shipped out to sugar mills, where they were quickly assembled on concrete pads. Each anhydrous distillery needed twelve columns, which, when set in place, loomed up over the sugar fields like blunt-topped silver spaceships. Behind the columns, huge vats with copper cooling coils fermented the alcohol. The sugar distilleries were fueled with the remains of the crushed sugarcane stalks—a fibrous material called bagasse—which was burned to produce heat for steam generation in long rows of brick ovens. All of this process steam could be generated by 75 percent of the cane processed through the distillery, leaving 25 percent of this biomass boiler fuel free for other industrial uses.*

Cane processing involved an initial washing step followed by crushing in giant rollers to extract the sugar juice. Once separated from the bagasse, the juice was filtered and sent to batch fermenting tanks. After one and a half to two days of fermenting, the dilute ethanol would be sent through fermenting towers,

---

*Figures on the ratios of bagasse necessary to fire sugarcane stills vary widely. Above ratios are from a paper by the Centro Tecicologia Promon. An article in *Agricultural Engineering*, meanwhile, said only 40 percent of the bagasse was needed to fire Brazilian distilleries.

where it was raised to 196-proof levels if destined for the 100-percent-ethanol-fueled vehicles or to the anhydrous level if it was to be used in blends.

The sugar mills sent their technicians to Austria to learn more scientific distilling techniques and put their agronomists to work developing new cane varieties with higher sugar yields. Mill trucks were modified to take advantage of this new fuel source, and visitors to the distilleries were apt to spot Chevrolet pickups jostling down the dirt roads with the words "Movido a Alcool" painted on the back.

By the end of 1978, ethanol fuel production levels had climbed to 1.5 billion liters and by 1980 to 4.1 billion liters per year. The fuel was enough to replace 33 percent of Brazil's total gasoline consumption for 1980, more than meeting the PNA production targets.[13]

As the new São Paulo distilleries entered production, an unexpected side effect of the program began causing some alarm. The watery byproduct of distillation, "slops" or "spent wash," was being poured into the state's river system. For every gallon of ethanol that was produced by a distillery, 12 to 17 gallons of slops were discarded.[14]

The Brazilian problem was somewhat, although not altogether, different from that facing U.S. distillers. In the first place, much of the spent wash from a corn distillery can be fed wet to cattle with the protein-rich grain byproduct of the fermenting tank, called "mash." Some of it can be recycled to the corn prior to cooking. And the remainder is not too salty to be spread over fields,* although Environmental Protection Agency regulations often stipulate that spent wash be kept in settling ponds until its biological oxygen demand is reduced.

But Dedini and other sugarcane refiners had found it convenient simply to pump the sugarcane slops straight into nearby rivers. The slops were filled with volatile solids and organic material that robbed the water of oxygen as they decomposed, killing fish and other aquatic life.

The effect of the slops on the muddy river which runs through the center of Piracicaba was clearly visible, as the remaining fish in the river slowly died. Despite the foul odors caused by the slops and other pollutants, children continued to bathe in its waters, but local fishermen were finding their catches sharply reduced. The government passed a law banning the dumping of slops, but found that it lacked the manpower necessary for effective enforcement.

Some attempts have been made to solve the problem. A number of distilleries trucked slops out to fields and spread them cautiously, since their high salt content and acidic nature could damage soil fertility if overused.

Dedini and other sugar industry groups were distressed enough by the mushrooming slop disposal problem to hire the Brazilian research institute, the Centro Tecnicologia Promon, to find solutions. CTP staffers set out from their headquarters in a Rio de Janeiro office building on a worldwide tour of the sugar industry. After an exhaustive period of study, CTP finally concluded that in nearly every instance the slops could be turned from an economic burden to a

---

*Highly acidic spent wash from a corn process must be applied with care to fields, despite its low salt content.

benefit. When refined, they could be used as liquid fertilizers since they had a high nitrogen and potassium content. The solids could be separated and used as cattle feed. The volatile solids would even form a good slurry for methane digestors, which, using an anaerobic process and different kinds of enzymes, could produce methane gas.* Local marketing options would determine the most economical use of the slops, CTP said, noting that considerable investments for processing equipment would be needed.[15] Despite assurances from the Brazilian government that the problem is under control, American engineers touring Brazil's alcohol fuels industry could not agree with the government's placid assurances. New, low-cost separation techniques, such as Columbia University professor Harry Gregor's ultrafiltration processes, may be one answer to the problem.

THE NATIONAL OIL COMPANY

The critical task of integrating ethanol into existing fuel delivery systems was entrusted to Petrobras, Brazil's national oil company. But the top executives at Petrobras were considerably less enthusiastic about PNA than the government ministers on the National Alcohol Commission.

While Petrobras had extensive experience in blending modest amounts of ethanol in 5- to 8-percent concentrations with gasoline, it had never tried to blend the 20-percent concentrations proposed in the PNA. To accomplish the goal would entail costly expansions and modifications of Petrobras's refining and distribution system, which would reduce the overall profitability of its corporate operation and direct revenues away from what was seen to be its primary responsibility—locating, producing, and refining oil. High revenues from $1.50-per-gallon gasoline in 1979 meant that almost $1.00 went to exploring for new oil offshore or in the Amazon or was otherwise added to Petrobras's profits. Ethanol, on the other hand, cost over $1.00 when purchased from distilleries and Petrobras realized substantially less profit from its sale.

Petrobras director Paulo Belotti repeatedly warned that government tax revenues and corporate profits would be gravely threatened by the expanded use of alcohol fuels.[16] But Petrobras President General Araken de Oliveira tried to downplay the oil company's antialcohol image by declaring, "Petrobras is more interested in the development and welfare of its people than in its own profits." And Brazilian Secretary of Commerce Jose Walter Bautista Vidal would boast at a symposium in Rio de Janeiro, "In Brazil, oil and alcohol are not enemies."[17]

Despite these public assurances, Petrobras moved slowly when the new alcohol program was first introduced. Top executives remained skeptical about the economics of the nation's plunge into alcohol production and wondered whether the $16-billion PNA budget might better be spent on oil exploration in the Amazon jungle. The general manager of Petrobras's research center outside

---

*An Indian engineer with a New Delhi chemical company, K. V. Malik, calculated that methane from slops could fire 75 percent of a distillery's needs.

Rio de Janeiro, Antonio Seabra Moggi, insisted that the PNA was based on "artificial" economics. Sugar cane would bring in more income, Moggi and others believed, than the oil replacement value of ethanol.

But Moggi came to appreciate the strategic significance of the PNA. When interviewed at Petrobras headquarters and asked about his statements on ethanol's "artifical" economics, Moggi joked: "Many things we do in life are not so natural as we think, such as drinking Coke . . . I used to hate Coke. Now I like it and spend a lot of money on it."

Moggi's major role in the PNA was to supervise the construction of twenty-six ethanol-gasoline blending centers for Petrobras. The original blending process was surprisingly simple. An empty tank truck would fill up 80 percent of its hold at a refinery gasoline outlet and then move on to another station where its tank would be topped off with ethanol. Once filled, the truck would jostle off down the highway, mixing the gasoline and ethanol together along the way. But trial-and-error experimentation showed that the truck needed to travel at least forty miles before the mixture was thoroughly sloshed together.

Petrobras officials found they were ill-prepared to cope with the rapid development of the national alcohol industry. To begin to take full advantage of ethanol's potential for reducing oil imports would involve a substantial revamping of the refinery system. This change would require extracting less gasoline from each barrel of oil and more diesel fuel, which was still desperately needed to fuel the trucks and buses transporting 80 percent of the country's goods. This refinery changeover was slow to come about, and Petrobras ironically found itself saddled with a surplus of gasoline which it had to export, often at cut-rate prices.

The situation prompted one Brazilian researcher, Sergio Trindade of the Centro de Tecnicologia Promon, to remark, "In Brazil we are at war with the oil barrel. The oil barrel has three armies: the diesel army, the fuel oil army, and the gasoline army. We have tackled the gasoline army and now we must challenge the other two."

In an attempt to "tackle the diesel army," Brazilian scientists began to experiment with the use of ethanol and vegetable-oil blends in diesel engines. Scientists working with the National Energy Council found that ethanol could be blended with diesel as long as fuel injector pumps were kept lubricated. Vegetable oils—peanut, soybean, sunflower, rape seed, and others—could be blended with diesel, blended with ethanol, or used straight in diesel engines, scientists found. After initial testing, the Agriculture Ministry proposed a plan to replace 6 percent of the diesel consumption by 1980 and 16 percent by 1985.[18] The economics looked marginal at best. A ton of soybean oil would bring $600 on the world market, enough to pay for two tons of imported diesel fuel. But, once again, the benefits of stimulating a domestic industry and reducing dependence on uncertain supplies of OPEC oil tipped the balance.

Proposals to reduce fuel oil consumption involved, for the most part, conversion of industries to abundant hydroelectric and coal supplies. One promising approach involving alcohol fuels was investigated by the Energy Company of São Paulo (CESP), a private Brazilian corporation. Construction of sixty-five wood methanol plants would eliminate the need for imported fuel oil by 1985, CESP said. The steam generators operating on fuel oil could be readily converted to operate on methanol, and the sixty-five plants would require wood from a 20,000-square-mile area in Brazil's dry interior which was useless for agriculture.[19]

Despite the appeal of the CESP proposals, the Brazilian government has not been enthusiastic about methanol development. The National Alcohol Commission, a backer of ethanol, has criticized the CESP methanol program, and has insisted that wood should be used for construction, and as a feedstock for ethanol.[20]

Much of the federal resistance to methanol stems from the fact that the Brazilian government and the powerful sugar industry have invested enormous sums of money in the ethanol program. And methanol could emerge as a major source of fuel not only for electrical and industrial power generation but also for transportation vehicles, thus fundamentally altering the direction of the Brazilian alcohol fuels program.

AN INTERNATIONAL CONTROVERSY

The rapid expansion of the Brazilian alcohol fuels industry in the late 1970s coincided with the phenomenal marketing success of gasohol in the United States. As demand for gasohol soared, U.S. distributors looked to Brazil not only as a source of technical assistance but also as a source of cheap anhydrous ethanol for blending. The distributors located in the eastern U.S. were particularly attracted to the Brazilian alcohol because they found it difficult to secure adequate supplies from the midwestern distilleries that dominated the market. It was soon apparent that these distributors could import Brazilian alcohol to eastern U.S. markets at less cost than it took to truck in midwestern alcohol. By the end of 1980, over 25 percent of the U.S. anhydrous ethanol market was supplied by Brazilian imports, presenting the U.S. gasohol industry with its first major internal dispute.[21]

The substitution of imported Brazilian alcohol for the homegrown midwestern product was bitterly denounced by Archer Daniels Midland and its farm lobby allies as a treacherous act which betrayed the founding principles of the homegrown gasohol movement. The East Coast distributors defended their action by insisting that the imports were just a temporary phenomenon which would end when more U.S. distilleries were built. In the meantime, they argued, Brazilian imports were necessary to "prime the pump," to keep consumer interest alive and demonstrate to skeptical financiers that markets existed.

The export issue also caused a sharp split among the Brazilians. Government officials gave little support to the nation's export industry, although some were angry later when the door was slammed in their faces. Most of the exported alcohol came from northern states which had more than could be consumed locally. Once internal alcohol distribution and storage systems were better established, they believed these regional imbalances would probably disappear.

And even if they continued, Camilio Penna, the Minister of Commerce and Industry, commented, "Brazil can't be turned into a giant sugarcane field to satisfy the world's fuel needs."

But while Penna was reassuring U.S. distillers with these comments, Brazilian Planning Ministry officials were preparing a proposal which would allow foreign investors to build new fuel distilleries in Brazil that would produce alcohol exclusively for export markets.

The midwestern distillers in the U.S. perceived this new policy initiative as a direct threat to their prosperity. For alcohol produced with the aid of Brazil's abundant cheap labor could seriously undercut market prices. In the spring of 1980, Dwayne Andreas began to lobby Congress to impose a tariff on imported Brazilian alcohol. This lobbying effort was supported by Robert Strauss, a key Carter administration official who helped negotiate numerous international trade agreements. The tariff was opposed by Alfred Kahn, the president's chief inflation fighter, who charged that the measure would seriously hurt American consumers. It was also resisted by Treasury Department officials who sided with the gasohol distributors in an in-house memo which concluded that the Brazilian imports, rather than hurting the domestic industry, "are currently being used by some future U.S. ethanol producers to prime the market and build up distribution networks until planned productive facilities can come on stream."

Andreas was able to recruit powerful supporters on Capitol Hill, including Kansas Sen. Robert Dole, who agreed to sponsor the Senate version of the tariff proposal, and Minnesota Congressman William Frenzel, who agreed to sponsor the measure in the House. Both Dole and Frenzel had received generous campaign contributions from both ADM and the Andreas family during past elections but both strongly denounced any suggestions that support for the tariff was in any way linked to these contributions.[22]

President Jimmy Carter's position on the tariff controversy remained murky until the final weeks of the 1980 presidential campaign, and few could guess whether he would side with Alfred Kahn and the Treasury Department or with his trusted political advisor, Robert Strauss. But five days before the election, Carter urged in a letter to Treasury Secretary William Miller that he protect the domestic alcohol industry, "immediately, by administrative means if possible, by legislation if necessary." In December, Carter signed into law a bill placing a 40-cent-per-gallon tariff on imported fuel alcohol—effectively offsetting the price differential between U.S. and Brazilian anhydrous ethanol. When Carter stepped down from the presidency in January, 1981, Strauss joined the ranks of ADM's corporate executives. The outcome of the high-stakes struggle had been a clear victory for Andreas and the midwestern gasohol lobby.*

*Although the bill was not proposed in Congress until the fall of 1980, the Brazilian delegation to the Interamerican Conference on Renewable Energy threatened, in the fall of 1979, to cut off all technology transfer to U.S. sugarcane growers and to withhold generous financing terms from the U.S. should the tariff pass. The Brazilians had learned of the behind-the-scenes machinations over a year before they became public.

THE AUTOMOBILE INDUSTRY

Even if some Brazilians were angry at the U.S. tariff action, most conceded that it had become a moot point, and that soon Brazil would have no surplus alcohol to export. By 1979, Brazil's farsighted program to create a new generation of automobiles fueled only by ethanol was beginning to pay off.

When President Geisel first announced the concept, many Brazilian auto executives, particularly those working for subsidiaries of U.S. automakers, refused to believe he was serious. They viewed his declaration as little more than an overblown political gesture which would result in a lot of talk but very little action. A major exception to this lackadaisical reaction was that of Volkswagen-Brazil, the nation's leading automaker, whose management quickly lobbied Volkswagen headquarters in Munich for funds to install research facilities.

Volkswagen had always managed to stay one step ahead of its competition in Brazil, and it was not about to be caught flatfooted when ethanol arrived at the pumps. In the early 1950s, when Brazil was still dependent on gas-guzzlers imported from the U.S., Volkswagen began staking out its claims on the Brazilian market. It set up the first Brazilian automotive production plant which turned out fuel-efficient cars that, by the mid-sixties, had captured the lion's share of the market. The VW "beetle" proved to be particularly popular. By 1979, eight out of ten cars speeding down the highway at any given time were apt to be beetles. They were perfect for the frenetic inner-city driving—filled with wild accelerations and abrupt halts—that Brazilian motorists appear to relish. In downtown Rio, beetles park bumper to bumper on top of the city's ornate mosaic sidewalks, forcing pedestrians to venture out into the streets. Yellow and black "mini-taxis" circulate through the downtown sections of São Paulo, and blue and white beetle taxis prowl the streets of Rio de Janeiro. Even more than the Volkswagen's German homeland, Brazil has become a nation of beetles.

Volkswagen built two manufacturing plants in Brazil. The larger, employing over 42,000 people, is located on the outskirts of São Paulo in a dreary industrial neighborhood. Since the Brazilian military dictatorship took power in 1964, few labor unions had dared to protest the low wages and inflationary spirals which were threatening the workers' standard of living. But in the spring of 1979, as Volkswagen prepared for assembly-line production of its first pure ethanol-powered automobiles, the industrial sector of São Paulo was swept by a wave of labor unrest.

A new Brazilian president, João Baptista Figuereido, had promised to democratize the nation in a preinaugural address, and the prospect of increased freedom galvanized workers' concerns about their conditions. The first open strike in fifteen years was staged by the metallurgical workers' union, and production was halted in almost all of the country's major industries.

Volkswagen was able to escape the first wave of strikes, and, partly as a result, was able to market the first 100-percent-ethanol cars. Volkswagen's intensive three-year research program that led up to the marketing of ethanol cars was headed by Dr. Herbert Heitland, an internationally known German scientist borrowed from the corporation's Munich headquarters. Heitland was "fascinated'' by what he heard about Brazil's ethanol experiment and was eager to put his years of theoretical studies and tests into practical application. In Germany, Heitland worked to develop an experimental methanol engine with Volkswagen and edited a book entitled *On the Track of New Fuels*, which dealt with hydrogen and alcohol.[23]

"The attitudes here are much different from those in Germany and the United States," Heitland explained in an interview in his São Paulo VW factory office, sipping a cup of strong Brazilian coffee. Behind his desk, a VW map of the world was posted on the wall indicating the prime sites for the production of biomass energy from renewable resources. "In the industrial countries, we always want to minimize the risk of everything," he said. "Here we are living like pioneers and can still experiment."

Accompanying Heitland to Brazil was a brilliant young Austrian engineer, George Pischinger, who would have direct responsibility for starting up assembly-line production of the new 100-percent-ethanol Volkswagens. Pischinger was somewhat of a loner who mixed uneasily with the Germans at Volkswagen and resented the aura of fierce nationalism which surrounded the Brazilian alcohol program. He felt the Brazilians were trying to claim too much credit for work largely developed in the German-based Volkswagen laboratories. "It's no great moral question to produce alcohol," he commented. "The Brazilians are not trying to do something great for the world. They just want to survive."

Although it had taken decades to perfect the gasoline-powered internal-combustion engine, Heitland and Pischinger were asked to develop the ethanol engine in only a few years' time. They found that a tremendous amount of research needed to be done. Plastic carburetor, fuel line, and gasket parts susceptible to the corrosive effects of ethanol had to be replaced. New fuel tanks, which would resist troublesome acid residues in the 196-proof ethanol, had to be developed. But the toughest problem was finding the right combination of temperature and fuel delivery to the new, higher-compression engines.

"During the early stages of our work, ethanol-powered engines were always destroyed when the intake air was heated excessively and engine knock occurred," Heitland told a November, 1977 symposium on alcohol fuels technology in Wolfsburg, West Germany. "The problem was solved when preheating was reduced, especially under full-load operation, and when a sufficiently safe margin from the knock limit was reached." Despite the problems, Heitland told engineers at the conference, Volkswagen was "certain that ethanol as motor vehicle fuel is a very attractive solution for the future."[24]

The final prototype VW Dasher averaged 18.5 miles to the gallon and housed an engine that delivered 10 percent more power than its gasoline equivalent. For several months prior to actual assembly-line production, Pischinger drove the prototype model around São Paulo, weaving through chaotic traffic snarls with the intensity of an Italian racing driver.

By spring, 1979 the major North American auto companies operating in Brazil had come to the realization that pure ethanol fuels were about to become a permanent fixture at Brazilian service station pumps. President Figuereido affirmed shortly after his inauguration the administration's commitment to the PNA. He set a new production goal of 10.8 billion liters of ethanol by 1985—enough fuel to replace two-thirds of the nation's gasoline consumption. Some Brazilian officials talked optimistically about a 20-billion-liter-per-year capacity by then, which would be enough to propel 60 percent of the nation's fleet of 12 million gasoline and diesel vehicles.[25] Later, the PNA goals were increased to 15 billion liters by 1985, although some officials doubted that the Brazilian bureaucracy could move that swiftly.[26]

As it became clear that Figuereido would give even greater priority to PNA than his predecessor, Ernesto Geisel, the North American automakers were roused into action. General Motors of Brazil, which had previously delivered one hundred experimental ethanol trucks to a group of São Paulo sugar growers, announced a $500-million investment program to revamp its diesel truck plant for the use of what it labeled "alternative combustibles."

Ford Motor Company of Brazil perfected a V-8 engine powered by 196-proof ethanol for its lavish Brazilian Landau model. Then, in an elaborate ceremony, Ford presented a new Landau to Figuereido. This was followed by a full-page ad placed in major Brazilian newspapers declaring: "Attentive to the needs of Brazil, the engineers at Ford are accelerating their projects to put at the disposal of all, in the shortest possible time, vehicles that run on alcohol. . . ."[27]

Brazil's first generation of ethanol-powered autos did not gain wide acceptance overnight. Models produced in 1979 were generally slower to start than gasoline-powered cars and contained some plastic parts that did not wear well. Many ethanol pumps installed at São Paulo service stations were often left hooded due to lack of use. But this lackluster reception gave way to an astounding surge in demand after the Iran-Iraq war cut off one-third of the country's oil supply. In an emergency measure to reduce demand, the price of gasoline was raised to $3.00 per gallon while the price of alcohol was kept at $1.50 per gallon. And while gasoline pumps were ordered closed on weekends, alcohol pumps were kept open on Saturdays.

Suddenly the Brazilian automakers could not manufacture the new cars fast enough to keep up with sales. September, 1980 proved to be the best month in Brazilian automotive sales history as some 30,000 ethanol vehicles were snatched up by eager buyers. The *New York Times* reported in October that "the rush for alcohol-powered cars occurred so swiftly . . . that the delay period for the purchase of one is already up to 60 days. The government-certified mechanics who perform the $250 conversion of gasoline engines to alcohol use are so swamped with orders that they are two months behind."[28]

Many motorists turned to unauthorized mechanics to make the gasoline-to-alcohol conversion. These unauthorized conversions cost only about $60 but were far less thorough than the certified conversions. In a twist of the popular name for the PNA, "proalcool", motorists who made the unauthorized conversions were dubbed "proalcapones." In the fall of 1980 federal officials were trying to persuade these motorists to convert their old cars back to gasoline in order to avoid unexpected alcohol fuel shortages and to make sure that auto manufacturers would not be discouraged from going ahead with increased production of the new ethanol-powered vehicles.[29]

By the Fall of 1981, the volatile market for ethanol automobiles had reversed itself once again as consumer confidence in the first generation prototypes waned, and the nation as a whole plunged into a deep recession. Many of the converted automobiles failed to operate satisfactorily, and even some of the assemblyline vehicles suffered from corrosion, cold-start problems, and increased fuel consumption. One official of the Rio de Janeiro Motor Rebuilders Association reported a 90 percent drop in the number of autos converted to operate on ethanol in the nine months following January, 1981. Sales of new ethanol vehicles also plummeted with reports surfacing that new car dealers had actually begun to convert ethanol cars in crowded showrooms to operate on gasoline. The technical problems that plagued the boom-bust birth of the ethanol automotive industry in Brazil were perhaps the inevitable result of the frenetic pace with which the nation attempted to make the transition from petroleum to renewable fuels. Many motorists still remained unconvinced that ethanol should be their nation's fuel of the future, and the government was forced to launch a new campaign to bolster confidence in the performance of ethanol.

THE RURAL POOR

The dissatisfaction of the motorists was but one of the problems facing the PNA. The motoring public, some 13 million strong, was really only a privileged minority in a nation of 120 million people. And it was far from clear that the nation's rural poor were benefiting from the PNA. As one cynical observer noted, "The government is more concerned with feeding the beast with four legs than the beast with two legs." For if sugarcane was to continue to be the sole agricultural feedstock for the PNA, its acreage would have to be more than doubled over 1980 levels in order to replace the OPEC oil imports. And this vast expansion of sugarcane acreage would increasingly put energy crops into direct competition with food crops for prime farm land, investment capital, water, fertilizer, farm management skills, farm to market transportation, and technical services.

Lester Brown, of the World Watch Institute, noted, "The decision to turn to energy crops to fuel the country's rapidly growing fleet of automobiles is certain to drive food prices upward, thus leading to more severe malnutrition among the poor. In effect, the more affluent one-fifth of the population, who own most of the automobiles, will dramatically increase their individual claims on crop land from roughly one to at least three acres, further squeezing the millions who are on the low end of the Brazilian economic ladder."[30]

William Saint, a rural sociologist with the Ford Foundation in Rio de Janeiro, was even more blunt in his assessment of the PNA. He stated that "in a country where an advantaged 20 percent of the population owns almost 90 percent of the automobiles and a disadvantaged 50 percent spends at least half their income on food, the policy decision—stated somewhat too simply—comes perilously close to choosing between allocating calories to cars or to people."[31]

The World Bank echoed these concerns in a September, 1980 report that backed up its decision to fund economically justified alcohol programs. "The possibility of large-scale biomass alcohol production has posed the question of whether, and to what extent, such a development is likely to compete for land and other agricultural resources that could otherwise be allocated for producing food. . . . The issue is complex and can sometimes be emotional. Basic considerations in assessing the extent of future competition for agricultural resources are the relative price movements for energy and food. As noted, on a global basis a sharper increase in energy prices than in food prices or most other agricultural products is plausible, at least over the next decade. Assuming this occurs, the potential land use conflict between food, export, and energy crops will increase as economic forces increasingly draw agricultural resources into energy production.[32]

Brazil's failure to develop food self-sufficiency along with its quest for energy self-sufficiency was starkly evident in 1979, when the nation's food import bill topped $1.5 billion. Although imports of beef, corn, rice, and beans had become a standard fixture of the Brazilian economy long before the PNA, due to its extensive system of sugarcane monocropping, the general fear was that the situation would steadily worsen with the increased demands of the PNA.

As the controversy intensified, the Commission Pastoral de Terra, an organization representing Catholic Church members, warned against "proletarianization" of poor Brazilians, and urged the government to adopt plans for large numbers of small-scale manioc distilleries instead of huge sugarcane plantations.[33] The Agriculture Minister, Amaury Stabile, tried to reassure critics by declaring that the government would not allow the country to run the risk of "having to import food in order to drive our cars." In 1979, the Agriculture Ministry began to cautiously implement a land-reform program aimed at penalizing owners of nonproductive farm lands. According to a May, 1979 issue of *Latin American Economic Report*, the main targets of this reform were "owners of latifundios, particularly in the northeast. The existence of vast unfarmed tracts of potentially productive land has meant that the northeast traditionally has bought large quantities of foodstuffs from the south, although the region could undoubtedly become self-sufficient."[34]

SECOND-GENERATION FEEDSTOCKS

From the onset of the PNA it was obvious to government policymakers that the social costs of an ethanol industry based solely on sugarcane would ultimately prove to be unacceptable. They hoped to rapidly develop second-generation energy crops which would provide a broader base for industry and bring more benefits to small farmers. Ideally, these new energy crops would be less demanding on soil fertility and be able to grow on marginal lands. These crops could then be cultivated in isolated interior regions and converted into fuel at small-scale distilleries similar to systems developed in the American Midwest.

Many second-generation feedstocks have been analyzed, including babassu nuts, which grow on a palm tree that thrives in swampy areas, and sweet sorghum, a crop which requires considerably less rainfall than sugarcane. The most important of these second-generation crops is thought to be manioc, a starchy root crop, often called cassava, which was cultivated for centuries by Indians on small plots of land in the Amazon. Today, manioc continues to be a staple food of Brazil's rural poor. The root is pulped, detoxified, and then made into a flour which is sprinkled on beans and rice. Sometimes, the flour is made into a heavy bread which lacks protein but helps to fill empty stomachs.

During the decade of Brazil's "economic miracle," as the living standards of the rural poor began to drop, many people turned to manioc to survive. By the end of 1977, small manioc plots took up some 2.1 million hectares of land—more than the total sugarcane acreage at the time.[35]

Manioc had three traits which endeared it to the government's agricultural research scientists. It was high in starch fibers which could be broken down readily into simple sugars for distillation into ethanol. It appeared to grow well in dry, poor soils which couldn't be used for sugarcane. And it could be harvested year-round, preventing the long shutdowns which plagued the sugarcane distilleries during the months when there was no cane to cut.

Hopes were initially high among the PNA's policymakers that manioc, grown on marginal lands, would eventually replace sugarcane as the nation's principal energy crop. But the sugarcane growers' lobby, whose power had increased markedly during the early years of the PNA, preferred to see the rapidly expanding ethanol industry remain dependent on sugarcane.

The technology of manioc distillation, except for the initial step of breaking the starches down into sugars, was not much different from that used in sugarcane distillation. "The problem," remarked one government official, "is not the technology but the know-how to apply it on a large industrial scale."

Petrobras was given the task of building the first modern manioc plant, which was engineered to produce 60,000 liters a day. Petrobras chose to locate the plant in the southern mining state of Minas Gerais, a depressed region where out-migration almost equaled that of the northeast. In the 1930s, Minas Gerais was the site of an earlier experiment involving manioc feedstocks for fuel. Between 1932 and 1934, a distillery in the town of Divonopolis produced 5.3 million liters of alcohol before being forced out of business by the return of cheap oil. The new distillery site was in the town of Curvelo, a few hundred miles from the old Divonopolis site in a desolate shrub-covered region of the *cerrado* badlands.

Brazilian scientists studied over five hundred different varieties of manioc before selecting a half-dozen types which they felt would be best suited for the Curvelo region. These varieties were nursed from seedlings in government greenhouses and then offered to small farmers in the area. Ideally, the revenues from the manioc energy industry would go directly to these farmers, giving them enough money to supplement their starchy manioc diets with more nutritious foods. But the American Carl Duisberg, in Brazil researching for his PhD dissertation, reported that "despite the presence of these nurseries . . . they [Petrobras] were unable to convince local small producers to increase their production to supply the distillery. *'Mineros neo se arriscam'*—'people from Minas Gerais don't take risks'—and farmers in the area showed no enthusiasm for, or confidence in, the project."

Petrobras abandoned its efforts to recruit small farmers and turned to agribusiness to produce the manioc it needed for the Curvelo distillery. By the summer of 1976 Petrobras had signed manioc production contracts with six large agribusiness corporations.

One of these firms—the Antunes Group—was partially controlled by Daniel Ludwig, an aging U.S. shipping magnate who had built a fleet of chartered tugboats into a multibillion dollar personal empire during a career that spanned four decades. Ludwig's efforts to develop a 3.5-million-acre tract of Amazon wilderness into a major world supplier of paper pulp had made him a highly controversial figure in Brazil. Starting in 1967, Ludwig replaced 250,000 acres of tropical forest with fast-growing paper pulp trees of melina and Honduran pine. A giant paper mill, prefabricated in Japan, was shipped to the project site by the Jari River, where Ludwig had gathered a work force of over 30,000 people. Ludwig took on this mammoth task shortly after his seventieth birthday, a fact that prompted *Fortune* magazine to note, in a lengthy personal profile, that "For a single man to undertake such an experiment in the final quarter of the 20th century is extraordinary to the point of eccentricity."[36]

The Brazilian government initially welcomed Ludwig to the Amazon, but, by the mid-1970s many government officials were openly critical of the Jari project over which they had so little control. There was concern that Ludwig's workforce would run roughshod over the delicate forest environment and wreak widespread destruction. "To allow a transnational into the Amazon region is the same as letting a goat into an orchard," declared Federal Deputy Modest da Silveria in June of 1979 during a face-to-face confrontation with Jari project directors.

But reporters such as Loren McIntyre of the *National Geographic* and several tropical foresters who visited the Jari project were impressed by the sensitivity of the Jari management to the ecology of the region and believed that these plantations stood a good chance of long-term success. Doubts still linger, however, about the ability of the fragile Amazon soils to sustain the intensive cultivation of paper pulp tree crops.[37]

Ludwig, who turned eighty-three in 1981, began the cultivation of manioc with the same intensity he displayed earlier in establishing the Jari project. His Antunes Group and five other corporations hired by Petrobras planted more manioc acreage than had ever before been attempted in one area. In experimental test plots, government agronomists achieved manioc yields of up to 60 tons per hectare. Petrobras agronomists expected yields of only about 15 tons per hectare. But even this figure, triple the amount of the average peasant's yield in the northeast, proved to be drastically over optimistic. The manioc plants did not take well to the intensive monoculture plantation farming methods the Antunes

Group used. Despite the addition of chemical fertilizers, the manioc yields averaged only half the 15-tons-per-hectare figure predicted by the Petrobras agronomists. Leopodtra, a leaf disease, attacked the plant, and a bacterial plague destroyed many of the roots. Drought added to Petrobras' problems. The diseases that struck the manioc plantations, agronomists noted, had rarely affected smaller plots of the crop, and Petrobras began contemplating a return to its original tactics.

The actual distillation process proved to be workable, despite failures in the first round of crop planting. The thick, tubular roots were sent to the Curvelo plant where they were weighed, washed, peeled, ground, cooked, fermented, and finally distilled. Wood from nearby stands of eucalyptus was used to provide the process energy for the plant. The Curvelo distillery's biggest problem during its first year of operation proved to be a simple lack of manioc supplies. Throughout 1978, it was forced to operate at only 30 to 40 percent of capacity.

After that first discouraging harvest, Petrobras tried once again to entice small farmers to produce manioc for the distillery. It helped them obtain the financing needed to increase their manioc production, removing some of the risk. Ludwig's agronomists, meanwhile, went back to the drawing board to try to develop a better large-scale manioc cultivation technique, and started a new experimental farm at the Jari project.[38]

Despite the disappointments of the early Curvelo experiments, the Brazilian government has continued to encourage the use of manioc for alcohol production in both large- and small-scale plants. A Japanese corporation announced plans in 1979 to open up a joint manioc venture in the Amazon basin, and a Brazilian firm proposed a manioc plant in the state of Mato Grosso. The government has also announced plans to build a series of experimental small-scale distilleries situated primarily in sparsely populated farming communities. Technical backup for the ventures is provided by Brazil's Institute for Small Technologies, which has developed a wooden distilling column that can be cheaply assembled by small farmers with inexpensive local materials.

TECHNOLOGY TRANSFER

Although Brazil's PNA is still in its infancy, it has already played a major role in stimulating renewed international interest in alcohol fuels. Despite the setbacks of the Minas Gerais experiment, many nations are interested in following the Brazilian lead in manioc fuels. In South Africa, "oil fields" of manioc have been planted in the arid Makatin Flats of Zululand. The Australians are experimenting with manioc on a large scale in their semidesert "outback" which covers much of the nation. Vast areas of Africa, Asia, and Latin America that now depend on manioc as a food crop may begin to see some manioc use as a fuel. The everpresent danger in this trend is that manioc destined for subsistence food crops may be diverted away to more lucrative energy markets. Most Third World farmers cannot afford the trucks or motorbikes that would be powered by manioc fuels, so their main interest in such projects would be to pocket additional cash that could be obtained by selling off occasional crop surpluses.

In Indonesia, the government is attempting to use alcohol fuels as an aid to a major resettlement program that will encourage people to migrate from densely populated islands to more remote outer islands. Fuel needs on these islands, according to the government plan, will be met largely by ethanol derived from sweet potatoes.

In Latin America, many nations have turned to Brazil for technical assistance in establishing their own alcohol fuels industries. Jamaica, Panama, the Dominican Republic, and Costa Rica all have pilot distilleries under construction, financed in part by the Brazilians. Paraguay is working with Dedini and the Brazilian government to develop small-scale sugarcane distilleries which can be built for about $20,000. And in neighboring Uruguay, a sugar growers' cooperative hopes to cultivate sweet sorghum as an alcohol feedstock.

The use of alcohol fuels transcends all political boundaries. Along with the interest shown by rightist-dominated nations such as Paraguay and Argentina, leftist nations such as Cuba and Nicaragua are counting on alcohol to ease their reliance on petroleum. Cuba, in fact, has been using alcohol fuels for decades, and when large sugar surpluses built up during U.S. boycotts of Cuban sugar, distilleries began producing fuel for the cars and trucks of this isolated communist nation.

Nicaragua's young revolutionaries, meanwhile, are planning on small-scale alcohol fuels plants to make the country's peasant population more self-sufficient. "We visualize a highly humanized life," an official of AGROINRA, a state agricultural organization, told the authors. "We need an alternative economic and social order that is humanized enough to avoid the flight to the cities. This is a strategic principle of the revolution. . . . [Thus] we need to pay particular attention to small-scale alcohol fuels development."

Meanwhile, as the Brazilian PNA advances in the 1980s, the ambitious production quotas sought by the Brazilian government appear to be well within reach. The future may yet bring the near total phase-out of imported oil in Brazil and the widespread use of alcohol and vegetable-oil fuels. But ultimately, the success or failure of the Brazilian PNA and the alcohol projects of other nations, will not be judged only by narrow technological yardsticks, but by their broader ability to improve the welfare of the population.

Brazil's victory over the OPEC oil barrel will be hollow indeed if it is achieved at the expense of the basic food needs of its people.

Chapter 8
Alcohol in Engines

In the fall of 1977, automotive research scientists from twenty different nations journeyed to Wolfsburg, West Germany to attend the second International Symposium on Alcohol Fuel Technology hosted by the Volkswagen motor company. There was a sense of excited anticipation among them, for the event marked a turning point in efforts to establish alcohol fuels as a realistic alternative to petroleum. Brazilian, Swedish, and German experiments with 100 percent-ethanol-powered engines had sparked international interest in the use of alcohol in both gasoline and diesel engines. And the most recent round of price increases by OPEC had dramatically improved alcohol production economics.

The 370 scientists attending the Wolfsburg conference spent three days listening to various speakers present some forty-five technical papers detailing their latest research findings. A joint conclusion at the end of the conference read: "Methanol and ethanol are the most attractive mid-term candidates as alternatives for current hydrocarbon fuels."

Despite disagreement between individual reports, the consensus at the Wolfsburg conference stood in stark contrast to research results published by U.S. automotive and oil industry scientists in the 1970s. The majority of these reports harped on numerous technical obstacles that might inhibit if not preclude the widespread use of methanol and ethanol fuels.

Controversy among engineers over the technical performance of alcohol fuels did not suddenly emerge in the 1970s; it had been simmering for decades. In 1949, S. J. W. Pleeth lamented in his book, *Alcohol: A Fuel For Internal Combustion Engines*:

> The bias aroused by the use of alcohol as a motor fuel has produced [research] results in different parts of the world that are incompatible with each other. . . . Countries with considerable oil deposits—such as the U.S.—or which control oil deposits of other lands—such as Holland—tend to produce reports antithetical to the use of fuels alternative to petrol; countries with little or no

indigenous oil tend to produce favorable reports. The contrast . . . is most marked. One can scarcely avoid the conclusion that the results arrived at are those best suited to the political or economic aims of the country concerned or industry sponsoring the research. We deplore this partisan use of science, while regretfully admitting its existence, even in the present writer.[1]

In a modern update on Pleeth's observation, Richard Pefley, Chairman of the University of Santa Clara's Department of Mechanical Engineering, commented in an interview with the authors that much of the research into the performance of alcohol fuels conducted by U.S. oil and automotive companies has been "defensive in nature," designed more to point out problems than to develop methods to overcome them.

Throughout the decade of the seventies Pefley had been on the cutting edge of automotive research into both methanol and ethanol fuels. He played a key role in organizing the Wolfsburg conference and in establishing an informal international network of university, industry, and government researchers who freely shared information about their respective efforts to improve the performance of alcohol engines. Thus, Pefley found himself in a unique position to assess the progress not only of his academic colleagues but also of many of his counterparts working with automotive and oil companies.

On close examination, the results of these research efforts indicated that both methanol and ethanol are surprisingly versatile fuels with the potential to substitute for many of the products refined from a barrel of crude oil—including diesel, electrical generating fuels, aircraft fuels, and petrochemical feedstocks. Although some questions still remain as to the long-term compatibility of certain engine components with methanol and, to a lesser extent, ethanol, the roadblocks which stand in the way of developing multiple end uses for alcohol fuels are less of a technical than an economic and political nature.

GASOHOL'S ROCKY ROAD TO RESPECTABILITY

The term *gasohol* was coined by a committee of the Nebraska State Senate, which in 1972 began an informal test of ethanol blends with two highway department trucks. The tests were directed by William Scheller, a chemical engineering professor at the state university, who recalled to the authors that "the petroleum industry was telling us that if we put this stuff in our cars, we were going to ruin the engines by washing the lubrication off the cylinder walls." When the trucks held up for 100,000 miles, Scheller said, "we just decided that this was a bunch of hot air coming out of the American Petroleum Institute."

The next step was to organize an expanded test program on forty-five state vehicles for the blend which soon earned the nickname of "gasohol," a term later trademarked by the state to define any 10 percent blend of renewable ethanol in gasoline.

The Nebraska gasohol tests found no problem with phase separation between the alcohol and the gasoline components of the fuel; the tests did show reduced fuel consumption and fewer pollutants in the engine exhaust. Fuel consumption was improved by 5.3 percent, Scheller reported. The report was greeted with skepticism by General Motors, Chevron, and the American Petroleum Institute, but since that time other reports—notably the U.S. Department of Energy's ongoing test programs at Bartlesville, Oklahoma—have shown that gasohol does improve mileage. One DOE test on Bell Telephone Company vehicles concluded that 4.48 percent mileage improvement was possible. The discrepancies in these test results were probably due to variations in the age and gross weight of the test vehicles, compression ratios, air-fuel mixtures, weather, and other factors that were better controlled in later tests.[2]

Skepticism also greeted the Nebraska finding that "engine cylinder wear measurements . . . made as a part of our Two-Million-Mile Road Test . . . indicate no unusual wear or deterioration of engines as a result of using gasohol fuel."[3] But the more extensive Brazilian experience with ethanol-gasoline blends has since borne out this finding. The 20 percent ethanol blends used routinely by motorists there have been successful "with little or even without engine modifications," according to Volkswagen researcher W. E. Bernhardt.[4] The principal adaptation that both U.S. and Brazilian engineers have, at times, found advisable is the replacement of clogged gas-line filters when gasohol is first used. This is necessary due to the ethanol's solvent action which will sometimes flush out carbon and grime in a fuel tank or fuel line. Occasionally, plastic and rubber fuel-line parts have also needed replacement.

Historically, blended fuels with over 20 percent ethanol in gasoline have been unpopular with motorists since they delivered poor mileage and frequently suffered from engine stumbles, surges, and other "driveability" problems as well as occasional phase separation. One team of Brazilian and U.S. researchers has managed to develop a system for adapting engines to operate efficiently on ethanol blends of up to 50 percent, however. In a paper published in 1980 by Carlos Luengo of the Brazilian University of Campinas and his U.S. associate Michael Leshner, a fuel control system was described that continuously adjusts air-to-fuel ratio through an electronic combustion control.[5] This constant adjustment results in only a few percentage points loss in fuel economy, even with 50 percent blends, they reported. Similar work with oxygen-sensing exhaust systems is underway with Toyota automobiles, and the hope appears to be that an "all fuels" car may be developed.

Converting gasoline engines to operate on 100 percent ethanol is a more complicated undertaking than making the minor adjustments sometimes necessary for alcohol blends. A thorough engine conversion involves redesigning the standard gasoline engine to take maximum advantage of alcohol's ability to transmit its heat energy into mechanical energy. Gasoline, for example, can transmit only about 25 percent of its potential heat energy into mechanical energy, while methanol can transmit about 35 percent and ethanol 32 percent. This heat energy is measured in British Thermal Units (Btu), and ethanol has only two-thirds as many Btu, gallon for gallon, as gasoline, while methanol has only half. Thus, gallon for gallon, fuel consumption is higher with alcohols than with gasoline, but the greater conversion of heat to mechanical energy—called thermal efficiency—can make up for this deficiency to some extent.

The two major modifications necessary to efficiently operate a 100-percent-ethanol-fueled engine involve an increase in the air-to-fuel ratio of the carburetion system and an increase in the cylinder compression ratios. The carburetor adjustment is required due to the different ratio of carbon and hydrogen molecules to oxygen in ethanol, which allows a cleaner burn, while the increase in compression ratio is carried out to take advantage of its higher octane rating compared to gasoline. By increasing the length of the piston, and thus of the compression inside the cylinder, more mechanical work is performed than in the lower-compression gasoline engines.

Another modification critical to the smooth operation of an ethanol-powered engine is the addition of a system to preheat the fuel-air mixture before it enters the engine cylinder. This step is necessary due to the fact that the heat of vaporization of ethanol is 3.5 times higher than that of gasoline. Automotive engineers in Brazil tackled this problem by using the heat radiating from the engine's exhaust manifold to warm the ethanol-air mixture. But such systems cannot be used when an alcohol car is first started up. One solution to this problem is to include a small auxiliary gasoline tank for cold starting. Another solution is to modify the fuel. A number of additives—such as isopentane, butane, or even gasoline—can help with cold starting. Volkswagen has recommended that the Brazilian alcohol pumps distribute a mixture of ethanol and ethyl ether, which would bring cold starting limits down to tolerable levels in the mild Brazilian climate.

One frequent point of controversy in the automotive engineering world has been the question of ethanol's long-term impact on engine wear. Ethanol is a "dry" fuel, containing less of the self-lubricating properties that gasoline provides, and it is also a powerful solvent. Mechanics who overhauled Brazilian ethanol engines have complained that valve wear is twice that of gasoline, while other reports have indicated that increased corrosion of aluminum, copper, and lead alloys in engines have resulted in pitted and cracked metal surfaces. This engine-wear problem, however, can be largely resolved by the addition of anti-corrosive agents and lubricants, and by replacing particularly susceptible metal parts in fuel pumps, carburetors, fuel tanks, and other critical contact points.[6]

Compared to the decades of intensive research that have been spent developing the modern gasoline engine, it would be unrealistic to expect that all problems with ethanol engines can be solved in a few years. However, the initial results of the Brazilian experience with ethanol engines are quite promising. Volkswagen noted in one report, "In extremely severe road tests covering more than 100,000 kilometers, no damage whatsoever was observed in engines that could be blamed on the use of alcohol."[7]

METHANOL AS A GASOLINE SUBSTITUTE

Although a small group of U.S. midwesterners and, eventually, the government of Brazil fought to establish ethanol as a gasoline alternative and blend, the majority of automotive engineering interest in alcohols has focused on methanol. This is principally because methanol could be produced from a wide variety of both renewable and nonrenewable feedstocks—wood, coal, and natural gas—at a cost that has been nearly competitive with petroleum throughout the 1970s.

One of the first researchers to experiment with methanol in the U.S. was Thomas Reed, a scientist who studied crystal growth at the Massachusetts Institute of Technology. The controversy sparked by his findings was indicative of the highly charged political atmosphere surrounding alcohol fuels research in the 1970s, and it raised once again the troubling question of the "partisan use of science" that Pleeth observed two decades earlier.

The soft-spoken Reed had been led to methanol blending experiments by reasoning that the hydrogen fuel projects on which he was working would have little immediate practical value. Reed was unaware of any other experiments on alcohol fuels as he crossed his fingers and poured a gallon of laboratory-grade methanol into his car's fuel tank one afternoon in May, 1973, but by December his experiments had attracted national attention when they were reported in *Science* magazine.[8] His article, coauthored by colleague Robert Lerner, appeared at the height of the Arab oil embargo and transformed him overnight into a celebrity.

One of Reed's many admirers was a Minnesota oil millionaire who offered him a $100,000 grant to carry on his research at MIT. The oilman, John Hawley, told Reed he had earned "a staggering amount of money from the American public in his lifetime, and it would make him feel good to know that after he was gone they would have some other fuel when the gas and oil ran out."

Reed turned Hawley's check over to David White, director of MIT's Energy Laboratory, who initially approved his methanol research plans. But this approval was abruptly rescinded several months into the program after a former Exxon oil company scientist, then working at MIT, sharply criticized Reed's research methods and professional qualifications.

White closed Reed's bank account and transferred him to other work. Reed was angered by the decision, believing that it stemmed more from Exxon's opposition to alcohol fuels development than any flaw in his program. Reed's belief was reinforced by the fact that the Energy Laboratory was the recipient of major grants from Exxon and Ford. "I thought we lived in a simple world where if you found something better, people would welcome it," Reed said. "When I first came across this 10 percent blend, I thought that the oil companies would love it, because it's a way of making a transition from the present fuel to some other fuel they could make out of coal. . . . But nobody wanted the better mousetrap."

The aura of intrigue surrounding the cancellation of Reed's methanol project increased with strong opposition to alcohol fuels from the American Petroleum Institute and General Motors in the mid-1970s. The strange sequence of events aroused the curiosity of *Science* magazine editor Allen Hammond, who published a lengthy account of the project's cancellation over a year after Reed was transferred to other work. Hammond noted in a sharply worded editorial that the incident raised "the spectre of universities adjusting their perspective and their research programs to mesh more smoothly with government and industry."[9]

Tom Reed was not the only academic researcher to lock horns with the U.S. oil industry over the question of methanol fuels use. During one 1979 conference in Monterey, California, Richard Pefley repeatedly clashed with members of the API alcohol fuels committee over the severity of the problems encountered in the development of alcohol-fueled engines. Pefley accused API of "abusing scientific evidence" in making its technical case against the use of methanol.* At another U.S. conference, Pefley found himself sharply at odds with research scientists from General Motors and Standard Oil of California (Chevron) when both strongly discounted alcohols as a reasonable alternative. "They in effect said, you guys are crazy even to consider this idea," Pefley recalls, "and we came away feeling that with giants like that making such statements, what chance do we have?"

The conflict between academic and oil industry engineers centered as much on the interpretation of experimental results as on the results themselves, which in many cases were not that far apart. Both academics and industry engineers agreed that methanol, in blends and straight fuels, presented some technical problems. Among these were vapor lock, phase separation from gasoline when water was present in blends, corrosion, and cold starting problems in 100-percent-methanol fuels. However, Pefley, Reed, and others were convinced that these technical problems could be overcome with sustained research, while oil industry researchers such as Robert Lindquist of Chevron viewed these problems as evidence that the U.S. should turn to more easily used synthetic gasoline fuels derived from coal and shale oil. General Motors researchers shared Lindquist's view, noting that pure methanol fuels were impractical because they couldn't start cars at temperatures below 41°F. This view was repeated in a 1979 paper, but as early as 1975 GM had found that certain additives solved the cold-start problem. The results of this research were reported to the U.S. Environmental Protection Agency (EPA) in what GM thought was a confidential report. Reports to the EPA are public documents, however, and GM researchers later noted that it was "unfortunate that the documents were made public."[10]

In Europe, Volkswagen was backed by the staunch support of the German government and committed its research labs to a sustained development effort, unlike its U.S. competitors. In the mid-1970s, VW began testing a fleet of forty-five vehicles fueled on a 15 percent blend of methanol in gasoline. After replacing

---

*Academics were not alone in their dismay at the API stance. Continental Oil Company researcher Robert Jackson, at the same conference, shouted that the API committee members were "liars" in portraying methanol as more toxic to marine life than oil.

plastic carburetor floats, solenoids, fuel filters, and fuel lines with more corrosion-resistant parts, VW was able to efficiently operate the vehicles on methanol blends. "Anticipated problems either failed to materialize or proved relatively easy to solve," said one VW paper that was specifically translated for the U.S. Department of Defense. The report noted that methanol blends in Germany "are already competitive with gasoline" economically and that problems with phase separation could be resolved with the use of additives, such as isopropyl alcohol, to stabilize fuel mixtures. The VW tests indicated that methanol—like ethanol—acted as an octane booster when added to gasoline and that a 15 percent methanol blend in gasoline could deliver as much as 6 percent more power than straight gasoline.[11]

Volkswagen also tested five vehicles for two years on pure methanol fuel. The engine conversion process was essentially the same as that required for pure ethanol engines; the air-fuel ratio of the carburetor was adjusted, the engine compression ratio increased, certain components of the fuel line replaced, and a vapor preheating system installed. A similar conversion process had long been practiced by race car mechanics who found that methanol-fueled engines delivered a 20 percent higher power output than low-compression gasoline engines, while consuming the same amount of fuel energy. VW researchers also used additives such as isopropyl alcohol, butane, pentane, and other compounds to improve the cold starting characteristics of the pure methanol engines.

ALCOHOL FUELS IN DIESEL ENGINES

The high-compression diesel engine is the workhorse of the transportation world, providing the power for buses, tractors, trucks, marine engines, and innumerable other machines required for heavy industry. Given the critical importance of diesel fuels to the smooth functioning of the industrial economies, it is not surprising that Brazil and other nations seeking independence from the OPEC oil barrel have turned to alcohol fuels as at least a partial replacement for diesel.

Diesel engines are quite different from gasoline engines. Where a typical gasoline engine will have an 8:1 compression ratio, and an ethanol a 10:1 or 12:1 ratio, diesel engines have a 17:1 compression ratio. They do not use spark plugs, but instead rely on the explosiveness of the fuel itself, at high compression ratios, to ignite at the proper moment. This explosiveness factor is measured by a diesel fuel's "cetane" rating. On a scale of 1 to 100, diesel fuel weighs in at a 40–45 cetane rating, while ethanol has a rating of 7 and methanol only 3. Thus, pure alcohol fuel injected into a diesel engine fails to ignite unless some form of fuel or mechanical change is made.

Along with ignition problems, diesel does not readily mix with ethanol and methanol, although higher-carbon alcohols (such as butanol) will mix with diesel. The problems are not insurmountable, but they are formidable.

Many of the tests on alcohols in diesel engines have involved two fuel tanks, two fuel lines, and some mixing either just before or during fuel injection. A carburetor system for ethanol on a fuel-injection diesel engine could accommodate 50 percent ethanol in the engine, a University of California researcher found. Mixing diesel and alcohols at the injection pump is another method of using blends; from 33 percent ethanol to 54 percent methanol can be used in such a system. Another system developed by VW engineer Franz Pischinger accommodated over three-quarters methanol before the engine began misfiring; it utilized a dual injection system in a specially designed combustion chamber.[12]

Volvo engineer E. Holmer noted that "a diesel engine can easily be converted to a dual-fuel engine and attain good performance and low emissions of smoke, hydrocarbons, and noise." Volvo's Brazilian subsidiary is in the final stages of perfecting a mass transit bus which will operate on varying blends of diesel and alcohol fuels. Holmer also said that hydrated ethanol or methanol (with as much as 15 percent water) can be used.[13]

Complete replacement of diesel with alcohol fuels is not a very practical direction. It involves either expensive cetane boosters that will bring ethanol and methanol up to higher explosivity, or major engine modifications that end up reducing overall efficiency. Brazilian engineers have found that a more promising approach to complete diesel replacement involves blending alcohol fuels with vegetable oils. These oils can be pressed from palm nuts, peanuts, soybeans, sunflowers, or tapped from certain types of tropical plants, much as maple syrup is tapped from maple trees.

Research results in Brazil, the U.S., and South Africa indicate that vegetable oils alone or in blends with diesel actually perform more efficiently than straight diesel fuels. A 1980 study by the Southwest Research Institute of San Antonio reported that a 60 percent blend of peanut oil in diesel used 11 percent less energy to accomplish a given task than did straight diesel fuel. South African engineers have reported that blends of 15 percent ethanol and sunflower oil gave greater thermal efficiency than diesel itself and produced 73 percent less smoke. Ethanol and other additives such as esters (part of an organic hydrocarbon family like the alcohol family, often used as flavorings) are used by the South Africans to cut the thickness of sunflower oil. A number of Brazilian experiments on nitric esters for cetane improvement of ethanol blends also have been reported.[14]

Another alcohol fuel which may be blended with diesel is butanol. With a high cetane rating, butanol blends much more readily than ethanol and methanol; concentrations of 50 percent were achieved by Colorado State University researchers before engine knock set in.

ALCOHOL FUELS IN AVIATION

In July of 1923, the *Washington Post* reported the results of extensive tests of a new high-octane aviation fuel which eliminated the knocking that plagued the performance of military aircraft engines. This new fuel, a blend of 30 percent ethanol in gasoline, had been tested at three different U.S. naval bases with "very satisfactory results." The newspapers, quoting an unnamed naval officer, concluded that this fuel ". . . will soon take the place of gasoline altogether for aircraft use."[15]

Alcohol fuels fell into disuse with the advent of leaded gasoline but returned in World War II when fighter planes and some bombers were equipped with "war emergency power" injection systems that pumped small quantities of 100-proof alcohol into cylinders to provide power boosts at critical moments. With the end of the era of cheap petroleum, interest in using alcohols in aviation fuels is again reviving.

Between 1973 and 1979 the pump price of aviation fuels skyrocketed; by the winter of 1981, a gallon of aviation fuel sold for $1.97 in some locations. This aviation energy crisis was aggravated as many petroleum refineries began to phase out production of the high-octane gasoline required by small aircraft. According to *Sport Aviation* magazine, "the major [gasoline] producers began phasing out [high-octane aviation fuel] before the 1974 crunch and there are airports east of the Rockies which haven't stocked a drop of the "red stuff" since 1971 or 1972. The companies still refining 80 [octane] are doing so with reluctance, constantly grumbling that the business—less than 1 percent of the total gasoline produced in the U.S.—is not worthy of their efforts to produce and distribute it."[16]

During the summer of 1979, when the gasoline crisis threatened to dry up all supplies of small-aircraft aviation fuels, amateur flying enthusiasts began to experiment with alcohol fuels. The Experimental Aviators Association, publishers of *Sport Aviation* magazine, worked with the staff of *Mother Earth News* to convert a Pober Pixie airplane to operate on pure ethanol. Engine modifications involved simply changing the engine's carburetor setting to accommodate ethanol. The first flight of the ethanol-powered Pixie occurred on July 26 at the Oshkosh, Wisconsin airport with Paul Poberenzy, president of the EAA, at the controls. *Sport Aviation* reported that "President Paul went through his checkout, taxied into position, and at 6:37 P.M. lifted off on the first homebrew-powered airplane flight we know anything about."

The flight of the Pixie was noted with interest by Gordon Cooper, the former astronaut who was catapulted into the national limelight in 1963 when his Mercury space capsule carried out a recordbreaking twenty-two orbits around the earth, and two years later when he made a second journey into space piloting the Gemini 5 space capsule. Cooper had long been interested in alcohol fuels; as a young man he had driven alcohol fueled racing cars and boats.

In the summer of 1980, a year after the flight of the Pixie, Cooper teamed up with Bill Paynter—a Sacramento commercial pilot who had once flown then-Governor Ronald Reagan around California—to begin experimenting with a Super Cub airplane operated on a formulated methanol fuel. Cooper and Paynter's first test flights were carried out in the midst of dire predictions from the local engineering community that the methanol-fueled craft would never get off the ground. After extensive ground tests on an engine redesigned by Sacramento entrepreneur Chuck Stone proved that methanol would fuel an aircraft engine, the local skeptics predicted that its fuel consumption would double since methanol had only half the Btu of gasoline.

One Sunday that summer, the Cub lifted off from the Sacramento airport.

Paynter and Cooper later carried out additional test flights, first reaching 4,000, then 8,000, and finally 15,000 feet in altitude. As the aircraft climbed through the atmosphere, Paynter and Cooper said that they found its fuel consumption dropping and exhaust gases sharply reduced.

Cooper, Paynter, and Stone believed the phenomenon was the result of the presence of oxygen in methanol's molecular structure, which allowed more complete combustion in the rarefied upper atmosphere. The theory has yet to gain wide acceptance among aeronautical engineers. Richard Rentz, a consultant to DOE, labels the theory "meaningless" and believes that improved efficiency at higher altitudes due to oxygenated fuel is "simply not possible." Eugene Ecklund, who reviewed the phenomenon with Rentz and the trio of Sacramento experimenters, noted in an internal U.S. Department of Energy memorandum that the response to the question of what research should be performed in this area "is an emphatic *none*." Ecklund wrote that the three Californians were "well-meaning enthusiasts" but "technological business neophytes."[17]

ALCOHOL FUELS AND ELECTRICAL GENERATION

The future of nuclear energy in the United States has been sharply called into question by the accident at the nuclear generating plant at Three Mile Island, Pennsylvania. Increasing public concern over the environmental and safety issues surrounding nuclear power plants, and the marginal economics of this source of energy have discouraged new investment capital and caused the federal government, even in the Reagan era, to reassess its once generous nuclear subsidy program. Repeated court clashes between utilities and environmentalists over the location of new plants and a lengthy federal nuclear licensing procedure, along with union strikes and licensing problems, have helped to slow the construction of new plants to a crawl. In the meantime, electrical rates have soared and many regions of the United States are threatened by brown-outs during peak use periods when utilities cannot generate enough power to meet demand.

One stop-gap measure to satisfy demand during these periods is the use of "peaking turbines" which utilize natural gas and fuel oil rather than coal or nuclear energy to produce electricity. By 1985 these peaking turbines are likely to require as much as 1.02 quads of energy annually, an amount equal to between 1 and 2 percent of the nation's total annual energy consumption. A series of natural-gas shortages during the winter of 1976–77, followed by a series of sharp OPEC oil price hikes, resulted in power cutbacks and economic chaos in the electrical utilities industry. At this time, several utility companies began to search for economical substitutes to both natural gas and fuel oil which could be utilized readily in already existing turbine systems. Many utility researchers concluded that methanol produced from coal represented one of the most promising alternatives for peaking turbines. When Westinghouse Electric Corporation engineer C. E. Seglem presented a paper on methanol's performance in turbines to a North Dakota conference in 1978, he prefaced his report by remarking, "It struck me that this is not so much a technical presentation of my . . . subject, but the more blatant statement of a snake-oil salesman. This is because it's hard to find negative things to say about methanol."[18]

Twenty years ago, when nuclear power was expected to provide an abundant supply of cheap electrical energy, the thought of turning to an alcohol fuel to generate electricity would have seemed to border on the ridiculous. However, Seglem's enthusiastic remarks are representative of the enthusiasm that many utility engineers now display for methanol. The cost of methanol produced from coal is already approaching economic competitiveness with some fuel oils, even though methanol contains only half as many Btu of energy per gallon. An added attraction is methanol's ready adaptability to turbine engines. One report prepared by Mueller Associates, a consultant to the U.S. Department of Energy, concluded that peaking turbine engines can be converted to methanol simply by "providing higher fuel flow rates to account for the lower heat content of alcohols (compared to conventional fuels) and supplementary lubrication for fuel pumps."[19]

Methanol may find another use in the utilities industry as a means of transporting coal to electrical power plants. Leonard Keller, a chemical engineer from Dallas, has developed a liquid slurry of finely ground coal and methanol which can be pumped through already existing oil and gas pipelines. This "methacoal" slurry concept is proposed as an alternative to the water-coal slurry pipelines that have been developed to transport western coal to eastern power plants. These slurry pipelines have been viewed as more economical for shipping the coal than the nation's already overburdened railway system, but they have been bitterly fought by environmentalists, who oppose having scarce western water resources diverted to this industrial use. Keller believes that his process is superior to the water slurry concept since it avoids the use of western water as a slurry medium and offers a more usable end product. The water-coal slurry must be dewatered before it can be put into a boiler, while the methacoal slurry can be fed directly into conventional natural-gas boilers or utilized in modified coal-fired boilers.[20]

If Keller's methacoal process is developed in the United States, methanol would be produced from wood, natural gas, or coal at refineries located next to western coal mines where it would be mixed with powdered coal and pumped to market. However, Keller's proposals aroused little enthusiasm among Carter administration energy officials, and it is still unclear whether the Reagan administration will commit any funds to developing this process. Keller's dealings with the private sector, at least outside the United States, have been somewhat more successful; Japanese, Swedish, Australian, and French companies have seriously considered using the methacoal technology.

Another concept in alcohol slurry technology involves the use of ethanol to transport pulverized wood chips to utility boilers. One of the major drawbacks of wood boiler fuels is their low energy content per ton when compared to such feedstocks as coal and fuel oil. This low energy content makes it uneconomical to transport wood boiler fuels over long distances by freight or barges. However, if pulverized wood could be mixed in suspension with ethanol and pumped like a fluid to electrical power plants, its potential boiler-fuel market area could be dramatically expanded. The ethanol in the slurry mixture could either be fed into the boilers along with wood chips or separated out and utilized as motor vehicle fuel.

ALCOHOL FUELS IN THE PETROCHEMICAL INDUSTRY

The postwar era has witnessed the rapid replacement of agricultural and coal-based chemical compounds with oil-based substitutes. The petrochemical industry has matured into one of the most dynamic sectors of the United States economy—with worldwide revenues of $200 billion in 1980—and it has attracted some of the nation's finest scientific talent to its laboratories. Almost no sector of the U.S. industrial economy has been left untouched by the petrochemical revolution. Hundreds of pharmaceutical products once produced from plants are now made from less costly oil substitutes. In the packaging industry, many wood, metal, and glass containers have been rendered obsolete by petrochemical plastics, while in the garment industry natural fibers must now compete with synthetic fibers for the consumer dollar.

A rough breakdown of the annual U.S. production of petrochemicals includes 10 million tons of synthetic rubber, 10 million tons of synthetic fibers, and 60 million tons of plastics and resins which are further refined into tens of thousands of different products. About 18 percent of each barrel of oil processed through U.S. refineries is funneled into the diversified petrochemical industry.[21]

Many of the technologies developed to produce petrochemicals utilize ethylene as a basic feedstock. It is a product normally produced from the naphtha refined from light crude oils. It can also be produced from ethanol, offering biomass the chance to play an important role in developing a renewable chemical industry.

Despite the dramatic oil price hikes of the past decade, the economics for producing ethylene from ethanol still remain somewhat marginal compared to petroleum-based ethylene. This cost disadvantage, however, has not discouraged some nations from beginning to lay the foundations for a biomass-based chemical industry. To date, Brazil and India have taken the lead in developing ethanol-to-ethylene plants. With severe unemployment problems, these nations favor the more labor-intensive ethanol-to-ethylene process over the more capital-intensive naphtha process. Research carried out by the Brazilian Centro Tecnicologico Promon Institute indicates that these ethanol plants, while producing a somewhat more costly end product, can be built on a smaller scale with relatively simple equipment.[22]

The Brazilian "alco-chemical" industry is expected to expand greatly during the 1980s. At the start of the decade the fledgling industry consumed some 35 million gallons of ethanol, and this figure, according to *Brasil Energy* newsletter, is expected to "increase nearly four fold by 1985 to 134.5 million gallons."[23]

Over one hundred different petrochemical feedstocks, in addition to ethylene, can be derived from ethanol. In India, where petrochemicals are generally more expensive than in the United States, ethanol is already competitive with seven of the eleven most commonly used petrochemical feedstocks and the future of the "alco-chemical" industry looks bright. Its future, however, remains uncertain in the United States, with most economic forecasters predicting that the industry will not be competitive with the petrochemical industry during the first half of the 1980s.

BREAKING INTO THE MARKET PLACE

The speed with which Brazil managed to introduce alcohol fuels—and the vehicles designed to utilize them—caught U.S. automotive and oil companies somewhat by surprise. Many of the companies' U.S.-based executives had been extremely skeptical of President Juan Baptista Figuereido's plan for a near complete phase-out of petroleum liquid fuels and were astonished by the sudden popularity of the "proalcool" program. They soon found it a matter of both political and economic survival in Brazil to assist in implementing the program. This put some companies in an awkward position in their public relations, however, as they promoted alcohol fuels in Brazil while fighting efforts of the midwestern "gasoholics" to introduce alcohol fuels in the United States. Exxon, for example, marketed a 20 percent gasohol blend in Rio de Janeiro under its Esso trademark while discouraging the efforts of its own U.S. retailers to market a 10 percent gasohol blend in New Jersey. Ford Motor Company was forced to vigorously promote its ethanol fuel vehicle to compete with Volkswagen for its share of the new-car market in Brazil. In the United States the company took a much less enthusiastic approach to the development of alcohol-powered engines. After demonstrating a 100-percent-ethanol-powered station wagon to U.S. congressmen in June of 1980, Ford issued a press release cautioning that a Brazil-style alcohol program in the United States "would not be practical since, in most of the country, we do not have the favorable climate or the ready availability of land on which to grow crops primarily for fuel."

While opposing a massive, sudden transition to alcohol-fueled automobiles as in Brazil, U.S. automotive companies are gradually becoming more receptive to the concept of developing a limited North American market for these vehicles. Such a move would be a radical shift for an industry which has spent the last eight decades perfecting petroleum-powered vehicles for U.S. motorists. However, for a brief period of time during the birth of the automotive industry, Henry Ford and other pioneers produced multifueled vehicles which could operate on kerosene, coal-derived benzene, alcohol, or whatever other fuels a motorist could secure. These versatile but somewhat primitive engines were phased out as gasoline and diesel fuels established themselves as the cheapest and most widely available fuels. Modern U.S. automobiles are designed to operate on a narrow range of petroleum fuels that have been developed by oil industry chemists and carefully introduced to the motorists.

The arrival of gasohol at midwestern service stations during the winter of 1978 represented a sharp departure from the traditional pattern of fuel development, for the first gasohol blends were not the carefully formulated fuels of oil industry chemists, but rather the hastily blended concoctions of a handful of midwestern gasoline distributors. These distributors, called "jobbers" in the industry lingo, originally practiced a technique called "stoplight blending," in which the fuel was partially mixed as gasoline and ethanol were pumped into the tank delivery truck, then further mixed as the truck lurched to its eventual destination. Such hit-or-miss blending techniques received a cool reception in Detroit, where the major auto manufacturers initially refused to extend new-car warranties to cover gasohol-powered vehicles.

The early gasohol blends were heavily promoted by such groups as the Nebraska Gasohol Commission, the Illinois Department of Agriculture, and the Iowa Development Commission, but the decision to begin marketing the fuel usually rested with the individual station owners.* Most of these owners were independents but a few were leaseholders with the major oil companies who were often hostile to the idea of gasohol suddenly showing up at stations which bore their markings. When gasohol first appeared at a Connecticut service station bearing Exxon markings, surprised officials from the company's New Jersey headquarters hurriedly told service station owner Wayne Konitshek to advise his customers that this was not an Exxon product. Konitshek promptly put up a sign to comply with this request but added a set of Mickey Mouse ears and the note: "Exxon will not sell gasohol to the American public." When gasohol began appearing at New Jersey service stations in January, 1980 Exxon notified the New Jersey Dealers Association that gasohol sales could not be charged to company credit cards and, furthermore, that dealers' franchises would be cancelled if pumps were not fully debranded. Exxon, however, was forced to tone down its antigasohol campaign as it rapidly turned into a public relations nightmare. The outspoken Wayne Konitshek charged that Exxon sales representatives were verbally intimidating dealers who tried to sell gasohol, and Al Ruth, president of the New Jersey Dealers Association, labeled the Exxon actions "illegal and unpatriotic" at a press conference held at a service station. These allegations triggered a probe by the U.S. Senate Antitrust Subcommittee to determine whether any of Exxon's actions represented an unfair restraint of trade. On January 11, Exxon's E. J. Hess, seeking to defuse the controversy, told the Senate subcommittee that the company's policy had been reversed and that credit card sales of gasohol would be allowed.[24]

Not all oil companies reacted as apprehensively as Exxon to the arrival of gasohol; some rapidly moved to develop their own marketing strategies. Texaco made no effort to discourage its service station leaseholders from marketing

---

*On the East Coast, several thousand service stations were supplied with gasohol through MARC-CAM Industries, a partnership between Mo Campbell and Buz Marcus. Campbell had been involved with Richard Merritt in many of the early gasohol lobbying forays, and was the first to bring gasohol to the East. When the gasohol market began drying up in the 1981 oil glut, MARC-CAM took the financial blow on the chin. Perhaps, said Campbell, the marketing strategy was wrong. "We were selling a champagne product at beer prices," he said. Instead of trying to compete directly with regular unleaded, Campbell believes that a better strategy would have been to promote gasohol as a premium fuel.

gasohol and soon developed its own gasohol blend which entertainer Bob Hope, standing in a cornfield, widely promoted in national media advertisements. And Chevron Oil, which had vigorously fought California legislative efforts to mandate methanol-gasoline blends, began cautiously to test-market gasohol in some midwestern and western states. As gasohol carved out a small share of the unleaded fuel market, mixing techniques became more professional and the distribution system was streamlined. Although gasohol sales suffered sharp drops in the eastern and southern states during the oil gluts of 1981, sales continued to hold steady throughout much of the Midwest. A principal reason for these steady sales was ethanol's increasing value as an octane booster in unleaded fuels.

THE U.S. AUTOMOBILE INDUSTRY AND THE ALCOHOL CARS

Despite the impressive performance of the first generation of Brazilian alcohol-fueled vehicles, the U.S. automobile industry has taken, as a whole, a decidedly go-slow approach to introducing similar vehicles in the United States. While European automotive companies such as Volkswagen looked upon Brazil as an exciting testing ground for a possible international shift towards alternative fuels, General Motors, the largest U.S. automobile manufacturer, was reluctant to commit its Brazilian subsidiary to a major research and development program in ethanol fuels utilization. It was only after Volkswagen neared assembly-line production of its first-generation ethanol vehicles that General Motors engineers received the go-ahead from Detroit headquarters to develop their own ethanol models. The slow pace of GM's research, according to some observers, is indicative of a deeper malaise which afflicts the giant corporation. John De Lorean, a former General Motors vice-president who quit his high-salaried post to form his own automobile company, complained that the corporation, despite the most expensive research facilities in the world, remained in "technical hibernation." *On A Clear Day You Can See General Motors*, a book by J. Patrick Wright detailing De Lorean's experiences, gives a picture of a GM leadership obsessed with superficial year-to-year market trends but slow to respond to major new automotive developments such as the arrival of subcompact imports in the 1970s.[25] A similar "technical hibernation" has been evident in the General Motors approach to alcohol fuels; its research scientists submitted only two papers to the four major international alcohol fuel symposiums held between 1976 and 1979 while Volkswagen submitted over fifteen papers.

Ford Motor Company has pursued alcohol fuels research and development somewhat more aggressively than General Motors but still lags far behind its foreign competition. Serge Gratch, a Ford vice-president, served on the U.S. National Alcohol Fuels Commission which recommended the rapid commercialization of alcohol-fueled vehicles, and a special alcohol fuels research division has been created at Ford's Detroit headquarters under the leadership of Roberta Nichols, a physicist who once held the women's world speed record for racing boats. A team of engineers led by Nichols has modified a fleet of Ford Escorts to operate on pure methanol. The tests performed on these vehicles, announced Ford's research vice-president D. W. Compton at a February, 1981 press conference, would provide "a real world demonstration of methanol use."[26]

By the time of Compton's press conference, Ford was not the only company that could provide vehicles for "real world demonstrations" of alcohol fuels. The lack of early activity in Detroit, coupled with the U.S. Department of Energy's inability to secure enough funds to finance large-scale fleet testing, has inspired a number of state legislatures, independent entrepreneurs, and foreign automotive companies to begin to test the U.S. market for alcohol-fueled vehicles. It was from this talent pool—and not its own in-house staff—that the corporation recruited Roberta Nichols.

One of the most impressive research efforts was funded in California. The research program carried out at the University of Santa Clara sparked the interest of the state legislature, which appropriated $44,000 in January, 1978 to convert a Ford Pinto to operate on methanol. The legislature chose Chuck Stone, a former NASA engineer who had worked with the Apollo project, to head a methanol research team which included Roberta Nichols. By adding a fuel injection system, installing longer pistons to increase the engine's compression ratio, and creating a manifold preheating system, the engineering team was able to convert the Pinto in less than six months to operate on a mixture of methanol blended with small amounts of proponol and butanol which Stone labeled "Methanol-X." Exhaust emissions from the methanol Pinto were so low, Stone claimed, that they could not be measured by standard pollution monitoring instruments. These low emissions became a point of contention as the California State Air Resources Board doubted Stone's claims and ruled that the Pinto could not be driven without the standard emissions-control equipment required for gasoline engines. Stone protested that his Pinto met all state emission standards without these controls and thus should not have to operate with them. By June, 1979 the controversy between Stone and the state board became so heated that the state legislature was forced to pass a bill exempting alcohol-fueled vehicles from the state pollution equipment laws if they otherwise met all federal and state standards. In the process of resolving this dispute with the state agency, Stone became increasingly frustrated at the resistance from the state bureaucracy and made several bitter attacks against politicians and lobbyists whom he suspected of slowing down the methanol program.

Stone was portrayed in much of the state press as a heroic figure, with one magazine dubbing him "Don Alcohote"—a modern knight bravely fighting the oil dragons.[27] However, Stone's outspoken ways alienated many of his colleagues and he was fired from his legislative position in June of 1979. His abrupt dismissal was due, according to *Sacramento Bee* columnist Herb Michaelson to the fact that Stone ". . . wanted too much too soon from bodies to slow to move in alternative directions. . . . Chuck Stone will not settle for anything less than 20 million cars in California fueled by methanol. By Saturday, sundown at the latest."[28]

Chuck Stone bounced back quickly from the loss of his state position and in the fall of 1979 formed his own automotive company which he named Future Fuels of America. The new company was generously funded by a Los Angeles Ford dealer and housed in an abandoned barn at his 100-acre homestead outside of Wilton, California. Many of Stone's Sacramento critics were extremely skeptical of the long-term survival prospects for Future Fuels of America and thought that his ego had perhaps overreached his engineering capabilities. However, in late 1979 Stone was able to persuade Merle Fisher, a vice-president with the California-based Bank of America, to test-drive a Mustang which the company had converted to operate on Methanol-X. Fisher was impressed by the vehicle's handling performance and subsequently asked Stone to convert first a van and then a fleet of Ford sedans and pickup trucks to operate on Methanol-X. It was a surprising move for the normally cautious Bank of America and, according to one source, it was originally resisted by a Chevron Oil official on the corporation's board of directors. However, after an extensive evaluation of the methanol fleet's performance by Carson and Associates, a Cambridge, Massachusetts systems-engineering company, the Bank of America decided to convert most of its vehicle fleet to operate on the methanol fuels. The primary reason for the bank's decision was economics: methanol produced from natural gas could be purchased for 85 cents a gallon from bulk distributors, substantially cheaper than the bulk purchase price of gasoline. Although Stone ultimately hopes to see the feedstock for his methanol fuel shift from natural gas to wood biomass, he took full advantage of the price advantage that methanol enjoys over gasoline to further the growth of his business. By December of 1980 Stone's Future Fuels of America had evolved into a multimillion dollar enterprise with new headquarters in an auto dealership.[29]

California's alcohol fuels testing program continued after Stone's departure, as the Sacramento-based Energy Commission received $2 million from the legislature to test three fleets of alcohol-fueled vehicles. One of these fleets, consisting of twenty-five Rabbits, was provided by Volkswagen of America, which produced the vehicles during a special production run at the company's Westmoreland, Pennsylvania assembly plant. Another fleet consisted of the Ford Escorts which were first unveiled by Ford vice-president Compton at his February, 1981 press conference.

While the California state legislature and Chuck Stone helped prod the U.S. automotive industry to quicken the pace of commercializing methanol vehicles, others sought to hasten the arrival of ethanol vehicles in the U.S. marketplace. Farm producers began offering 180–190 proof ethanol as their plants generated small amounts of surplus fuel, and organizations such as the American Agriculture Foundation compiled listings of these producers for interested motorists. In Nebraska and Arkansas this network has evolved one step further as two service stations have actually opened up pumps offering 190-proof fuel for sale, as well as technical assistance in converting vehicles to operate on this fuel.

It was this small but growing fuel supply that led Jim Floyd, a young Alabama entrepeneur who owned an ethanol plant, to attempt to import 100 percent ethanol cars into the U.S. Floyd turned to the Brazilian automakers, and when General Motors' Brazilian subsidiary expressed interest in a deal, Floyd and business partner Michael Pete lost no time arranging a meeting with Joseph Sanchez, president of GM in Brazil. Sanchez agreed to export 1,000 Brazilian-made Opalo Commodoros as long as certain conditions—including exemption from all liability claims against GM of Brazil and its parent in Detroit—were met.[30]

Two alcohol cars arrived in early 1981, one of which was sold at cost to the Governor of Alabama. Floyd took to the road in the other, obtaining 23 orders by April. But in May, 1981, GM of Brazil told Floyd he would have to deal with GM's Detroit-based legal department, and no other orders for the alcohol cars would be accepted without their OK. A series of complex legal entanglements ensued, and the 23 cars were not delivered. By November, an angry Floyd had filed an $80 million suit against GM, alleging that the automaker had broken the contract and violated the Sherman Anti-Trust Act. "GM did not want a car in this country for which a person could manufacture his own fuel," Floyd said.

The U.S. automotive industry will face stiff competition from its foreign competitors when it finally enters the alcohol fuels market. Volkswagen expects to make a limited number of ethanol-fueled vehicles available to U.S. motorists in 1982. Toyota has contracted with University of Santa Clara professor Richard Pefley to study the performance of sophisticated multifueled engines. Pefley, who is discouraged by what he calls a lack of "initiation and innovation" in Detroit, glumly predicts that the U.S. alcohol-powered cars, when they finally arrive, will fail to measure up to the foreign imports.

Chapter 9
The Environment

When oil replaced alcohol as a lamp fuel in the 1860s, few of its early boosters stopped to ponder the environmental consequences of developing this combustible black liquid. It would have taken a leap of imagination worthy of Jules Verne to conceive of a modern world in which thick photochemical smog would enshroud cities and oil spills from ocean tankers would send tar balls washing up on coastal beaches.

It would have been equally difficult to conceive of a more subtle yet potentially more dangerous environmental impact of oil development: the dramatic rise in the level of carbon dioxide in the earth's atmosphere, which scientists attribute (at least in part) to the combustion of petroleum- and coal-derived fuels. This carbon dioxide has increased the amount of infrared solar radiation trapped in the earth's atmosphere, resulting in a "greenhouse effect" which scientists have warned could trigger a global warming trend with far-reaching climatic consequences.

Tanker spills, smog problems, and the greenhouse effect are all well-known impacts of oil development. Until recently, however, little was known about the impacts of substituting alcohol for petroleum.

The development of an alcohol industry—as Brazil's Proalcool program has already demonstrated—will also pose environmental risks. Liquid waste from alcohol distilleries creates serious water pollution problems when dumped untreated into rivers and bays. Alcohol exhaust emissions also produce objectionable air pollutants—though, on the whole, considerably less than gasoline or diesel. And an unplanned rush to use forest or other biomass resources could trigger increased soil erosion and ecological disruption.

Although the environmental risks involved in the increased use of renewable alcohol fuels are not to be taken lightly, neither can they be considered in a vacuum. Instead, they must be weighed against the risks of continued global reliance on petroleum liquid fuels or a sudden shift to synthetic fuels derived from coal and oil shale.

ALCOHOL FUELS AND WATER QUALITY

The 1978 decision by U.S. gasohol marketers to import Brazilian ethanol marked the birth of the oceangoing-tanker trade in alcohol fuels. While defended by marketers as a necessary expedient until U.S. production could come on line, the imports were seen as increasing U.S. dependence on foreign energy sources. (The development would have been denounced in an earlier era as well, by chemurgists such as William Hale, as a departure from the all-important principle of national energy self-sufficiency.)

While this alcohol tanker traffic pales in comparison to the 20-million-barrel-per-day global oil traffic, it is likely to grow substantially during the next decade as major markets are established for these fuels. Much of this trade will involve methanol refined from natural gas—a fuel which usually has been shipped overseas in frozen liquid form.

One dramatic example of the risks involved in LNG (Liquid Natural Gas) occurred in Cleveland, Ohio in 1944, when a tank containing the mixture exploded with a force that killed 130 people and injured another 200. The tank which ruptured in the Cleveland explosion was thirty-five times smaller than those in use today, which, if they were ever to rupture, would result in explosions of catastrophic proportions. One U.S. Federal Power Commission report concluded that over 100,000 people might be injured or killed if an LNG explosion were to occur at the Staten Island, New York shipping terminal.[1]

As safety concerns mount over the LNG tanker trade, necessitating increasingly costly tankers to ship the fuel, methanol has begun to stand out as an attractive form in which to ship natural gas. Davy McKee, a large international engineering firm, concluded in a 1978 study that methanol, rather than LNG, was the most profitable fuel to produce from offshort natural gas that must be shipped, rather than piped, to market.

"Ten years ago, before the development of newer technology, that might not have been the case," said one Davy McKee official. "However, today . . . both logic and economics favor the LCF (liquid chemical fuel, methanol) rather than the LNG route."[2] Announcement of this conclusion coincided with an agreement between Davy McKee and the Soviet Union to construct two of the world's largest methanol plants in remote Siberian gas fields. About half the fuel from the two 2,750-metric-ton-per-day plants will be shipped by tanker to Great Britain.*

Other nations may soon be following the Soviet lead in methanol development. A major portion of Saudi Arabia's long-term development plan involves the production of methanol from natural gas which has been simply flared off (burned) at the well head. Most of this methanol will then be shipped by tanker to Japanese markets.[3]

As the methanol trade expands in the years ahead, the chances of an accidental spill become increasingly likely. Scientists have only recently begun to

---

*Much of this methanol will be used in a relatively new process for making single-cell protein feed supplements for livestock.

study the effects of alcohol spills on the marine environment, but the early evidence shows that such spills would have far less impact than comparable oil spills.

According to a report from the Argonne National Laboratory, a federal laboratory, the impact from a major alcohol spill "would be relatively short and would be limited to a relatively small area. The length of time alcohol would be present in the water at toxic concentrations could be measured in hours compared to years for crude oil or gasoline." Argonne concluded that the impacts would be "dramatically less" than those from oil, in part because ethanol and methanol alcohols mix readily with water, rapidly diluting to low concentrations rather than forming a slick on the ocean surface.[4]

Dr. Peter D'Eliscu, a biologist at the University of Santa Clara, has found that many marine organisms display a high tolerance for low levels of methanol, which may be due to the fact that small amounts of methanol are naturally produced by the actions of some ocean bacteria. D'Eliscu also discovered that marine life—unlike humans—demonstrated a greater tolerance for methanol than for ethanol. One octopus he studied became gray and sickly when exposed to a low level of ethanol, but became bright red, alert, and combative when exposed to an equivalent level of methanol in his tank. D'Eliscu has concluded from these studies that "the consequences of aquatic spills of methanol appear to be minimal and of short duration."[5]

The most important contribution which the alcohol fuels industry, as a whole, can make to improving the ocean environment is in reducing the global demand for petroleum fuel which supports the international oil trade. This trade results in oil spills with impacts on marine life which can scarcely be described as either "minimal" or "of short duration." At least 1.5 million tons of oil are spilled from oceangoing tankers each year with another 3 million tons attributed to offshore drilling and industrial discharges into coastal waters.

French oceanographer Jacques-Yves Cousteau reports that signs of oil pollution are now evident in even isolated Pacific regions—thousands of miles away from the well-traveled tanker routes. The National Academy of Sciences, in a cautiously worded yet disturbing report on the possible impacts of oil pollution, concluded that:

A basic question . . . remains unanswered. . . . At what level of petroleum hydrocarbon input to the ocean might we find irreversible damage occurring? The sea is an enormously complex system about which our knowledge is very imperfect. The ocean may be able to accommodate petroleum hydrocarbon inputs far above those occurring today. On the other hand, the damage level may be within an order of magnitude of present inputs to the sea. Until we can come close to answering this basic question, it seems wisest to continue our efforts in the control of inputs, and to push forward to reduce our current level of uncertainty."[6]

Another assessment of the potential danger was made by Noel Mostert in *Supership*, his pathbreaking book on the dangers of the giant oil tankers. He noted: "Considering how far the Atlantic plankton already appears to be polluted, the general diminution of life . . . might be far more advanced than we could suspect. If it is, then it would represent, surely, one of man's single most calamitous acts."[7]

STILLAGE SLOPS—A WATERY DISPOSAL PROBLEM

One of the most serious threats to water quality posed by the development of the alcohol fuels industry lies in the enormous volume of stillage which will be generated. Stillage, also called slops or vinessa, is the liquid leftover of the fermented mash after the alcohol has been distilled off and the solids separated for livestock feed. Slops have from 1 to 10 percent dissolved minerals and organic material. The organic compounds rapidly decompose when dumped into waterways, robbing the water of oxygen required by fish and other aquatic life for survival. Even after all the oxygen in a waterway has been dissolved, a second-stage anaerobic (oxygen-free) decomposition process is set in motion, producing a noxious-smelling hydrogen sulfide gas.

Brazilian water quality has already suffered greatly from the rapid expansion of Proalcool, as 12 to 17 gallons of slops are produced for each gallon of sugar-cane alcohol. The 1980 Brazilian production level of 4.1 billion gallons of fuel is a gargantuan pollution problem. Even worse, the slops are usually discharged into rivers and streams during the six dry months of the year in which sugar cane is harvested. At this time, shrunken streams and rivers are least able to assimilate the discharges. Carl Duisberg reported in his doctoral dissertation on the Proalcool program that several large river systems in the state of São Paulo, including the Piracicaba, Mogi-Guassu, and Pardo "have been virtually poisoned to death by stillage."[8]

Many alternatives to direct discharge of the slops have been proposed by the Brazilian environmental agencies seeking to curb the water pollution. But most of these alternatives have entailed costs which the sugar industry has been reluctant to shoulder. The traditional way of treating the slops has been to discharge them into large lagoons where they slowly decompose prior to entering the waterways. However, a number of new methods for dealing with the stillage are now under development. Hopefully, they will provide profitable alternatives to lagoon treatment.

One of the most promising involves returning the organic and mineral nutrients of the slops to the land. Their nitrogen and potassium content makes them a good fertilizer, but since they are so dilute, the economic breakeven point for pumping the slops out onto fields is estimated at about 3 kilometers. After that, it is less expensive to simply buy fertilizers than to pump slops. Thus, researchers have sought efficient ways to dewater stillage so that it can be profitably transported over longer distances. A number of dewatering options have been examined, including the use of molecular sieves or ultrafiltration membranes, large evaporation ponds, or evaporation by bagasse heat.

Another option for using the watery wastes involves culturing single-cell protein in the liquid. This protein can serve as a valuable additive to livestock feed or even be consumed directly by the malnourished segment of the Brazilian population. A Peruvian distillery already has experimented with a process which provides enough protein from the stillage to provide partial rations for a herd of 6,000 cattle.[9] The nutrients in the stillage also could be used in fish farming, and some research in the U.S. is proceeding in that direction.

Still another option involves producing methane gas from the anaerobic decomposition of organic materials. While the dilute solution of organic materials is probably only a marginal feedstock for methane gas production, some researchers estimate that perhaps 40 percent of the process energy requirements of the average Brazilian sugar cane distillery could be met by generating methane from slops.[10]

Methane production from stillage wastes has also been carefully scrutinized in the U.S. as a potential process energy source for distilleries. The U.S. alcohol fuels industry faces a costly disposal problem in dealing with its corn stillage slops. The average 20-million-gallon-per-year corn distillery produces a stillage output equivalent to the sewage of a city of one million people. Water-quality laws are more strictly enforced in the United States than in Brazil, and companies that try to ignore them usually face stiff fines. For the most part, standard waste processing techniques, using aerobic decomposition, have been employed.[11]

OTHER WATER POLLUTANTS: GRIST FOR THE ALCOHOL MILL

Along with the potential for producing water pollutants, the alcohol fuels industry also has the potential to turn some water pollutants into energy and byproduct livestock feed. The most oustanding example of the industry's ability to help resolve—rather than create—liquid waste disposal problems can be found in Wisconsin, where a distillery owned by the Milbrew Corporation profitably turns cheese whey into ethanol fuel. Whey, a liquid byproduct of cheese production, has long been a major headache for the dairy industry. While some of the whey has been given to farmers as a cattle feed supplement, much of it, prior to the enactment of federal clean water standards in the 1960s, was simply dumped into rivers or streams. Whey has about 93 percent water and 7 percent organic material, and has had much the same damaging impact in U.S. waterways as sugar distillery slops in Brazil. During the 1970s, the Environmental Protection Agency began to crack down on whey dumping and forced cheese producers to use costly sewage treatment processes to dispose of it.

In 1973, as the full impact of new federal water quality standards became clear, Milbrew opened its first experimental distillery to convert the lactose sugars present in cheese whey to ethanol for the beverage industry. By 1977 Milbrew had expanded the distillery to serve the newly created gasohol market, supplying ethanol for some of the first gasohol stations that opened in the Midwest. The distillery used a partly condensed form of whey which was picked up from a number of Wisconsin cheese factories eager to find a new use for the burdensome waste.[12]

Milbrew found that it was able to market not only the whey-derived ethanol but also a high-protein byproduct which the baking industry now uses in pastries and breads. The success of the patented Milbrew technology has caught the eye of other major cheese producers eager to turn an economic burden into a benefit. However, the economics of whey-derived ethanol are fairly site-specific. High capital costs are required to finance the sophisticated technology used in the Milbrew process, and thus, a high concentration of whey producers in a relatively small geographical area is needed to insure the profitability of such a venture. A study by the Rocket Research Corporation concluded that the Washington State cheese industry has the potential to produce approximately 1.75 million gallons of ethanol per year from whey. But due to the widely scattered locations of the major cheese producers, the new fuel distilleries would be faced with high shipping costs which might diminish profits. But when faced with the choice between installing waste treatment systems and distilleries, the distillery might prove the more economic option.[13]

Cheese producers are not the only industry that produces waste streams suitable for ethanol production. Food processors that market frozen and canned fruits and vegetables also produce a multitude of byproducts which may be suitable for ethanol production. For example, each week the Campbell's Soup Company bulldozes several thousand cases of products with expired shelf-life into special landfills. Total food and agricultural wastes for the U.S. have been put at 400–550 million tons per year.[14] Using these wastes may prove difficult, however, since they share a problem with cheese whey: their high water content makes them expensive to transport to a distillery. The most efficient kind of use would involve a distillery set up adjacent to the processing plant.*

The wood-processing industry also produces waste streams which can be converted to ethanol. Georgia Pacific, one of the nation's largest paper and wood products manufacturers, converts waste streams from the paper process into ethanol in a Bellingham, Washington distillery. The plant was built during the 1940s as part of the crash effort to produce ethanol for the synthetic rubber industry. The waste sulfite liquors were commonly dumped into Puget Sound prior to the emergency distillery construction.

Costs for a wood-liquor-to-ethanol plant have skyrocketed in the four decades since the Bellingham plant was built, and Georgia Pacific officials caution that the economics of the process are wholly dependent on obtaining a reasonable price for paper products as well as for ethanol. The production process is also energy inefficient, requiring 150,000 Btu of fuel oil and natural gas to produce the 75,000 Btu of energy in a gallon of ethanol. However, one Canadian firm, Tembec Forest Products, is attempting to update the Bellingham process at a Quebec paper plant where they are constructing a 6-million-gallon-a-year waste-wood-pulp-to-ethanol plant.

---

*University of Oklahoma researchers have developed immobilized yeast reactors that will accept raw dilute feedstocks such as cheese whey and other sugary fluids, thus shortcutting the condensation step and making dilute feedstocks easier to handle.

AIR POLLUTION: GETTING THE LEAD OUT WITH GASOHOL

One of the biggest culprits in the poor quality of urban air is the octane-boosting lead compound used in regular gasoline. This compound—tetraethyl lead—has been identified by medical researchers as one of the most dangerous and pervasive air pollutants of the twentieth century, with a variety of damaging effects on human health. Although leaded gasoline was known to be dangerous when General Motors and Standard Oil introduced it in the 1920s, a government-mandated phase-down only began in the 1970s. Due to the low cost and ready availability of lead, it has long been favored by oil and auto industries over safer octane boosters.

Alcohols, in blends with gasoline, have been recognized as good octane boosters and an alternative to lead and other dangerous additives for most of the century. Used in gasohol blends of 10 percent, ethanol can boost octane from three to five points, depending on the grade of base gasoline. A gram of tetraethyl lead will raise octane by about the same amount.

Octane rating is an important measure of the antiknock quality of gasoline. Low-octane gasoline must be used in low-compression engines to prevent knocking, which can damage spark plugs, valves, and rod bearings. Low-compression engines are less efficient than higher-compression engines and deliver reduced mileage per gallon.

The early gasolines that fired up horseless carriages at the turn of the century were crude forerunners to the complex mix of hydrocarbons in use today. They had an octane rating* of 50–60, which meant that compression ratios had to be kept low (about 4:1) in early engines. In contrast, today's gasolines range in octane rating from 85–95, and auto engines usually have from 8:1 to 10:1 compression ratios.

The incessant knock in early automobile engines led researchers far and wide to search for economical antiknock compounds. While blends of alcohol and/or benzene were often used with gasoline, the discovery of a cheap lead additive completely overshadowed their development. Made by the General Motors research labs in Dayton, Ohio in December 1921, this discovery was hailed by some as "the beginning of high-compression fuels leading to the conservation of resources."[15]

Charles Kettering and his top researcher, Thomas Midgely, were regarded as heroes in many engineering circles, and their creation—ethyl gasoline—was marketed by the Ethyl Corporation set up by Standard Oil of New Jersey and GM in 1924. But the accolades were abruptly silenced by the cracking of a large chemical retort at the Standard refineries in Elizabeth, New Jersey in October of 1924. Forty-nine workers were exposed to the escaping lead fumes and had to be rushed to the hospital. Five died, "violently insane," according to a front page article by the *New York Times*, from lead poisoning.[16]

---

*Actually, octane rating systems were not developed until the late 1920s and early 1930s; thus it would be safer to say they had what would later be called an octane rating of 50–60.

The accident touched off a campaign to ban tetraethyl lead, and a daily tabloid of the era, the *New York World*, dubbed ethyl gasoline "loony gas." Reporters digging into the story found that two researchers had died in Kettering's labs and that nine more had been killed at DuPont labs developing leaded gasoline. Soon after the Elizabeth accident, public health authorities in New York told Ethyl Corporation to take lead off the market until a full inquiry could be made. While lead was pulled off the national market almost immediately, officials showed little remorse for their employees' deaths.

When a *New York Times* reporter asked what would be the reaction of Standard Oil to benefit claims by families of the men who died, a company official responded that "the daily physical examinations, constant admonition to wear rubber gloves and use gas masks, to not wear away from the plant clothing worn during work hours, should have been sufficient indication to every man in the plant that he was engaged in 'a man's undertaking.' "[17]

Standard Oil officials tried desperately to calm what they viewed as a hysterical overreaction to the New Jersey accident by pointing out that those unfortunate workers were exposed to far more concentrated doses of lead than was present in automotive exhaust fumes. Frank Howard, research director for Standard, adamantly maintained at a hearing convened by the U.S. Surgeon General that "Present day civilization rests on oil and motors. . . . We do not feel justified in giving up what has come to the industry like a gift from heaven on the possibility that a hazard may be involved in it."[18]

The Kettering laboratory even went so far as to hold a press conference in which Thomas Midgely rubbed his hands with tetraethyl lead to demonstrate its safety, and declared: "So far as science knows at present, tetraethyl lead is the only material which can bring about these [octane-boosting] results which are vital to the continued economic use . . . of automobiles. . . . Unless a grave and inescapable hazard exists, its abandonment cannot be justified."

Midgely's insistence that no alternative to tetraethyl lead existed was extremely misleading, since the ability of ethanol to raise the octane rating of gasoline had already been well documented in the United States and Europe. Standard Oil had even test-marketed a 25 percent ethanol blend in Baltimore in 1925.[19]

The thick red liquid Midgely thought safe enough to wash over his hands, this "gift from heaven," was viewed with considerably less enthusiasm by the medical community of that earlier era. A study carried out in 1925 by Dr. Alice Hamilton indicated that lead workers ran a high risk of suffering neurological damage and that, because of the presence of lead in street dust, this risk might also be shared by city dwellers. Dr. Wendell Henderson, a physiology professor at Yale University, called leaded gasoline pollution "the greatest single question in the field of public health that has ever faced the American public."[20]

Henderson warned against slow poisoning from fumes of leaded gasoline. "If leaded gasoline kills enough people soon enough to impress the public," he said in 1925, "we may get from Congress a much-needed law and appropriation for the control of harmful substances. But it seems more likely that conditions will grow worse so gradually—for this is the nature of the disease—that leaded gasoline will be in nearly universal use, and large numbers of cars will have been sold that can run only on that fuel before the public and government awaken.''

The U.S. Surgeon General remained unswayed by these medical findings, and issued a report in 1926 saying ''there are no good grounds for prohibiting the use of tetraethyl lead in gasoline.'' While tests for the Surgeon General showed increased levels of lead in the blood of garage workers exposed to the exhaust of leaded gasoline, the exhaust did not cause symptoms of lead poisoning. Ethyl Corporation immediately announced plans to remarket leaded gasoline.[21]

For the next four decades leaded gasoline would remain the oil and auto industries' fuel of choice for U.S. motorists. While chemurgy-movement leader William Hale denounced the use of this ''poison-spreading'' compound in the 1930s, the oil industry was not much interested in using more expensive alcohol octane boosters. Although Henry Ford confidently predicted to the *Christian Science Monitor*, in an interview at the height of the loony-gas controversy, that alcohol was the ''fuel of the future,'' he was content to sell a byproduct of his industrial complex—benzene—as the octane-boosting additive of the 1920s.[22]

By the time Congress passed the Clean Air Act of 1970, which ordered the gradual phase-out of leaded gasoline, scientists were able to detect lead levels in the typical urban air which averaged four to five thousand times higher than natural lead concentrations. Even more alarming, studies showed that the lead level in the blood of an average urban adult measured one hundred times the natural blood levels of lead.[23] Lead measured in children's blood, medical researchers reported, was high enough to cause physiological disturbances which could lead to serious learning disabilities. Dr. Herbert Needleman of Harvard Medical School studied the scholastic performance of children with both high and low lead levels. He found that 26 percent of the children with high lead content in their blood were rated as unable to follow a sequence of directions by their teachers, while only 8 percent of the low-lead group had similar problems. Teachers also noted that the high-lead group was much more disorganized, hyperactive, impulsive, and easily frustrated.[24]

Another study by Dr. Morris Wessel of Yale Medical School concluded that 98 percent of the lead in children's blood is caused by gasoline emissions, and estimated that over 400,000 children a year become ill from lead poisoning, some suffering permanent damage. Federal health officials have set the ''safe'' level for lead at .30 micrograms per 100 cc. of blood, but Wessel estimated that between 10 and 50 percent of urban children maintained lead levels of over .40 micrograms per 100 cc.[25]

In the face of such substantial evidence, the oil industry has been forced to begin a long-delayed effort to replace lead with safer forms of octane boosters. Even so, many oil companies continued delaying tactics to slow the lead phase-down, including a series of unsuccessful court actions.

But the lead phase-down proceeded in two directions. First, cars made after 1975 could use only unleaded gasoline, which meant that by 1980 almost half of the gasoline sold was unleaded, and by 1985 almost three-quarters will be unleaded. Secondly, what lead there is in regular gasoline was reduced according to an EPA schedule. This lead reduction for regular has been frozen since 1979, when a shortage of refinery capacity resulted in what came to be known as the "octane crisis."

AROMATICS, GASOHOL, AND THE OCTANE CRISIS OF 1979

The summer of 1979 was a memorable one for American motorists, as interminable lines at the gas pumps combined with a wilting heat wave. Tempers rose with the temperature, occasionally flaring into violence, and providing—perhaps—a taste of things to come.

While the cut-off of Iranian oil after the fall of the Shah was partly responsible for the gasoline shortage, government and oil industry officials were more concerned about the shortages produced by a lack of refinery capacity than by a lack of crude oil on world markets.* Although nearly ten years had passed since the government served notice on the oil industry that lead octane boosters would be phased out, American refineries were still unprepared to meet the demand for increased octane in unleaded gasoline.

There are basically two ways to increase octane in gasoline. By employing an additive, such as lead or alcohol, a refiner can boost octane above the level that comes out of the basic refining process. Or, by using a more expensive type of refining process, a refiner can produce gasoline with a higher basic octane level.

As lead was phased down in the mid-1970s, refiners turned to a new additive, MMT, a manganese-based compound developed by the Ethyl Corporation in the 1950s. As the demand for unleaded gasoline grew, refiners relied on MMT in about 40 percent of the new unleaded. Ethyl sold about 10 million pounds per year to the booming new market. But, in late 1977, tests run by the major auto companies and the EPA showed that MMT damaged the new cars' catalytic converters, and the substance was banned.[26]

As a result of the MMT ban, refiners turned to the more expensive refining process to upgrade octane levels. This process, known as "severe reforming," involves greatly increased temperature and pressure. It uses from 6 to 8 percent more of the original barrel of crude oil just for process energy, and involves substantial long-term investments in new equipment.

The main octane boosting compounds produced during the severe-reforming process are called aromatics, and they make up 25 to 40 percent of most unleaded gasoline. They are composed primarily of toluene, benzene, and xylene compounds. Wayne Barney, an environmental research scientist with the Argonne National Laboratory believes that these compounds are potent carcinogens which

---

*Crude oil production was up 5 percent in the first quarter of 1979, according to oil industry figures, and the Iranian shortfall was made up by increased Saudi Arabian production.

may ultimately prove to be as hazardous to human health as lead. "We're saving the inner city kids from lead poisoning only to kill them with cancer," said Barney in an interview with the authors.

By 1979 many of the aromatic compounds once utilized by the petrochemical industry to produce plastic and other synthetic compounds were being retained by refineries to boost octane via the severe refining process. This process reduced the overall U.S. daily refining capacity by about 300,000 barrels of gasoline (from a total daily output of seven million barrels.) While this reduction in refining capacity was little noticed during 1978, it became readily apparent when supplies of crude oil tightened up in 1979. By the summer of 1979, oil refiners complained of an "octane crisis," and asked EPA to slow the phasing out of leaded gasoline and to allow temporary use of MMT in an effort to boost production.[27]

The 1979 gasoline shortage sent shock waves of panic through the petrochemical industry as the octane demand for aromatics began to compete sharply with that of the plastic and synthetic fiber industries. A group of the major petrochemical industries—including DuPont, Dow, and Monsanto—began an intensive lobbying campaign for an easing of the phasing out of leaded gasoline, claiming that the aromatic shortage, "threatens the jobs of the 14 million Americans directly dependent and the 29 million Americans indirectly dependent on the petrochemical industry for employment."[28]

The intense pressure from the chemical industry was evident at a June, 1979 hearing convened by the Environmental Protection Agency. Monsanto's Ralph Kienker went so far as to declare that "refiners have no means of increasing octane without lead and MMT until new refining equipment can be brought on line."[29]

Kienker's testimony brought a sharp rebuttal from the gasohol lobby's Richard Merritt, since Kienker, like Midgely of Ethyl Corporation, neglected to mention that alcohols were also useful octane boosters. While Merritt found environmentalists such as Barry Commoner eager to support his campaign to promote alcohols as octane boosters, the EPA reluctantly gave in to the chemical industry and halted the lead phase-down.

Two other major options for octane-boosting additives also were noted at the EPA hearing. Tertiary Butyl Alcohol (TBA—a form of butanol) has been extensively explored by Sunoco (Sun Oil Company) and Arco (Atlantic Richfield Corporation), and has been marketed as an effective octane booster in 5 to 7 percent concentrations for many years. But the supply of TBA hinges on petrochemical byproducts, and building plants to manufacture TBA does not appear to be an economically attractive option at present.[30]

Methyl ethers—such as MTBE or TAME—are also feasible octane boosters researched by Gulf Oil and others. But Gulf researchers estimate that, due to the limited supply of isobutylene feedstock for MTBE, only 3 percent of the U.S. gasoline pool could be brought up to adequate octane levels. "No one should be fooled into thinking MTBE is an answer to the unleaded crisis," Gulf's Spencer Sheldon said. While TAME may be more favorable in terms of potential supply, it will require the development of natural gas or coal-based methanol facilities for any substantial impact.

Any of the alcohols or ethers would be preferable as octane boosters to aromatics. Benzene is a known carcinogen, strictly regulated in the workplace by the Occupational Safety and Health Administration (OSHA), although no standards have been set for public exposure to gasoline. Toluene and xylene are suspected carcinogens.

Richard Pefley noted that the problem with aromatics may increase in the future. "As we start putting these (aromatics) into gasoline for tetraethyl lead, we're trading one bad component for another. . . . [Alcohols] avoid that, and I just keep arguing, let's make this whole assessment before the nation commits itself to synthetic gasolines* in preference to methanol."

But the air-quality arguments which favor the use of alcohols over aromatic octane boosters are unlikely to make much of an impact under the Reagan administration, whose outlook is marked by an avowed distaste for forcing industry to comply with federal regulations.

Even with all the air-quality arguments aside, there is a strong economic case for alcohols as octane boosters. DOE analysts note that it takes 12.2 gallons of raw, low-octane gasoline to make 10 gallons of regular unleaded, while it takes only 10.6 gallons of the same basic low-octane gasoline to make the 9 gallons of lower-octane gasoline needed for blending into gasohol. Thus, "refinery operations could reduce the degree of reforming and increase the yield of gasoline by about 20 percent" if full advantage was taken of gasohol blending, DOE said in a 1980 report. In other words, a gallon of ethanol used for gasohol blending replaces not just one gallon of gasoline, but two.[31]

The promising economics of ethanol as an octane booster go a long way towards explaining the interest of major oil companies in marketing gasohol blends and entering into joint ventures with grain processors for new distilleries. Texaco, for example, announced its widely publicized gasohol program in the fall of 1979, several months after major breakdowns in refinery systems hindered the flow of unleaded gasoline. And Chevron (Standard of California) holds a majority ownership in a Kentucky corporation which received one of the first federal loan guarantees approved by the Reagan administration for ethanol plants.

A number of oil companies, including Chevron, have experimented with ethanol as an octane booster without labeling their pumps "gasohol," according to the Renewable Fuels Association's David Haulberg. And the Ethyl Corporation released a paper in June, 1981, saying that there was "little question" that

---

*The synthetic gasoline Pefley refers to is the refined synfuel from coal-based processes such as Gulf's SRC II or Exxon's EDS. The aromatic content of such fuels is high, and they are so liable to environmental challenge that one oil company, Conoco, announced in 1980 that it would abandon the synfuel route in favor of methanol.

alcohols would be used as fuels and fuel components. G. H. Unzelman of Ethyl, denouncing the impositions foisted on industry by the Clean Air Act, noted that the crude oil losses to refiners who make unleaded gasoline would approach 110 million barrels per year in 1981.[32]

Yet gasohol, somewhat paradoxically, still represents a distinct threat to the major oil companies, especially in periods of oil glut. For the industry as a whole has made tremendous, multibillion dollar investments to develop the severe-reforming capacity required to produce the new generation of aromatic-laced "premium unleaded" fuels which were unveiled in service stations across the country in 1980–81. To recoup these enormous investments, the oil industry will have to convince the motoring public that aromatic unleaded gasoline—and not the more environmentally benign gasohol—is the best premium fuel for the 1980s.

AIR POLLUTION: HOW ALCOHOL REDUCES TAILPIPE EMISSIONS

In recent years the citizens of São Paulo, Brazil, have noticed a distinctly new odor lingering in the air of their smog-choked city. Carl Duisberg, on one of his many trips to Brazil, found this sweet, pungent odor—the result of an increase in alcohol-powered vehicles—"not necessarily offensive or harmful" but rather a sharp reminder that man's every step on earth leaves a mark. Duisberg notes that the São Paulo experience with alcohol fuels demonstrates that they are not, as some proponents have claimed, "pollution-free fuels," but more accurately speaking, they are pollution-reducing fuels as compared to petroleum.

During the 1960s, as concern about air pollution mounted, three major types of pollutants were identified: carbon monoxide (CO), hydrocarbons (HC), and nitrogen oxides (NOx). These compounds, along with lead, came under federal regulation with the Clean Air Act of 1970. Another type of emission problem, evaporation from fuel tanks and carburetors of automobiles, was also included in regulations by the mid-1970s.

The regulated air pollutants are found in the exhaust of gasohol-powered cars and pure-alcohol-powered cars, but at generally lower levels. Also, another class of emissions, aldehydes, are a source of some concern to automotive engineers studying the emissions of alcohol-fueled autos. Emission changes (compared to gasoline) from use of alcohol fuels and blends were studied by the U.S. Office of Technology Assessment for a report to the United States Congress in 1980. The results of the study are recorded in table 9-1.

Despite a great deal of variation and controversy over emissions tests performed in the mid-1970s, a general consensus that gasohol and pure alcohol fuels produce fewer overall emissions problems than gasoline has been reached among automotive engineers.

Fuel researchers have consistently found CO and HC to be reduced with

TABLE 9-1

Emission Changes (compared to gasoline) from Use of Alcohol Fuels and Blends (Source: U.S. Office of Technology Assessment: *Energy From Biological Processes*, U.S. Congress, 1980, Volume II, p. 211)

Pollutant/fuel	Methanol	Methanol/gasoline	Ethanol	Ethanol/gasoline "gasohol"
Hydrocarbon or unburned fuels	About the same or slightly higher, bu much less photochemically reactive, and virtual elimination of PNAs; can be catalytically controlled	May go up or down in unmodified vehicles, unchanged when Φ remains constant. Composition changes, tho, and PNAs go down. Can be controlled. Higher evaporative emissions	Not very much data, should be about the same or higher but less reactive. Expected reduction in PNA	May go up or down in unmodified vehicles, about the same when Φ remains constant; composition may change, expected reduction in PNAs. Evaporative emissions up
Carbon monoxide	About the same, slightly less for rich mixtures; can be catalytically controlled; primarily a function of Φ*	Essentially unchanged if Φ remains constant, lower if leaning is allowed to occur	About the same, can be controlled; primarily a function of Φ	Decrease in unmodified vehicles (i.e., leaning occurs), about the same when Φ remains constant
Nitrogen oxides	1/3 to 2/3 less at same A/F ratio, can be lowered further by going very lean; can be controlled	Mixed; decreases when Φ is held constant, but may increase from fuel "leaning" effect in unmodified vehicles	Lower, but not as low as with methanol; can be controlled	Slight effect, small decrease when Φ is held constant, but may increase or decrease further from fuel "leaning" effect in unmodified vehicles
Oxygenated compounds	Much higher aldehydes, particularly significant with precatalyst vehicles	Aldehydes increase somewhat, most significant in precatalyst vehicles	Much higher aldehydes, particularly significant with precatalyst vehicles	Aldehydes increase, most significant in precatalyst vehicles
Particulates	Virtually none	Little data	Expected to be near zero	Little data, no significant change expected
Other	No sulfur compounds, no HCN or ammonia	Unknown	No sulfur compounds	No data

*The symbol, Φ, represents the air-to-fuel ratio.

gasohol and greatly reduced with pure alcohol fuels,* although NOx is often slightly higher or slightly lower than gasoline in many tests. Aldehydes are usually higher with alcohols than gasoline, but are not as significant a problem as the other emissions on a volume basis and can be controlled with the aid of catalytic converters.**

To understand why alcohols burn cleaner than gasoline, automotive engineers studied the more complete and cooler combustion and "leaner" burning properties of alcohol. Pefley and his associates noted that an increase of air in the air-fuel mixture flowing through the carburetor would reduce emissions from straight gasoline. Blends of ethanol or methanol, they found, could achieve this same leaning effect without adjusting the carburetor.[33]

In a gasoline-fueled car, leaning the mixture decreases HC and CO, but it also raises engine temperature, which increases NOx.† Pefley's research demonstrated that the cooler burning property of alcohol blends and straight alcohols could keep NOx low, even at leaner air-fuel mixtures.

The air-fuel mixture proved to be a major parameter in measuring the emissions for alcohol fuels and blends. In one comparison between methanol and gasoline, Pefley found CO dropping from 17 to 2 grams per mile when gasoline mixtures were leaned by 20 percent, while methanol dropped from 13 to 1.7 grams per mile when leaned by a theoretically similar amount. Thus, carburetor air-fuel mixtures had far more to do with the exhaust-emissions level than the type of fuel, but alcohol proved to be more suited to the lean burn that could keep emissions low.

That a minor carburetor adjustment could create such a wide variation in emissions results is indicative of how widespread disagreement over the effect of alcohols came about among automotive engineers. But as the impact of such adjustments became a greater source of concern among fuel scientists in the late 1970s, a greater degree of agreement was seen about the emissions tests.

*HC emissions have also been reported up in some tests, when all emissions are measured, but automotive engineers and EPA specialists note that the hydrocarbon emission—which is also known as unburned fuel—consists mostly of ethanol or methanol in cars powered by either of those two alcohols. Methanol is not photochemically reactive, while ethanol is only slightly photochemically reactive, and thus neither is considered a cause of smog. A realistic measurement of HC takes this fact into account. One DOE report noted that HC emissions from a pure methanol car were up by a factor of three, but that 98 percent of these were nonreactive methanol or ethanol compounds. This same question of realistic measurement applies to evaporative emissions for gasohol. Although the EPA notes that "evaps" are up 100 percent with gasohol, specialists in Ann Arbor note that most of these emissions involve the alcohol component of the gasohol, which is not reactive.

**While aldehydes can be controlled in the United States with catalytic converters, they are a source of concern in Europe and Brazil, where catalytic converters are not used. On balance, though, Volkswagen engineers note that increased aldehyde emissions are "outweighed considerably" by gasoline's emissions of polynuclear aromatics (PNAs), which are "severely carcinogenic" according to VW. PNAs are uncontrolled in Europe and are the most troubling emission from aromatic octane boosters there. The EPA believes the problem is mostly controlled in the United States with catalytic converters.

†The lean air-fuel mixture was the prime pollution-reducing strategy for U.S. automakers between 1970 and 1975; the introduction of catalytic converters made possible a shift back to richer air-fuel mixtures which, for pure gasoline, give better performance. If gasoline is burned at too lean a mixture, it approaches the "lean limit," and engines hesitate and stall. Alcohol's ability to move back the lean limit is another plus that Pefley and other automotive engineers considered.

The one major class of emissions that increased consistently, in comparison after comparison with gasoline, proved to be aldehydes. These emissions—which include formaldehyde, a strong eye irritant and suspected carcinogen—were relatively low in comparison to the total emissions of other types, but their presence did cause some concern. This concern was allayed somewhat by a 1980 report by the Argonne National Laboratory which concluded that the new generation of post-1980 emissions-control systems ". . . are effective in eliminating . . . increased emissions of unburned alcohol (HC) and aldehydes over the long term. In the near term, the aldehyde emissions which will accompany the utilization of alcohol blends are not expected to cause any significant problems."[34]

While alcohols are not a perfect fuel, they are better than gasoline. Positive health benefits from substituting alcohol for lead and aromatics in blends, and bringing down total emissions in pure-alcohol-fueled cars, are anticipated. On balance, a major shift to the use of alcohol fuels holds the promise of improving urban air quality, rather than contributing to its further deterioration.

PUBLIC SAFETY: HAPPY HOUR AT THE GAS STATION?

Newspaper cartoonists have had a field day depicting the arrival of alcohol fuel pumps in the United States. Inevitably, these cartoons picture an intoxicated motorist lurching away from a pump in which he has filled up his own tank—as well as his car's—with a potent brew. While the humorous intent is clear, there are some individuals who take the threat of increased drunkenness seriously. One congressman on the House Agriculture Committee protested that if Congress helped the alcohol fuels movement "there won't be a straight furrow plowed in Kansas."

But neither the pure ethanol nor the methanol that are beginning to make their appearance at fuel pumps around the country can be considered safe for human consumption. The strong denaturants that are added to ordinarily potable ethanol when used as fuel make it unpalatable and painful to swallow. And stiff law enforcement by the Treasury's Bureau of Alcohol, Tobacco and Firearms place any would-be moonshiners seeking to avoid beverage taxes and regulations at considerable risk. Proper labeling—for both denatured alcohol and methanol, which is naturally toxic—should provide ample warning to most people.

Despite these precautions, the long, sad history of alcohol abuse leads one inevitably to conclude that some unfortunate soul—no matter how clearly these fuels are labeled as toxic—will try to drink them instead of drive with them. The painful results of such experimentation should succeed in detering all but the most desperate of alcoholics from a second drink.

Alcohol vapors may present a workplace hazard, but a 1977 DOE paper noted "gasoline is much more toxic on inhalation than methanol, and gasoline vapors, which contain aromatics, are carcinogenic." Methanol is also considered more toxic as a drink than gasoline, DOE also noted, "unless the benzene content is high."[35]

Argonne labs noted that skin exposure to gasoline "is considered to pose a greater public and occupational health risk than either methanol or ethanol." While skin exposure to methanol can be highly toxic, "unwitting dermal exposure to methanol is unlikely because when in contact with the skin, methanol and methanol blends cause a burning sensation. . . . By comparison, gasoline contains proven and suspected carcinogens which are readily absorbed by the skin when contact is made, or inhaled when exposed to alcohol vapors."[36]

Alcohol fires are generally far less hazardous than gasoline fires since they can be extinguished with water. It is this safety feature as well as its clean combustion that makes alcohol a favored engine fuel on racing tracks. But pure alcohol fires are also nearly invisible, and there have been recommendations that flame colorants be used for 100 percent-alcohol fuels.

Gasohol fires may be slightly more difficult to extinguish than gasoline fires, although experiments conducted at Iowa State University showed the potential hazard to be small. The alcohol portion of the gasohol will break down small amounts of foam as it is applied initially, but there was "no observable difference in extinguishment of gasohol," university spokesmen said.[37]

THE GREENHOUSE EFFECT: THE ULTIMATE ENVIRONMENTAL DILEMMA

Carbon dioxide—the gas that bubbles out of a healthy batch of fermenting mash—is released into the earth's atmosphere whenever a carbon-containing material burns or decays. It is drawn back out of the atmosphere during the photosynthesis process in which plants absorb carbon dioxide (CO_2).

In nature, a rough balance is maintained between the carbon dioxide released through decomposition and the amount fixed by plant life through photosynthesis. But this natural equilibrium has been upset by combustion of fossil fuels—petroleum, natural gas, coal, tar sands, and oil shale which contain the fossilized carbon of sedimentary materials formed millions of years ago.

Scientists now believe that the sudden increase in fossil-fuel use during the past 120 years of the industrial age has resulted in a massive buildup of CO_2 in the earth's atmosphere, which one DOE report labels "the most important environmental issue facing mankind."[38] This atmospheric CO_2 gas reflects infrared radiation normally dispersed into outer space and reflects it back down to earth; it retains heat, like the panes of glass in a greenhouse, hence the term "greenhouse effect."

Climatologists with DOE have predicted, on the basis of computer simulations of the greenhouse effect, that the average world temperature could rise as much as 2 to 3 degrees centigrade during the first half of the twenty-first century. Such a global warming trend—greater in magnitude than cooling trends in ice ages past—would have far-reaching impacts. Rainfall patterns would probably shift, probably to the detriment of the major U.S. grain-producing regions, and the ice sheets at the poles could melt, flooding most of the world's coastal cities.

A two-year study of the greenhouse effect by the prestigious National Academy of Sciences has concluded that "the primary limiting factor on energy production from fossil fuels over the next few centuries may turn out to be the climatic effects of the release of carbon dioxide."[39] If the greenhouse effect does prove serious enough to force a global shift away from fossil fuels, renewable alcohol fuels may be forced to supply a much greater portion of the world's liquid fuel needs. But it is not at all certain that a warning could be sounded in time to make a transition.

The use of renewable alcohol fuels presents an alternative to fossil fuels in that they need not be net contributors to the CO_2 buildup. The CO_2 released as plants and trees are harvested, refined into fuel, and finally burned can then be fixed back into carbon by the next generation of biomass crops. On the other hand, alcohol fuels made from fossil fuels, such as methanol from coal, would contribute as much CO_2 as most other types of coal fuel.

The key to maintaining the balance with renewable fuels would lie in establishing a rough balance between the amount of biomass harvested for energy production each year and the amount of new biomass growth. If the forests and fields are simply "mined" for biomass feedstocks and the erosive forces of nature destroy the soil's fertility, then renewable fuels could also contribute to the CO_2 buildup. In fact, some scientists fear that the overharvesting of many of the world's forests and the steady advance of deserts may already be helping to trigger the greenhouse effect.

The steady buildup of atmospheric CO_2 was first documented at a Hawaiian weather observatory dramatically situated on the side of a still-active volcano which rises starkly out of the Pacific Ocean. The observatory was built in 1949 by the U.S. Weather Bureau at a time when few scientists had any inkling of the danger it would find. The observatory was designed simply to make a series of routine observations to help climatologists understand weather systems.

Charles Keeling, a climatologist with the Scripps Institute of Oceanography, was one of the first scientists to seriously investigate the greenhouse effect. In 1958 Keeling went to the observatory to set up equipment to monitor CO_2 levels in the atmosphere. After four years of collecting data, Keeling found a steady upward climb in CO_2 levels amid the seasonal variations. He concluded that "approximately half of the carbon dioxide from fossil fuel [combustion] was accumulating in the air."[40]

In the early 1970s the significance of the data began to be widely debated. Scientists who had worked at the observatory wrote several landmark articles detailing their research for professional journals. In Washington, Dr. Lester Machta, a distinguished climatologist, was chosen to lead a CO_2 monitoring program at the National Oceanic and Atmospheric Administration (NOAA). Machta moved quickly, and new monitoring stations were set up in Barrow, Alaska, American Samoa, and Antarctica to further document the CO_2 buildup.

Despite the stepped-up level of research, scientists working for NOAA have found it difficult to make any firm predictions as to the precise nature and magnitude of climatic impacts from the greenhouse effect. One major stumbling block results from their limited knowledge about the basic mechanisms of the carbon cycle. Scientists have been unable to pin down, for example, the amount of CO_2 absorbed by ocean plankton and at what point the oceans may become saturated with CO_2, dramatically accelerating the rate of CO_2 buildup.

One effort to predict the impacts of the greenhouse effect was undertaken by the Stanford Research Institute (SRI), a private think tank which caters to an elite group of corporate and government clientele. SRI analyzed available data and then developed a "hypothetical error band" or "envelope of uncertainty" for every major potential impact of the greenhouse effect. The report was submitted to the DOE forerunner, the Energy Research and Development Administration (ERDA) but was never publicly released. It concluded that the CO_2 buildup presented a "high probability of causing large, persistent fluctuations of global food supplies within fifty years" and a low but significant probability of "disruption of the U.S. economic system due to chronic water shortfalls, below levels needed to sustain energy technologies and agriculture."

The report also noted a "low but significant probability of climate-related collapse of selected webs in regional economies," and "widespread concern and political dissension about prevention of atmospheric CO_2 buildup." Its conclusion mirrored the view of the National Academy of Science when it declared that "world fossil fuel resources cannot be exploited at maximum feasible rates without incurring CO_2-related impacts that may be defined as intolerable. This would necessitate an earlier transition to inexhaustible energy resources than would otherwise be the case."[41]

Although the scenarios laid out by SRI read like an excerpt from a script outline for a Hollywood disaster movie, evidence was uncovered by Dewey MacLean, a geology professor at Virginia Polytechnic University, indicating that a slight increase in atmospheric carbon during a bygone era caused major environmental upheavals. "An examination of rock and life records at the Mesozoic-Cenozoic time interface, which was a time of worldwide faunal extinction," MacLean wrote in the August, 1978 issue of *Science* magazine, "spells out the potential effects of a global temperature elevation in living organisms and suggests that human alterations of the carbon cycle could trigger conditions resembling those of the so-called 'Time of Great Dying' that terminated the Mesozoic Era."[42]

The strongest public warning about the hazards of CO_2 buildup was issued by a team of climatologists led by Princeton University professor Gordon MacDonald, who authored a 1979 report for the White House Council on Environmental Quality (CEQ). The report concluded that "the carbon-dioxide-induced warming trend will probably be conspicuous within the next twenty years. If this trend is allowed to continue, climatic zones will shift and agriculture will be displaced. Such a series of changes would have far-reaching implications for human welfare in an ever more crowded world, would threaten the stability of food supplies, and would present a further set of intractable problems to organized societies. The best estimates suggest that there would be the least change in temperature in the tropics but polar regions would grow substantially warmer. With sufficient high-latitude warming, the ice cap in the western part of the Antarctic continent would disappear in a period as short as two centuries, causing a 20-foot rise in sea level with resulting inundation of low-lying coastal zones. Enlightened policy in the management of fossil fuels and forests can delay or avoid these changes but the time for implementing these policies is fast fading.''[43]

Despite the stark tone of the report which MacDonald delivered to CEQ, it made little impression on the energy policymakers of the Carter administration. The warnings put forth by the nation's scientific community, however ominous, were still filled with too many "*if*'s, *and*'s, and *but*'s" to convince either James Schlesinger or his successor at DOE, Charles Duncan, to seriously alter the direction of federal energy policy. Both Schlesinger and Duncan were more concerned with trying to steer the nation safely through the energy shortages of the eighties and nineties than with heading off still-uncertain climatic disasters which could arise in the twenty-first century.

A cornerstone of the Carter administration's energy plan was the creation of a synthetic fuels industry that would, by 1992, have the capacity to produce 2 million barrels a day of "synfuels" from coal, tar sands, and oil shale. While dubiously conceived on other environmental and economic grounds, the potential impact of such a synfuels industry was frightening to scientists concerned with the greenhouse effect—frightening enough to send MacDonald, and a number of others, through a round of Senate subcommittee hearings in 1979 that injected them squarely into the middle of the energy debate.

MacDonald estimated that the doubling of the earth's present atmospheric carbon-dioxide concentration—the level expected to trigger the greenhouse effect—would begin in the year 2030 if the current mix and rate of fossil-fuel use continued, but could begin as early as 2010 if a crash program for synfuels were to be developed. He denounced DOE's plan to lend federal money for a synfuels industry and testified: "Before we commit ourselves to the construction of a major synthetic fuels infrastructure, involving the investment of hundreds of billions of dollars, we should make every effort to understand not only the short-term benefits and costs, but the longer-term consequences to the generations that must live with decisions taken today."[44] MacDonald recommended that the government discourage all private-sector investments in synfuel processes based on coal and oil shale, because these fuels release three times more carbon dioxide per Btu than petroleum. He urged that the synthetic fuels program be scrapped in favor of increased conservation efforts and the development of natural gas resources and renewable fuels.

Although the surprising declines in U.S. oil consumption during 1980 and 1981 compared to 1978 may slow the greenhouse effect, it will require a truly global effort to reduce the 20 billion tons of carbon dioxide released annually throughout the world by the burning of fossil fuels.

Whatever the percentage of liquid-fuel needs it eventually replaces, alcohol promises to be much less harmful to our atmosphere than the nonrenewable, hydrocarbon sources that we continue to deplete at a massive, even if slowed, rate.

Chapter 10
Alcohol Future

The thorny mesquite tree was a valuable ally to the Indians who once roamed the arid southwestern United States, providing them with wood for fuel and sugary pods for nourishment. But to the ranchers who now populate the region, the mesquite constitutes a hostile presence whose thick foliage shades out scarce grasses which their cattle need for sustenance. In Texas alone, some 56,244,800 acres of pasture lands are listed as "infested" with mesquite, prompting the federal government to initiate a major eradication program.[1]

In the decades ahead, Texas ranchers may take a somewhat more kindly view of the lowly mesquite since its sugar-rich pods are a valuable feedstock for ethanol production. Tree crops, such as the mesquite pod which can be grown on marginal lands without triggering soil erosion, together with wood and aquatic energy crops, form the foundation for a second generation of renewable feedstocks which are likely to play an important role in the alcohol fuels industry of the future. A broad use of these feedstocks will make possible major boosts in alcohol fuels production without harvesting grain crops from prime farm lands or mining the earth for coal. Thus, the potential food-versus-fuel conflict associated with a large grain ethanol industry and the environmental hazards associated with coal-to-methanol production can be largely avoided.

The technologies required to utilize the "second generation" feedstocks are already under development at academic and private research centers both in the United States and abroad. The use of energy crops that can be grown on marginal lands or in ponds and marshes holds a particular attraction to many Third World nations whose hungry populations can scarcely afford the luxury of putting vast tracts of fertile farm lands into sugarcane and corn fuel crops. Some of the second-generation feedstocks are already on the verge of commercialization—offering the tantalizing possibility of eventually providing enough alcohol to totally eliminate the need for imported gasoline fuels—while others may require at least another decade of concerted development efforts.

The mesquite tree stands out as one of the most intriguing of the second-generation feedstocks. It already thrives on some 72 million acres of semidesert lands in the United States which receive a miserly rainfall of only 15 to 20 inches a year. These lands can produce few profitable food crops without extensive irrigation and their value as pasture land is minimal—in some areas offering a net return of less than $10 per acre.[2]

A key to mesquite's ability to thrive in the arid Southwest is its extensive root system which may reach down over 40 feet into the ground and extend laterally from its base by as much as 50 feet. This root system helps to anchor soil in place, protecting the land from erosion, and enables the tree to survive periods of prolonged drought by reaching deep in the earth to collect subterranean water. On the tips of this root system tiny bacterial nodules help the mesquite to provide its own growth nutrients by taking nitrogen from the air during the photosynthesis process and fixing it in the soil.[3]

The cultivation of mesquite as an energy crop involves a minimal disruption of fragile soils since only the pod—and not the entire tree—is harvested for fuel. These pods contain only small amounts of water and high concentration of fermentable sugars and carbohydrates. Dr. Peter Felker, a professor at Texas A&I University, has carried out a series of field trials which indicate that a 6.8-square-mile tract of southwestern lands planted in mesquite could provide enough pods to sustain a 42,000-gallon-a-day, 12-million-gallon-a-year ethanol distillery. Each acre of land could be expected to produce some 4,000 pounds of pod per year which could be converted to 190 gallons of ethanol plus byproducts of cattle feed and a seed gum which can be used in a variety of different industrial, cosmetic, and food products. Different varieties of mesquite have been examined by Felker at a field site in California's Imperial Valley to learn more about their growth cycles. In one planting of some 1,300 trees, six trees displayed the surprising ability to produce their first pods only six months after transplanting, with the potential to produce commercially attractive quantities of pods in less than five years. A number of different harvesting techniques developed for the walnut and pecan industry have been examined to see if they might be adapted to mesquite, but as yet no fully commercial pod harvesting system has been fully developed.[4]

Felker's mesquite pod research has been modestly funded with grants from the U.S. Department of Energy but he hopes to attract private capital to develop a pilot plant. The process energy for this plant would be provided by mesquite wood which would be harvested from about 12 percent of the total land area of the 6.8-mile tree farm. One oil company has discussed the possibility of funding the first mesquite-pod-to-ethanol distillery with Felker but has yet to make any firm financial commitments. Felker now spends most of his time on basic research in mesquite tree cultivation techniques near the Kingsville campus of Texas A&I University.

On the international front, mesquite has aroused interest not only for its potential as an ethanol tree crop but also due to its ability to both combat erosion and increase soil fertility through nitrogen fixation. In East Africa, mesquite trees have been planted to help fight the advance of the desert into grazing lands;

in Asia, the trees have been intercropped among food grains such as millet to boost the nitrogen content of marginal soils.

Another promising tree crop for ethanol production is the common honey locust which, like the mesquite, produces a sugar-rich seed pod. Honey locust appear able to thrive in over 90 percent of the terrain within the continental United States, including marginal pasture lands. Certain varieties can be planted to form a light, open canopy over the ground which does not shade out pasture grasses. Thus, an acre of pasture planted in honey locust could also support cattle grazing, providing a farmer with two different ways to simultaneously earn income from the same piece of land.[5]

The potential of the honey locust to provide an economic boon to U.S. farmers was first noted by J. R. Smith in his classic work, *Tree Crops: A Permanent Agriculture*.[6] In this book, first published in 1929, Smith recommended crop-yielding trees such as the honey locust, rather than erosion-prone row crops, as the "best medium for extending agriculture to hills, to steep places, to rocky places, and to the lands where rainfall is deficient." During the 1930s, when the farm chemurgic movement reached its zenith, the Tennessee Valley Authority (TVA) conducted an extensive research program to determine the best pod-producing varieties of honey locust and their growth characteristics. The TVA project compiled a mass of information which has contributed greatly to current efforts to develop the honey locust.

An acre of honey locust, according to more recent research carried out by the TVA and the Center for the Study of Biological Systems in Saint Louis, will produce enough pods each year to yield an average of 81 gallons of ethanol.[7] While this yield is considerably less than the 250 gallons produced from the 100 bushels of corn harvested from a typical acre of midwestern farm land, it is achieved on marginal lands without the extensive use of petrochemical fertilizers and pesticides or the soil loss that often accompanies row-crop production.

The commercial cropping of honey locust—like its dryland counterpart, the mesquite—suffers from a dearth of efficient pod harvesting equipment. The honey locust pods do not drop off the tree all at once but gradually over a year's time. Once this harvesting problem is resolved, preliminary studies indicate that honey locust ethanol could be produced for as little as $1.14 per gallon if a cattle feed market is found for the pod's high-protein byproduct stillage.[8]

POPLAR, EUCALYPTUS, AND WOOD ENERGY FARMS

Another major area of interest in the scientific community is in the development of technologies to cultivate fast-growing trees whose wood could then be harvested as an alcohol feedstock. These wood energy farms, like tree crop farms, could be planted primarily in marginal forest and farm lands or possibly even used to help reclaim strip-mined coal sites. Most of the trees considered for energy cropping have the ability to "coppice," or resprout from cut stumps, thus alleviating the need for replanting after each harvest.

During the Carter administration, a broad-ranging if somewhat disorganized effort was launched to establish experimental wood energy at various sites across the nation. The blueprints for these farms were contained in a detailed six-volume study compiled by the Mitre Corporation for the U.S. Department of Energy in 1977. However the report, according to its principal author, Robert Inman, was highly speculative in nature since it was difficult to assemble much reliable, first-hand data on the economics and technology of intensive wood energy cropping.[9]

The report explored the possibility of creating what Mitre termed "energy plantations" on up to 30 million acres of the 324 million acres of land classified by the U.S. Soil Conservation Service as forest, pasture, and range, with from 20,000 to 50,000 acres of wood crops cultivated on each plantation. These lands, while generally not suitable for intensive, traditional agriculture, were arable and could generally support biomass energy farming.

Production costs for methanol from these plantations were estimated at between 48 cents and 75 cents a gallon with ethanol production costs ranging from $1.60 to $2.20 a gallon. The critical variables in these cost estimates were the relative productivity of the energy plantations and the efficiency of the processing plants. On the basis of these initial cost estimates, Mitre predicted the commercialization of wood-to-methanol production in the early 1980s, with wood-to-ethanol production not coming on line until sometime after 1990. In the interim, the Mitre report recommended that the federal government sponsor a number of basic research projects to investigate cultivation and processing techniques for promising wood energy crops such as poplar, eucalyptus, black locust, sycamore, alder, and salt cedar, a tree which can be grown in salt-laden coastal lands.

Mitre's timetable for wood-to-alcohol commercialization was thrown askew by major advances in wood-to-ethanol technologies which promise to substantially lower production costs. Until recently, scientists had been severely hampered in their efforts to convert wood to ethanol by the difficulty of efficiently separating the cellulosic and hemicellulosic ethanol feedstocks from the protective lignin material which largely encapsulated these materials. This step proved to be nearly as formidable a task as developing the enzyme and acid technologies necessary to break down these feedstocks into simple sugars required for fermentation. Yet without the means for economically achieving this separation, the cellulose-to-ethanol technology would remain confined largely to preprocessed feedstocks, such as shredded paper, and avoid any direct use of unrefined wood.

One of the most successful efforts to achieve this critical separation step was carried out by a team of researchers led by Dr. Kendall Pye at the University of Pennsylvania in Philadelphia. Pye utilized an alcohol-based solvent to separate out cellulose and hemicellulose materials from a fast-growing strain of hybrid poplar. These materials were then converted into ethanol via an enzyme process while the remaining lignin remained behind for possible use as a boiler fuel. This high-powered research project was jointly funded by the U.S. Department of Energy and a General Electric research division headquartered in Philadelphia. In the meantime, Pye worked with Miles Fry and Son, a Pennsylvania nursery, to determine the potential yield per acre from the fastgrowing hybrid poplars. The results of a four-year study indicated poplar growth rates that would produce enough wood to yield an average of 900 gallons of ethanol per acre, with future yields possible of a 1,000 gallons per acre.[10]

Once these promising test results were reviewed by General Electric's corporate chiefs, the company quickly moved to hire Pye to direct a new subsidiary, christened the Biological Energy Company, which was created with the express purpose of developing the poplar-to-ethanol technology. But as the new venture began to take shape, Pye opted for a more cautious strategy. He decided to forego ethanol production over the short term in favor of commercializing an intermediate process, completing only a first-stage conversion of the poplar to cellulose, lignin, and molasses. In the light of the oil glut of the early 1980s, he concluded the new subsidiary could make more money marketing these products than by carrying out the full ethanol conversion process.[10]

General Electric's strategy for developing poplar trees as an ethanol feedstock does not call for any massive corporate takeover of marginal farm and forest lands to plant vast energy plantations of poplar trees. Instead, as Pye explained in an interview with the authors, General Electric would rely primarily on midwestern farmers who would be willing to plant narrow shelterbelts of poplar trees around their prime farm lands. Pye is sensitive to the soil erosion problems which might be exacerbated by a major grain-based expansion of the alcohol fuels industry. "Farmers have got to move away from their reliance on annual crops toward a greater use of perennial tree crops," he commented. "The Red River Valley area of South Dakota underwent incredible erosion during the 1930s and there was a vast program to plant shelterbelts to protect the land from the harsh prairie winds. Now, as economic pressures mount, there is a tendency to cut down the shelterbelts, so you can see that soil erosion is increasing dramatically. We are encouraging farmers to replant the shelterbelts around the fields in rows of poplars which are twelve trees wide. Then every year we will go and harvest one-third of the row."

Pye predicts that the first poplar will be ready to harvest four years after the first seedlings are planted in shelterbelt areas, with possible annual yields of up to 1,000 gallons of ethanol per acre. The poplar can be harvested by a mechanical system which snips the tree off at the trunk; the stump soon sprouts new limbs and is ready for harvesting again in another two years.[11]

The leaves of the poplar trees, which constitute what Pye terms "a superb cattle feed," could be separated out from the chips prior to processing. Pye believes that these leaves, when taken together with the byproduct stillage feed, could produce more protein on an acre of marginal lands than is possible with most forage crops such as alfalfa.

Before Pye left his post at the University of Pennsylvania to enter the corporate world, he predicted that poplar ethanol could be produced for between 65 cents and 80 cents a gallon—less than half the costs projected in the 1977

Mitre report. In his position as chief corporate officer of the Biological Energy Company, he takes a somewhat more guarded approach in discussing production economics, confirming only that "we'll be selling this alcohol at just below the market price of anhydrous corn ethanol—because that's the name of the game. But we're quite sure we can make a reasonable profit doing so."

The sweet-smelling eucalyptus is another promising wood energy crop in frost-free areas of the South and West. The eucalyptus sprouts from seed in only seven days and can reach heights of over 14 feet in only seven months.[12] New shoots rapidly resprout from its harvested stump, while its extensive root system seeks out water in arid soils. Due to the rough, uneven grain of eucalyptus wood, it has little commercial value to the timber industry, although the Japanese and Brazilian paper industries have purchased eucalyptus chips for pulp. For several decades the Brazilian government has provided tax incentives to reforest logged off areas of the nation's dry interior with eucalyptus trees. The Energy Company of São Paulo, in recent years, has come to view these mature eucalyptus stands as prime candidates for methanol production. A study completed by this corporation concluded that the cost of this eucalyptus methanol would be well below the Brazilian cost of liquid fuel refined from imported crude oil and called for the construction of sixty-five methanol plants at a total capital cost of $15 billion. The study concluded that these Brazilian plants, having the advantage of greatly reduced labor costs in comparison to U.S. plants, could produce methanol at a cost of 37.8 cents per gallon compared to ethanol production costs of $1.09 per gallon. The plants would harvest eucalyptus from a 20,000-square-mile area of marginal Brazilian lands unsuitable for traditional agriculture.[13]

In Japan, Saurzo Takeda, a researcher at Mie University, has looked into the intriguing possibility of refining an oil similar to petroleum fuel from the leaves of the eucalyptus. Takeda's research has been supported by the Sekisu Company, a Japanese firm which has planted a small eucalyptus plantation which it hopes will produce up to 330 gallons of eucalyptus leaf fuel per year at a reported cost of $1.75 per gallon.[14]

In the United States, the railroads were among the first to realize the energy potential of the eucalyptus tree. During the nineteenth century, several U.S. railroads imported the tree from Australia and planted them in large stands near train depots. When the steam-driven locomotives of that era stopped to refuel and let passengers off, their boilers were filled with eucalyptus wood which would then power the trains on to their next stop.

The most ambitious modern effort to develop the energy potential of eucalyptus in the United States has been undertaken by C. Brewer and Company, a Hawaiian corporation founded in 1826 by a New England sailing captain named James Hunnewell. For the present, C. Brewer and Company has shown more interest in using eucalyptus wood as a boiler fuel to supplement the sugarcane stalks that now power its electrical generating plants on the island of Hawaii. These trees had long grown in dense forest stands on the volcanic slopes of the island, but they had never been put under intensive cultivation in carefully managed energy plantations. In these plantations, spacing between trees and rows would be carefully controlled and petrochemical fertilizers and herbicides would be used to increase growth rates.

The first plantation was started by sprouting some 100,000 seedlings from seed in two greenhouses equipped with sophisticated trickle-down irrigation systems to optimize their germination. Then in March of 1979 they were taken to their planting site, a 41-acre tract of land on the upper slopes of an abandoned sugarcane field. Seven months later some of these seedlings had shot up to surprising heights of over 17 feet, and Tommy Crabb, the project's director, spoke confidently of making the plantation a commercial success. When asked about the economics of the venture, Crabb, casting an admiring glance at one of the tallest trees, replied, "It's looking better every day."[15]

WOOD ENERGY CROPPING AND SOIL FERTILITY

Although many forestry experts are excited about the prospects of moving into the new frontiers of wood energy farming, others are skeptical of the impact which the intensified cultivation practices will have on soil fertility. Conventional silviculture to raise tree crops for plywood, pulp, and paper have traditionally planted seedlings at spacing densities ranging from 600 to 2,000 stems per acre and allowed for growing periods of thirty to eighty years or more. The energy cropping concept calls for planting from 12,000 to 77,000 stems per acre and harvesting them on four- to twelve-year rotations. Moreover, many of these harvesting techniques would remove all of the above-ground biomass from the plantation site, leaving behind almost no residues to decompose and recycle nutrients to the next crop. This type of all-inclusive harvesting differs radically from conventional timber industry practices, which leave behind between 9 and 24 dry tons of wood residues per acre.[16]

The Mitre report proposed resolving the problem of soil nutrient depletion in much the same way the midwestern corn farmer would—with intensive use of petrochemical fertilizers. However, other foresters, wary of the rising costs and decreasing efficacy of many petrochemicals in boosting grain production, favor alternative approaches such as the introduction of nitrogen-fixing trees such as alder and mesquite into poplar and eucalyptus plantations in order to bolster soil fertility.[17]

Earle Barnhart, a researcher at the New Alchemy Institute in East Falmouth, Massachusetts, writes that ". . . monocultures are kept ecologically simple by frequent applications of pesticides, fungicides, and herbicides. Recent ecological theory of pest management . . . indicates that epidemics of pests and diseases can be avoided by including many different plant species and a number of genetic varieties of each species."[18] Barnhart proposes that the tree plantations of the future produce a multitude of both food and fuel crops, thus maintaining a greater degree of ecological stability than could be achieved by planting vast monocultures of a single energy crop. This development strategy would reduce the risk of a disastrous blight or insect infestation that could wipe out an entire plantation and bankrupt its owners.

Barnhart's strategy for developing tree plantations may be particularly useful to Third World nations, whose citizens are often in desperate need of both food and fuel. In the impoverished island nation of Haiti, for example, remaining forests lands are being destroyed at a rapid rate as peasant farmers scavenge for miles in search of a few saplings to stoke cooking fires. Elsewhere in the Third World, traditional slash-burn agricultural systems that allowed marginal lands to lie fallow and return to forest over ten-year periods before replanting in food crops are breaking down under growing population pressures. As the fallow periods are reduced in an effort to grow more food in shorter periods of time, the fragile ecological balance which allowed the land to regain its natural fertility before replanting is destroyed. In such desperate situations, any efforts to develop wood plantations must focus on reforestation of denuded or damaged lands with both food and fuel crops rather than on any kind of monoculture planting.

In the United States, remaining forest lands are subject to somewhat less intensive pressures than those in the Third World. Taken as a whole, forest lands are not being rapidly decimated by logging and encroaching urbanization but rather are slowly increasing in size. The history of U.S. forests was summarized by Marion Clawson in *Science* magazine: "When the first European settlers arrived in the eastern United States, some 950 million acres of forests covered the nation. By 1920, after large sections of land had been heavily logged and cleared for farms, only 615 million acres of forest remained." However, in recent years, Clawson noted, abandoned farms have returned to timber and large areas of forest land have fallen under federal protection, with the result that forest acreage in the United States has actually been increasing at an annual rate of some 2.2 billion cubic feet per year.[19]

Nearly half of the most promising lands for wood energy production identified in the Mitre report are located in once-logged areas of the southern United States, with most of the remaining sites in the Great Lakes states, Appalachia, the Northeast, and the Pacific region. The Mitre report found the Great Lakes states to be "particularly attractive" since their severe winter climate which limits food production potential, would have little impact on wood energy farming.

Some foresters see a more intensive management of existing forest lands as a logical first step toward development of wood energy farms in the United States. This intensive management need not translate into a wholesale logging off of vast tracts of forest lands; rather it would mean a selective thinning of dead and diseased wood designed to increase the long-term productivity of these timber stands. Much of the potential for such stand improvement lies in the northeastern and southern states, whose forests have largely been stripped of high-quality, first-growth timber. In these forests, a gradual harvesting of low-value timber may coincide with a gradual replanting of more valuable species both for energy production and, when possible, for commercial use as timber.

Another intensive management strategy calls for increased harvesting of the wood residues left over after the commercial logging of prime timber sites. This strategy is somewhat controversial, however, since it carries with it the same dangers presented by harvesting farm crop residues, mainly that harvesting the wood residues deprives the soil of valuable nutrients and can contribute to soil erosion in logged-off areas. Some foresters argue that the greatest value for most of these residues lies in their ability to enhance the growth of the next generation of trees through on-site decomposition rather than removal from the land for use as an alcohol feedstock.

A third strategy for harvesting alcohol feedstocks from commercial forests involves a partial use of the 74 million dry tons a year of bark, sawdust, and wood chips generated by milling and paper pulp operations.[20] These wastes have the potential of producing as much as 8 billion gallons of ethanol a year if they are all diverted to distilleries. However, in reality, most of these wastes are already finding markets as replacements for fuel oil and natural gas in industrial and residential heating systems, commanding a market price which many ethanol distilleries would be hard pressed to match.

AQUACULTURE—A NEW FRONTIER

In recent years a major new branch of research and development in alcohol fuels technology has emerged as some scientists have turned away from land-based crops to explore the energy potential of aquatic crops growing in wetlands, lakes, bays, and oceans. They have found that many of these plants demonstrate a surprising capacity to produce sugars, starches, and cellulose during their seasonal growth cycles, with one federal supply concluding that they are often "capable of higher rates of photosynthesis" than land plants.[21]

Experiments with aquatic farming first began during the 1950s at a time when scientists working at the Carnegie Institution sought to cultivate microalgae as a source of cheap protein. Progress to date toward this goal has been slow and painstaking, with certain strains of algae only recently beginning to appear as a specialty item in some health food stores. However, as scientists studied the growth cycles of algae, they discovered that many strains showed considerable promise in helping to purify waste waters in sewage lagoons. The cultivation of algae helps put oxygen into the water, which promotes the decomposition of wastes and the removal of excess nitrogen and phosphorous. A seven-year research and development program carried out in Israel indicated that some 30 dry tons of algae could be harvested from a single acre of land covered by waste water. After the waste water is purified by the algae, the Israelis found that it is usually clean enough for the irrigation of agriculture crops or for recharging depleted underground water supplies.[22]

Gedaliah Shelef, a researcher at the Israeli Institute of Technology, has developed a sophisticated dewatering system to harvest the algae which involves a series of centrifuge processes to dry the tiny plants to the point where they can be used as cattle and fish feed. The algae strains which Shelef has worked with are approximately 50 percent protein with lesser amounts of fermentable starches and sugars. In the future, with the aid of genetic engineering, it may be possible to create new strains of high-sugar algae specifically for ethanol production.

Researchers believe that algae may eventually be able to produce more food and fuel per acre than most land-based crops. Already one private firm has received limited financial backing to construct a network of canals in Florida to produce algae for methanol production. However, the most economical algae-to-ethanol processing is likely to occur in conjunction with already existing projects which utilize algae as part of integrated waste treatment systems. Costs for algae production are still much higher than for most land-based crops.

A promising aquatic energy crop that has been extensively investigated by the National Aeronautics and Space Administration (NASA) for possible use in future space stations is the water hyacinth, which, like the common water lily, grows on top of the water and rapidly expands into open areas in ponds and lakes. NASA scientists have identified the water hyacinth as "the primary candidate for aquatic energy farming" in the subtropical United States.[23]

Water hyacinth thrives in some 2.2 million acres of southern waterways and in some areas must be continually cleared from vital shipping channels and reservoirs. In many tropical Third World nations, the water hyacinth thrives in small ponds and waterways that have been polluted with farm and urban wastes.

NASA had funded the development of four waste water treatment systems in southern Mississippi, utilizing a combination of water hyacinths and two other aquatic plants to achieve what one report labels "a high degree of waste treatment and water reclamation."[24] The water hyacinth purifies water in essentially the same fashion as algae, drawing up excess nutrients during its growth cycle and adding oxygen to aid in the decomposition process of organic materials.

To date, most of the research into the energy processing of water hyacinths has involved their anaerobic conversion to methane gas. Its high water content and relatively low percentage of fermentable sugars and starches make it a rather unattractive feedstock for ethanol production. However, the water hyacinth may find a vital niche in integrated ethanol production systems: if cultivated in the stillage lagoons of distilleries, it would perform the dual service of purifying the slops and providing biomass material to generate methane gas for use as a boiler fuel. In such a system, the water hyacinths would help to cut the costs of pollution control and process energy.

A third aquatic plant considered to be a prime candidate for energy production is the cattail, which thrives in marshlands and lakes throughout North America. Cattails were once highly prized by native American Indians, who cooked the reed's thick brown head of grain and wove its stalks into bedding and baskets. Douglas Pratt, of the University of Minnesota's Botany Department, launched an extensive study of the cattail's energy potential after he discovered the plant to be one of nature's most efficient converters of solar energy to carbohydrate energy. In some wetland areas it is not uncommon for cattails to produce over 40 tons per hectare of biomass growth a year. Pratt believes that, even in its natural state, the cattail has the potential to produce over a 1,000 gallons of alcohol per acre, with yield improvements possible through genetic engineering.

The question of turning to the nation's delicate wetland areas, which shelter an abundance of wildlife, to provide energy is a controversial one which rightfully provokes a strong reaction from environmentalists. Pratt is not insensitive to the environmental devastation which might result from a wholesale conversion of marshlands into energy farms, and he does not advocate such a drastic approach. Rather, he proposes a selective cutting of some cattail stands, which could actually improve certain marshland animal habitats. In Agassiz National Wildlife Refuge in Minnesota, for example, huge stands of cattails are periodically mowed down by the U.S. Interior Department in order to maintain the ratio of half-open to half-reedy areas that federal studies indicate waterfowl prefer for breeding areas.

Outside the marshlands, there are large areas of peat-based soils which receive enough rain for the cattail to grow in. These soils can be harvested to form a low-grade fuel which can either be directly combusted for heat or sent through a gasification process. As energy prices have risen during the past decade, there has been increasing corporate interest in the large-scale mining of major peat deposits in Minnesota. Pratt has cautioned against such mining efforts, arguing that

> It may be more desirable . . . to use peat as a substrate for the production of a renewable energy crop rather than as a consumable commodity which, if heavily exploited, would not last long. If a decision is made to harvest some of Minnesota's peat resources as energy, cattail or other wetland plants might serve as tools for the reclamation of lands from which peat has been removed.[25]

The greatest, yet in many respects the most elusive, potential for aquatic energy farming lies in the oceans, where floating kelp farms could one day produce both protein and vast quantities of methane gas which could possibly be synthesized into methanol and shipped to shore. The most ambitious effort to develop kelp-based fuels has been sponsored by General Electric with the aid of major grants from the U.S. Department of Energy and the Gas Research Institute. The project began in 1973, when plans for the first experimental kelp farm were developed for the California coastal waters.

The kelp variety chosen for the energy farm has been known to grow as much as 2 feet per day in the open oceans, but scientists are uncertain how it will perform in the more controlled environment of the artificially cultivated floating farms. A key component of any such farm would be a massive pumping system to bring nutrients contained in the colder waters of the ocean depths up to the surface where the kelp grows. This system might be combined with another promising aquatic energy system called Ocean Thermal Energy Conversion, or OTEC for short. The OTEC system makes use of the temperature difference between warm surface waters and colder deep waters to generate electricity. In an integrated OTEC-kelp farm system, the same pumping system could carry out the dual role of nutrient enrichment of the warm upper waters where the kelp grows and the condensation of vapors in the electrical turbine.

The first small ocean energy farm was established in the coastal waters off Corona del Mar, California, consisting of a large floating structure which enclosed 103 kelp plants. Initial research efforts have been severely hampered by numerous equipment failures caused by rough weather during winter storms. Nevertheless, preliminary cost estimates compiled by General Electric indicate that gas produced from kelp farms can be competitive with most other biologically produced methane gases, although it is still considerably more expensive than natural gas captured from oil wells. If current project development scenarios prove accurate, General Electric will have its first commercial sized kelp farm in operation by 1992. This farm is being designed to occupy a 100-square-mile area of ocean and yield from 25 to 70 dry tons of biomass material per acre each year.[26]

Some 1.5 billion acres of bay and ocean are within the 200-mile coastal zone of U.S. waters, offering a huge area for possible development of aquatic farming. To develop about 4 percent of this area, or some 100 million acres of ocean kelp farms, would be a massive undertaking which might cause extensive conflicts with commercial shippers, fishermen, and other major users of the coastal waters. If 100 million acres of kelp farms were consolidated into a single, vast floating mass, they would occupy a square 335 miles long on each of its four sides. One federal report by the Congressional Office of Technology concluded: "The dedication of large areas of open ocean surface for a single commercial purpose such as this is unprecedented. It would require complex special regulations after review of current local, national and international laws."[27]

HOW MUCH ALCOHOL—HOW SOON

Piecing together estimates on the total amount of alcohol that may be available in the future from renewable resources has proved difficult for resource planners and federal energy officials, who usually hedge any such predictions with numerous if 's, and's, and but's. The single most important factor influencing the development of the alcohol fuels industry will be the price and availability of oil. For example, a decision by the Saudi Arabian government to drastically cut current oil production levels would soon send OPEC prices on another upward spiral. If these production cuts resulted in a sustained period of oil shortage, they would trigger a rapid shift towards renewable alcohol fuels. Moderate estimates of alcohol's land-based potential indicate that some 47 million acres of marginal lands might eventually be brought under cultivation if a concerted effort was made to develop energy farms. The annual ethanol potential of these lands could range from 81 gallons an acre for land planted in honey locust pods, to an average of 800 gallons an acre for land planted in fast-growing wood crops. Thus, ethanol production levels could vary from a low of 3.8 billion to a high of over 37.6 billion gallons a year, depending on the mix of energy crops that evolves. Methanol production from these same lands, assuming an average yield of 960 gallons per acre per year, could total over 55.1 billion gallons of fuel annually.[28]

Perhaps the most optimistic estimate of alcohol's future potential comes, not surprisingly, from General Electric's Dr. Kendall Pye and is based largely on his research in converting poplar hybrids into ethanol. Pye predicts that it will ultimately prove feasible to put at least 70 million acres of marginal lands into wood crops (an area roughly twice the size of Pennsylvania), and that this land mass will produce an average yield of up to 1,000 gallons per acre. Thus, Pye believes, it will one day be possible to produce from 70 to 100 billion gallons of ethanol a year. This would provide the nation with enough liquid fuel to replace roughly from 60 to 90 percent of current U.S. gasoline consumption.[29]

In the absence of another oil shortage, wood crops, tree crops, and aquatic energy crops are likely to be developed at a relatively moderate pace. Some technologies, such as the poplar-to-ethanol process developed by Kendall Pye, may show substantial short-term growth if the end product is price competitive with gasoline. As long as surpluses depress grain markets, corn and sorghum will continue to play a role in the midwestern gasohol. And as production technologies improve, more and more municipal garbage, food processing wastes, cheese whey, and below-grade cull crops will be funneled into ethanol production. These technologies need not await the arrival of $2.00-a-gallon gasoline; they can find major industrial markets for their end products as octane boosters in premium unleaded gasolines. Thus, even if oil prices remain relatively stable throughout this decade, a modest expansion of the alcohol fuels industry, to the point where it replaces 5 to 10 percent of current U.S. gasoline consumption, appears possible. During the 1990s as the potential of methanol begins to be realized, and new technological advances in ethanol production are more fully refined, alcohol fuels may begin to undercut the cost of gasoline at the service station pumps, increasing the demand considerably.

A second major factor affecting the future of alcohol fuels lies in the manner in which the federal government implements energy policy. If the laissez-faire economics of the Reagan administration predominates during the next decade, the nation may once again begin to import increasing amounts of foreign oil to satisfy its liquid fuel needs. This path of action (or inaction) carries with it the everpresent risk of a supply cut-off which could lead to armed conflict or—even in the absence of an abrupt cut-off—an acceleration of the already massive transfer of wealth from U.S. consumers to oil-producing nations.

Increasing domestic liquid fuel supplies involves a choice between the centralized technologies required for synthetic gasoline and methanol production from fossil fuels and the more decentralized technologies available for producing renewable alcohol fuels. If a decision is made to move ahead with the exclusive development of synthetic fuels, the nation risks not only the immediate environmental problems of air pollution, depletion of western water supplies, and loss of land to coal mining, but also the uncertain long-term risks associated with a global climatic change triggered by the greenhouse effect.

A more promising path to follow for the future involves a limited development of a broad base of renewable feedstocks for alcohol fuels which can be combined with and complement other efforts to develop biomass, solar, co-generation, and conservation energy resources. These feedstocks can be used to produce methanol and ethanol fuels which will both be able to find niches in the transportation, energy, and petrochemical industries of the future.

However, even if synthetic liquid fuels from coal and oil shale take a back seat to renewable alcohol fuels development, the road ahead will not be without dangers to the environment, as was summed up by Dr. Thomas Reed in an address to his colleagues at the International Solar Energy Society:

> Man has demonstrated through the centuries that the overharvesting of biomass for food, wood, and energy has the potential for destruction of large areas of the earth's lands. If now we are to turn even more to biomass as a significant source of energy, we must realize the implicit potential for the destruction of the land base.[30]

It is this vast but ultimately fragile land base which first inspired the chemurgists to dream of an alcohol-powered America at a time when gasoline sold for less than a dime a gallon. Today, in an era when motorists retain only fond memories of a $1.00 a gallon fuel, the alcohol fuels industry, which William Hale fought to establish, is a reality. The major oil companies, which so often maligned the Agrol blends of the chemurgists, routinely market gasohol to urban motorists. Hundreds of farm-scale distilleries now form the skeletal backbone of an alternative fuel network in rural America which is entirely independent of the major oil companies. The integrated food-processing greenhouse-distillery complexes built by Archer Daniels Midlands rival any of the most sophisticated systems which Leo Christensen may have contemplated when establishing the first Agrol distillery in Atchison, Kansas. After a five-decade hiatus, the automotive industry is once again marketing vehicles specifically designed for alcohol fuels. The OPEC oil price hikes of the past decade have finally lifted the long prohibition on Henry Ford's farm fuel.

Exhibit 10-1
ESTIMATES OF ETHANOL-FUEL PRODUCTION IN THE UNITED STATES AND IMPORTS (In Millions of Gallons)

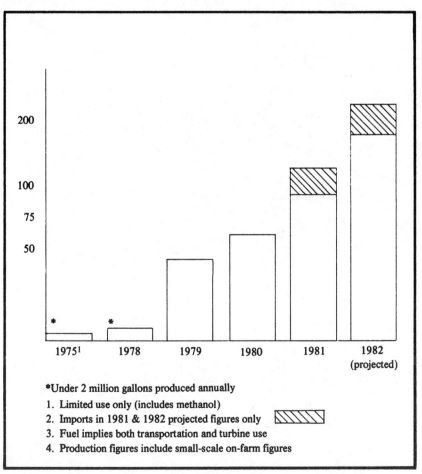

*Under 2 million gallons produced annually
1. Limited use only (includes methanol)
2. Imports in 1981 & 1982 projected figures only
3. Fuel implies both transportation and turbine use
4. Production figures include small-scale on-farm figures

Appendix A
Chemistry And
Production Processes
of Alcohol

WRITTEN BY CLIFF BRADLEY AND KEN RUNNION

EDITED BY STEVE MARLENS

ILLUSTRATED BY HANS HAUMBERGER AND KRIS ELLINGSEN

The material in this appendix first appeared in *Fuel Alcohol: Answers To Common Questions*, published by the National Center for Appropriate Technology in Butte, Montana. It is reprinted here with permission. The National Center for Appropriate Technology (NCAT) is a nonprofit organization established to develop and implement small-scale technologies that promote self-reliance and help alleviate problems of low-income Americans. Principal funding for NCAT is provided by the U.S. Community Services Administration.

I. WHAT DOES "ALCOHOL FUEL" MEAN?

The term "alcohol fuel" generally refers to any or all of three kinds of alcohol which can be blended with or substituted for gasoline or diesel fuel. These three are ethanol, methanol, and butanol. Each fuel alcohol has a different chemical composition and is produced in a different way from the others (see table A-1 and exhibit A-1.)

II. HOW IS ETHANOL MADE AND WHAT DOES IT TAKE TO PRODUCE IT?

Ethanol is made in a multistep process (see exhibit A-2). It is extremely important that anyone considering small-scale ethanol production understand the process, since each step must work well for an ethanol plant to be economical.

A. FEEDSTOCK

Ethanol can be produced from any sugar or starch feedstock (a few are listed in Table 1). However, not all feedstocks are economical. Whether a feedstock is economical depends on: 1) its initial cost, 2) how much sugar it yields for conversion to ethanol, and 3) whether its byproducts can be used or sold.

The feedstock is usually the most expensive ingredient in ethanol production. In general, agricultural crops are economical sources of fuel ethanol only when the high-protein byproduct of production is used for animal feed or when the crop has a very low

TABLE A-1
A COMPARISON OF THE THREE COMMON ALCOHOL FUELS

CHEMICAL	METHYL ALCOHOL	ETHYL ALCOHOL	BUTYL ALCOHOL
COMMON NAMES	METHANOL WOOD ALCOHOL	ETHANOL GRAIN ALCOHOL	BUTANOL
CHEMICAL FORMULA	CH_3OH	CH_3CH_2OH	$CH_3CH_2CH_2CH_2OH$
FEEDSTOCKS USED IN PRODUCTION	NATURAL GAS (methane) wood, coal, garbage	STARCH (corn, barley, potatoes, and other grains) SUGAR (sugar beets, sugarcane, etc.)	STARCH (corn, barley, potatoes, and other grains) SUGAR (sugar beets, sugarcane, etc.)
PRODUCTION PROCESS	CHEMICAL CONVERSION AT HIGH TEMPERATURES AND PRESSURE	FERMENTATION USING YEAST	FERMENTATION USING BACTERIA

EXHIBIT A–2
CHEMICAL STRUCTURE OF ETHANOL, METHANOL, AND BUTANOL

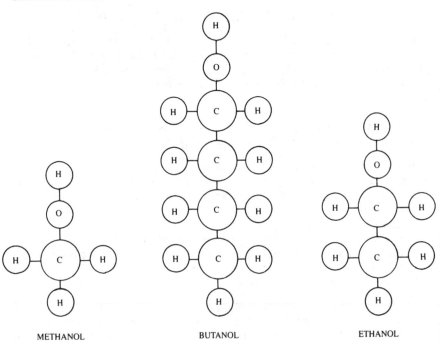

METHANOL   BUTANOL   ETHANOL

value—as with cull potatoes or distressed grain. Starch- or sugar-containing wastes such as cheese whey or bakery waste often make economical feedstocks because they are inexpensive and require minimal preparation. Factors other than cost which affect choice of feedstock are availability (seasonal, year-round) and the amount of space or type of structure needed to store the feedstock.

B. PREPARING THE FEEDSTOCK

In order to make the starch or sugar available for conversion, high-moisture feedstocks such as sugar beets and sugarcane are washed and crushed, and low-moisture feedstocks such as grain are ground. Water is added to the low-moisture feedstocks to form a "mash" solution; high-moisture feedstocks often can be used without additional water. Many farms already have the hammermill, conveyors, augers, and storage facility necessary for feedstock preparation. If this equipment must be purchased, plant costs increase significantly.

C. CONVERTING THE STARCH TO SUGAR (Starch feedstocks only)

Yeast cannot ferment starch: starch must be converted to sugar before fermentation can occur. This starch-to-sugar conversion is called "cooking." Cooking requires a tank equipped for heating, cooling, and stirring. The prepared starch feedstock is augered or poured into this tank and heated while enzymes are added. The conversion process must be well monitored, since these enzymes only work under specific, controlled conditions.

Cooking is time-consuming—it generally takes three to four hours for each tankful of starch solution. The process must also be efficient. If starch remains after cooking, the operator is wasting costly feedstock and adding to his expense. Other expenses of starch conversion include the equipment, the enzymes, and the energy (heating fuel) used to cook the solution.

D. FERMENTATION

During fermentation, the sugar in the mash is converted to ethanol by yeast. Yeast are one-celled organisms, which, under the right conditions, produce ethanol and carbon dioxide from sugar. (Most of us are familiar with yeast used in baking and in alcoholic beverage production.) Successful fermentation requires a good understanding of the biological processes taking place in the conversion of sugar to ethanol. Poor fermentation, the biggest problem in small-scale ethanol production, is often caused by bacterial contamination or slow yeast growth.

Fermentation requires one or more tanks sized to the overall system. These tanks must be equipped with stirring and cooling systems, and must be designed for ease of cleaning. The major cost of fermentation is the tank and associated equipment.

E. DISTILLATION

Fermentation produces a mixture called "beer" of about 10 percent ethanol and 90 percent water. Distillation separates the ethanol from most of the water to make the ethanol fuel-grade—generally about 80 to 95 percent pure. This is accomplished in a distillation column (exhibit A-3).

Although the distillation column and associated equipment are expensive, the major ongoing cost of distillation is the energy (fuel) used to boil the ethanol/water mixture in the distillation column.

Exhibit A–2
FUEL ALCOHOL PRODUCTION PROCESS

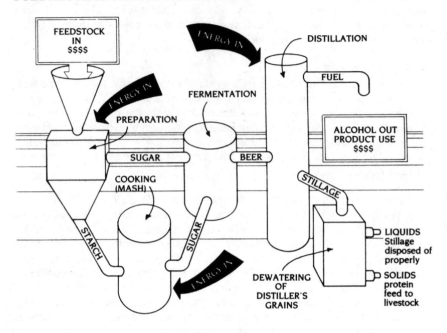

Exhibit A–3
EQUIPMENT OVERVIEW OF A FUEL PRODUCTION PLANT

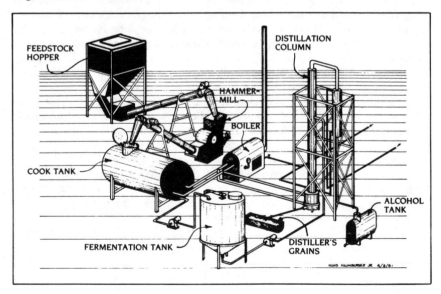

Appendix B
Economics of Ethanol

CHRIS HURT, WALLY TYNER, AND OTTO DOERING

This article, published in 2006 by Purdue Extension Center and reprinted here with permission, replaces the original appendix B, "Some Economic Issues in Ethanol Production." It provides a more up-to-date analysis of current ethanol economics; however, its assumptions about corn prices and ethanol profitability proved to be short-lived as evidenced by market conditions in 2008–9.

Farming for fuel is a relatively new concept for U.S. agriculture. Biofuels include both ethanol (corn) and biodiesel (soybean oil), but ethanol is far in the lead. Production capacity across the country is expected to exceed 8.0 billion gallons by early 2008 and substitute for approximately 5 percent of U.S. gasoline consumption. Some hope that biofuel production can eventually substitute for as much as 25 percent of the country's gasoline over the next twenty to thirty years. The ultimate importance of biofuels will be determined by events that are still to unfold.

The drivers are expected to be found in energy prices, state and federal energy policy (Doering 2006), and technology, particularly the improvement of the process to produce ethanol from cellulose (plant material) (Mosier 2006). Why is there such startling interest in fuels from farms? The nearly "gold rush" status is driven by powerful profitability, especially for ethanol. The federal subsidy of fifty-one cents per gallon of ethanol was established when crude oil was less than thirty dollars per barrel. At that price of crude oil, the subsidy was necessary to make ethanol profitable. However, with the price of crude oil much higher, ethanol has shifted from being just profitable to being highly profitable, and thus major investment in the sector has been stimulated. The value of ethanol can be thought of as coming from three components: (1) the energy value as a replacement for gasoline; (2) the value of subsidies and policy incentives provided to ethanol; and (3) the value of ethanol as an additive that is primarily an oxygenate (to produce cleaner-burning fuel) and octane enhancer for gasoline.

ENERGY VALUE

The energy value in a gallon of ethanol is less than in a gallon of gasoline. While exact difference in gas mileage will probably vary somewhat, it is expected that a gallon of ethanol will only do about 70 percent of the work of a gallon of gasoline. Therefore, we would expect the energy value of ethanol to be about 70 percent of the wholesale price of gasoline.

SUBSIDY-POLICY INCENTIVE VALUE

Federal government policy is to stimulate ethanol production and thus provides a $0.51 per gallon subsidy to blenders of ethanol. This $0.51 per gallon is about $1.35 per bushel of corn used. There are other federal ethanol subsidies primarily targeted at initial production years and smaller plants. The national energy bill passed in the summer of 2005 mandated the use of at least 7.5 billion gallons of biofuels by 2012, a level that will be exceeded in 2007.

Some states also have a state subsidy for ethanol production, and still other states provide financial incentives to ethanol producers such as support for infrastructure development and job training assistance. Finally, more states are passing their own state renewable fuels standards. Minnesota, for example, mandates all gasoline sold in the state must be at least 20 percent renewable.

ADDITIVE VALUE

Ethanol tends to trade at a premium price even above its value of energy and the subsidies. Twenty-five states have either restricted or outlawed the use of MTBE (methyl-tertiary-butyl-ether) as a gasoline oxygenate because it is highly toxic and has been found in groundwater. The 2005 federal energy legislation ended the federal requirement for specific oxygen levels in gasoline. Oil companies are now free to meet the clean air requirements in any way they choose. Thus in May 2006, when the oxygen requirements ended, oil companies were no longer required by the government to add a certain level of oxygen, and most companies feared legal liability if they continued to use MTBE. For most blenders the best way to meet the emissions standards in the Clean Air Act is now to use ethanol to blend with their

Exhibit B-1

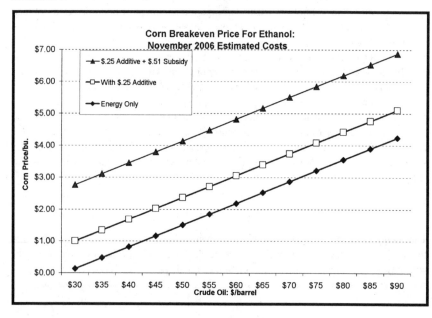

gasoline. The largest part of this premium is related to the value of ethanol to replace MTBE as an oxygenate. Also, ethanol has an octane of 106 compared to 87 for gasoline, so it has value to enhance octane. Beyond these technical values, some drivers will pay premiums to use ethanol blends over straight gasoline. There is also a strong national interest in reducing the dependence on foreign oil, which helps enhance ethanol demand as well.

ECONOMIC BOTTOM LINE

Figure 1 illustrates the economics of ethanol, depicting the relationship of crude oil prices and the estimated breakeven price per bushel that an ethanol plant could pay for corn. Breakeven corn prices still allow the plant to be paid off in fifteen years and for equity investors to receive 12 percent per year return on their investment dollars. Construction and operating costs similar to November 2006 are assumed. The three lines relate to the three component values for ethanol. (The appendix explains the complete set of assumptions behind the relationships in figure 1.)

The bottom line in figure 1 represents the value of the energy in ethanol based on 70 percent of the value of gasoline. As an example, with $60 per barrel crude oil an ethanol plant could pay $2.19 per bushel for corn. The middle line represents the corn breakeven price when the value of the $0.51 per gallon federal subsidy is added, and at $60 oil this is $3.96 per bushel. Finally, when an oxygenate premium of $0.25 per gallon is added, this raises the estimated breakeven price an ethanol plant could pay to $4.82 per bushel. During some periods, the oxygenate premium has been considerably higher than the $0.25 per gallon assumed here.

Given these assumptions, if a plant can buy corn at less than $4.82 per bushel, the owners will get a higher return than 12 percent and/or a quicker payback than fifteen years. We should note also that the capital cost component of ethanol production cost is about $0.30 per gallon, or $0.80 per bushel. This means that existing plants with capital costs already recovered could potentially pay $0.80 more per bushel, or about $5.60.

This summarizes some of the great opportunities in ethanol but also highlights some of the extreme vulnerabilities. One vulnerability is the oxygenate premium. As the supply of ethanol increases to meet the amount needed to replace MTBE, the oxygenate premium could drop sharply. We have not experienced the situation in which ethanol production exceeds oxygenate demand, so there is considerable uncertainty regarding ethanol market value once we reach that threshold. Without the oxygenate premium, the ethanol industry will be operating on the middle line in figure 1. One can see that lower crude oil prices could make ethanol profits vulnerable as well. The corn breakeven on the middle line with $50 oil as an example is a bit over $3 per bushel. The high demand to build ethanol plants is bidding up construction and processing costs, which also make ethanol profits vulnerable. Another major vulnerability is that as more ethanol capacity comes on line, the increasing demand for corn results in higher corn prices, thus narrowing ethanol producers' margins. Finally, the federal subsidy is very large and could be subject to change, as higher corn prices have adverse impacts on livestock producers and ultimately on livestock product consumers (Tyner and Quear 2006).

CONCLUSION

The future direction of ethanol will be highly dependent on state and federal governmental policy and on energy and corn markets. If all factors were to stay as they are today, the exponential expansion of ethanol plants would continue until corn prices were bid up to near their breakeven level. It is much more likely, however, that policy and energy prices will also be dynamic, that corn prices will rise, and that other constraints will dramatically slow the growth of the industry after 2007. It is clear that the ethanol industry cannot continue to grow at the current rate based on the use of the corn seed as a feedstock source without hitting major constraints, including extreme competition for corn to be used for feed, exports, and food. After 2007 the industry will have to grow at a much slower rate, probably keeping pace with corn production increases. The hope is that cellulose-based ethanol can then emerge by 2010 to 2012. However, as long as corn based ethanol is profitable, investors will probably prefer the more certain technology to the still uncertain cellulose technology.

REFERENCES

Doering, O. 2006. "Ethanol and Energy Policy." ID-340. Purdue University Cooperative Extension Service. http://www.ces.purdue.edu/bioenergy.

Mosier, N. 2006. "Cellulosic Ethanol—Biofuel Beyond Corn." ID-335. Purdue University Cooperative Extension Service. http://www.ces.purdue.edu/bioenergy.

Tiffany, D., and V. Eidman. 2003. "Factors Associated with Success of Fuel Ethanol Producers." Staff paper series, P03-7, Department of Applied Economics, University of Minnesota.

Tyner, W. E., and J. Quear. 2006. "Comparison of a Fixed and Variable Corn Ethanol Subsidy." *Choices* 21, no. 3: 199–202 of PDF. http://www.choicesmagazine.org/2006-3/grabbag/index.htm.

ARTICLE APPENDIX

The link between crude oil price and breakeven corn price requires numerous assumptions. Following are the most important assumptions updated to November 2006:

1) Relationship between crude oil price and gasoline price. This relationship is given by the equation below:

Wholesale gasoline price ($/gal.) = 0.3064 + 0.03038 * crude oil price ($/bbl.)

The data for this equation were monthly data 2000–6 from EIA/DOE. However, longer and shorter time periods were tested, and the results are remarkably stable. The adjusted R2 for the equation is 0.93, meaning that 93 percent of the variability in gasoline price over time is explained by changes in the crude oil price.

2) Relationship between gasoline price and ethanol price. The energy equivalent price of ethanol is assumed to be 70 percent of the gasoline price. That is slightly higher than the pure energy equivalence.

3) Relationship between corn price and DDGS price—DDGS price is a function of the prices of corn and soybean meal as follows:

DDGS price ($/ton) = 1.52 + 0.205 * soybean meal price ($/ton) + 21.98 * corn price ($/bu.)

Substituting a price for soybean meal of $200/ton into this equation yields the equation used in the model:

DDGS price ($/ton) = 42.52 + 21.98 * corn price ($/bu.)

All data are from USDA, monthly 2003–6. Illinois prices were used for corn and soybean meal, and Lawrenceburg, IN, for DDGS. It is assumed that 18 pounds of DDGS is produced per bushel of corn used.

4) Ethanol yield per bushel of corn is assumed to be 2.65 gallons. Newer plants may have higher yield, but this figure is close to the industry average.

5) Capital cost for the plant is assumed to be $1.80 per gallon of capacity. Older plants had considerably lower capital cost, and much of the capital probably has already been paid off. The plant is assumed to operate at full capacity.

6) Financial assumptions: The plant is 40 percent equity and 60 percent debt finance. The debt interest rate is 8 percent, and the equity return is 12 percent.

7) No value was assigned to the CO_2 produced.

8) Energy costs:

Natural gas	$9/mil. BTU
LP	$1.20/gal.
Electricity	$0.06/KWH
Total energy	$0.383/gal. of ethanol

9) Other costs: Chemical and enzyme costs $0.182/gal. of ethanol. Other processing costs $0.297/gal. of ethanol. Given these assumed relationships and values, the Tiffany and Eidman (2003, University of Minnesota) spreadsheet model of a dry-mill ethanol plant was used to calculate profitability and thus derive the breakeven prices. Breakeven was assumed to be the point of zero economic profit; that is, it includes the payment of debt and stipulated return on equity. Clearly, any of these assumptions and values could be modified in the future as conditions change.

Appendix C
An International Historical Survey of Alcohol Fuel Programs: 1910-1960

Practically every industrial nation has had to substitute part of its oil supply at one time or another. The alcohol fuel option has been well known and practical, if somewhat expensive. Many nations made fuel marketers blend fixed amounts of alcohol into gasoline, while others encouraged power alcohol markets with tax breaks.

European power alcohol programs peaked in 1936, and began slowing down with crop failures in 1937. World War II disrupted most programs, except in unaffected nations such as Switzerland and Sweden.

Meanwhile, the war's effect on oil shipments necessitated alcohol fuel programs in China, Latin America, Africa and India.

The following country-by-country description of alcohol fuel programs from 1910 to 1960 was developed by the authors for the International Institute for Environment and Development as part of a briefing paper for a United Nations conference on renewable energy.

ANGOLA–A decree from Lisbon, Portugal made two types of alcohol-gasoline blending mandatory in 1933. One was "Alcoolaina" with 25% alcohol, 75% gasoline and a blending agent. Another was "Gasalcool," a 75% alcohol/25% gasoline blend. Imported gasoline could not clear customs until oil companies bought proportionate amounts of alcohol.

ARGENTINA–An extended survey of grain and gasoline markets resulted in a 1931 report recommending a 30% blend of alcohol, since it could be produced at less cost than importing gasoline. Some marketing was carried out but no record of an official decision is available.

AUSTRALIA–A two million gallon per year ethanol plant was built in 1926 that used primarily molasses, along with cassava (manioc) and sweet potatos grown in rotation with sugarcane. "Shellkol" fuel was marketed by Shell Oil Co. at 15%-35% concentrations. Power alcohol was mostly confined to Queensland, and the province required that all gasoline distributors purchase alcohol from the National Power Alcohol Distillers at 1.5% of their gasoline sales volume.

AUSTRIA–A 1931 law authorized the Minister of Finance to mandate 25% blending as long as the price of alcohol did not rise above the wholesale price of gasoline. The program did not succeed in bringing alcohol to market and in 1934 gasoline importers were required to buy alcohol at 2% of their sales volume while domestic gasoline producers were required to buy at 3.75% of their sales volumes. The alcohol was sold in 20–40% blends. About 35,000 barrels (5.5 million liters) were blended with gasoline in 1936.

BRAZIL–Gasoline importers were required by a 1931 law to buy alcohol in volumes of 5% of their imports under the supervision of the Minister of Agriculture. The law was inspired by French blending laws that were studied by young agriculture ministry officials who later influenced the Proalcool program of the 1970s. Sales tax exemptions for blends and reductions on taxes of high compression motors were also instituted. Much of the interest in alcohol fuels came from sugarcane planters, who often used pure alcohol vehicles. Blends of various proportions were marketed, including a 90% alcohol 10% gasoline blend, a 70% alcohol and 30% di-ethyl ether blend, and a 12% alcohol 88% gasoline blend. The Instituto do Assucar e do Alcool was establshed in 1933 to promote alcohol fuels and lend technical assistance. It was not until about 1937 that alcohol production reached 7% of the nation's fuel consumption, about 51.5 million liters. Blending continued intermittently through the 1950s as an outlet for sugar surpluses.

CHILE–An August 1931 law required blending of more than 10% and less than 25% alcohol in all gasoline, but was dropped to 9% and later 6% as the law became effective. The ratio varied from year to year. In 1938, 3% alcohol was blended with gasoline sold primarily in Santiago, Aconcagua and Valpariso. Some 1.7 million liters were blended in 1938, mostly produced from imported Peruvian molasses.

CHINA–"Benzolite" a mixture of 55% alcohol, 40% benzene and 5% kerosene was sold in the 1930s, and annual consumption was around 500,000 liters of alcohol. Blends and straight fuels were used extensively during WWII by civilians, the Japanese, the British and American armies. Fragmentary data does not reveal the extent of alcohol use, but a spokesman for the Chia Yee Solvent Works noted at a United Nations conference on power alcohol in 1952 that "the shortage of gasoline was so acute it became impossible for civilians to get any amount of gas. At that time the use of alcohol was no longer a question of costs or efficiency, but of necessity." There is at least one report of a diesel-powered bus that ran on a number of vegetable oils, including those extracted from peanuts, tea leaves, poppy seeds, tung, cotton seed and cabbage seed. Many American soldiers in China in WWII remember the potent, potable alcohol that doubled as a fuel for their jeeps and generators.

CUBA–A blend of 80% gasoline and 20% alcohol called "Espiritu" was sold in the early 1920s, apparently without much success. Several laws to mandate alcohol use were proposed but failed in the 1930s, but blends fared better on the open market. "Nofuco" was said to be the most popular, containing 64% ethanol, 22% gasoline and benzene, gas oil and other additives. About 2% of Cuba's total fuel consumption in 1937 was alcohol, or about 12 million liters. The government adopted a mandatory blend shortly after the begining of WWII. Official formula was 35% gasoline, 65% alcohol, but it was changed in 1943 to 75% alcohol and 95% was also frequently used. Minor adjustments were made to automobiles, such as replacing cork floats in carburetors, and mileage suffered.

CZECHOSLOVAKIA–"Dynakol," a mixture of about 50% alcohol, 20% benzene and 30% gasoline was used exclusively by government agencies between 1923 and 1932, when a law making 20% alcohol blending compulsory was passed. Some 66 million liters, or 12% of the total fuel supply, were sold in 1936. The government subsidized low-cost power alcohol with a tax on beverage alcohol.

FINLAND AND NORWAY produced about 20 million liters per year from waste sulfite liquors from the paper pulping process in the 1930s. Sale of the fuel byproduct made Scandinavian paper commercially competitive with Canadian paper in the U.S., as paper costs were reduced by 20%. U.N. officials estimated that 75% of Finland's fuel in the 1940s was supplied by gasogen units that consumed only three kilograms of wood for a liter of gasoline equivalent compared to about 8 kilograms of wood for a liter of ethanol. By 1952, a 20% blend was still on the market and about 1.5 million liters were sold.

FRANCE–A bureau set up to encourage agricultural reconstruction after WWI encouraged research into and production of alcohol fuels, and in 1921 a full committee was set up to investigate power alcohol. Their report in 1923 recommended a 40%-50% blending level, and on Feb. 28, 1923, a law was passed requiring gasoline importers to buy alcohol for 10% blends from the State Alcohol Service. The law had a far-reaching impact as many other nations were influenced to enact similar laws based on the French and German programs. The French blending program was not without problems, however. As the level of blending rose to 50%, motorists complained about stalling, hesitating and starting difficulties. In 1928 the blend level was revised to 25%. Several brands were marketed: "Carburant Pois Lourds" (lorry fuel), "Tourisme" and "Supercarburant," By 1931 blending stocks had passed 87 million liters and the Alcohol Service acquired even more alcohol from the glutted wine industry. Power Alcohol use peaked in 1935 at 406 million liters, accounting for over 7% of all fuel use, and declined by 1937 due to poor harvests to 194 million liters.

During WWII, the Michelin Co. marketed two blends, one a 62% hydrated alcohol with 20% gasoline, 10% benzene and 8% acetone, and another of 80% hydrated alcohol, 5% benzene and 15% gasoline.

FRENCH INDOCHINA (NOW VIETNAM, CAMBODIA AND LAOS)–The Societe Francaise des Distilleries de l'Indochine produced about 100,000 liters of anhydrous alcohol for blending in 10% mixtures with gasoline during the 1930s. Blend percentages began rising in 1941, reaching 50% by July. About 450,000 liters were used in blends and straight by 1942, but war damage decreased production in subsequent years as U.S., French and British bombers knocked out Japanese-occupied distilleries along with other fuel facilities. Work stopped at all distilleries in 1945, and when most reopened in 1950 they turned to beverage production.

FRENCH WEST AFRICA (NOW MALI)–The Centre Experimental des Carburants Vegetaux in Segou tested a variety of vegetable oils in the 1930s, including a 50% hydrated ethanol, 35% benzene and 15% alcohol blend that was used at the Centre. Tests of corn and sorghum crops for improved distillation and numerous vegetable oils from various oilseed wastes were also performed.

GERMANY–As the nation recovered from WWI, the agriculture and the distilling industry struggled to regain some of its fuel market. Power alcohol production grew from about 800,000 liters in 1923 to 67 million liters in 1930. "Monopolin," a 25% alcohol blend, was sold in competition with gasoline and endorsed by airplane pilots and famous auto racers. In August, 1930, the government required all gasoline importers to buy 2.5% of the volume of their imports from the German Alcohol Monopoly, and the ratio was increased to 6% and then 10% by 1932. Estimates of alcohol used in 1932 vary from 44 million liters to 174 million liters. By one industry account, some 36,000 small farm alcohol stills, owned by the monopoly, were in operation in the 1930s.

A 10% ethanol blend was compulsory in 1932, but increasing production of methanol from coal led to some use of it, too, in the "Kraftspirit" fuel. By 1938, Germany was producing about 267 million liters of ethanol, about two thirds from potatoes and the rest from grain, wood sulfite liquors and beets. Some 89 million liters of methanol were produced from coal, while other synthetic fuels included 550 million liters of benzene and over one billion liters of synthetic gasoline. All told, 54% of the pre-war German fuel production was derived from non-petroleum sources, of which 8% was ethanol from renewable sources.

GREAT BRITAIN–"Koolmotor" marketed by the Cities Service Corp., "Cleveland Discoll" marketed by Standard Oil Co.'s subsidiary, and "National Benzol" marketed by the National Benzol Association were well known fuels in Great Britain during the 1930s. Various acts of Parliament encouraged use of power alcohol, although blending was never made compulsory. Use of alcohol in blends increased from 800,000 liters in 1931 to 20 million liters by 1937. Import duties on alcohol were lifted in 1931 but reimposed in 1938, probably causing a decline in sales. The fact that Standard Oil advertised their blends as having greater power, less knock, no starting or phase separation problems in Great Britain while attacking alcohol blending proposals in the U.S. on the same technical grounds was a source of heated controversy in the U.S. in 1936.

HUNGARY–"Moltaco," a blend of 20% ethanol, was made compulsory by royal decree in 1929, and almost 12 million liters of ethanol made up 11% of the nation's total fuel consumption in 1936.

INDIA–The central government was opposed to alcohol blending in the 1930s, although a few sugar refining plants converted inedible molasses to alcohol for blending in the Mysore province. In 1938 a Power Alcohol and Molasses Committee of Uttar Pradesh and Bihar provinces recommended that the many sugar mills begin turning molasses into alcohol instead of dumping it on land as fertilizer. India's molasses feed market was rather small in comparison to most other countries, since vegetarianism is an important part of the Hindu religion. In 1940, Uttar Pradesh passed a 20% power alcohol blending law, and several other provinces followed suit. Pure alcohol fuels were widely used during WWII, as an emergency measure, since by 1943 transport came to a complete halt from lack of petrol. About 8 million liters were used in 1946, increasing to 9 million at the peak use in 1948. Another 20 million liters were used in blends, out of about 1 billion liters of gasoline used in 1951. Indian leaders were conscious of possible "food or fuel" conflicts and prohibited the use of grains and root crops as feedstocks, but also felt that the power alcohol industry had to be protected from petroleum interests. The 1948 "Indian Alcohol Act" provided for 20% blending.

ITALY–A royal decree in 1926 set fuel composition at 30% alcohol and 70% gasoline, and eleven alcohol manufacturers were producing for the fuel market by 1931. A controversy over the technical feasibility of alcohol blending erupted between the government and Standard Oil in 1926, as Standard and other gasoline importers refused to blend alcohol, claiming it was impossible to produce a satisfactory fuel. A report from the University of Iowa said of this controversy: "The malice and intention to sabotage the young industry were very apparent." Government pressures prevailed, however, and blends called "Benzalcool," "Elsocina" and "Robur" were marketed in the late 1920s and 1930s. "Robur" was 30% ethanol, 22% methanol, 40% gasoline and other additives, while "Benzalcool" had 20% ethanol and 10% benzene with gasoline. Supplies of alcohol never exceeded 9.5 million liters per year, and production levels never rose above 4% of the nation's fuel supply.

JAPAN–Compulsory blending of 5% ethanol became effective in 1938. One 1936 report noted that 166,000 liters of ethanol and 125,000 liters of methanol were used. Much of the Army, Navy and Air Corps was using alcohol during WWII, due to scarcity of petroleum supplies. At the begining of the war about 365 million liters of ethanol were used as a fuel compared to 5.4 billion liters of gasoline, coal fuels and shale oil.

LATVIA (NOW U.S.S.R.)–Compulsory mixtures of 25% ethanol were set in 1931, increased to 50% and 66% by 1936, when the legislation was lifted. Production from about 50 distilleries peaked at 13.5 million liters in 1935.

LITHUANIA (NOW U.S.S.R.)–An April 1936 law set national fuel levels at 25% alcohol. About 1.5 million liters of ethanol were sold in a blend called "Motorin," and gasoline imports dropped by 12% to 5.8 million liters by 1937.

MALAYA–Citing the example of France and Czechoslovakia, the government developed a 500 acre nipah palm plantation for power alcohol in the 1920s. Malayan officials said "the view of the petroleum experts was against the proposal as not being technically feasible," and the Nipah Distilleries Inc. which had already built the distillery near the development, went bankrupt before it marketed a liter of alcohol. The facilities were reorganized for sugar production but that business, also, went bankrupt, and the distillery equipment was sold to Thailand. In 1926, a number of wood distilleries were built as part of the same program in Krambit, Phang and the Collieries, and apparently produced wood ethanol for the fuel market.

PANAMA–A 1934 law exempted gasoline imported for blending with alcohol from import duties, and a number of blends were sold in competition with regular gasoline. One of these, "Alcolina," was popular when it sold for 5¢ less than gasoline, but a "price war" kept local distilleries from expanding for the new market. Sales fell off when oil companies dropped their prices.

PHILIPPINES–"Gasonol" (spelled with an "n") blends of 20% ethanol and 5% kerosene were introduced in these sugarcane-producing islands in 1931, and by 1937 alcohol consumption rose to 75 million liters. More than any other nation at the time, the Philippines had the most experience with pure alcohol fuels. Autos, buses, trucks, railway locomotive and mining operations commonly used the abundant fuel. Studebaker, McCormack, GM and International Harvester sold pure alcohol-fueled cars and trucks, advertising them as "more economical . . . (and) free from carbon." Three large bus companies including Manila's Batangas Transportation Co. were running their buses on 100% ethanol, while buses and trucks on Negros and Panay also used pure alcohol as a common fuel. No compulsory blending or tax advantages were given alcohol fuels in the Philippines, but one U.S. Commerce Dept. official commented, "The sugar interests have felt reasonably well satisfied." When ethanol fuel use reached 90 million liters in 1939, according to the Philippine delegate to a 1952 United Nations power alcohol conference, "The use of blended motor fuel was abandoned, for the simple reason that the gasoline interests fought hard to kill it. After such a very sad experience, we fully realize that proper legislation similar to that in India should be adopted in the Philippines." Power alcohol production reached a standstill in WWII, but climbed back to 30 million liters by 1950. A four-year plan was then in place to produce 120 million liters, or 20% of the nation's fuel supply, but it was abandoned as new sources of cheap oil became available.

POLAND–Along with about 1,500 small scale alcohol plants serving farm communities, a major alcohol factory was built at the Kutno Chemical Works in 1927. This plant alone produced 54,000 liters per day (as much as 20 million liters per year) by 1939. Polish sources reported that oil interests were helpful in starting the power alcohol industry there, after compulsory mixtures at 25% were decreed in 1932. Standard Oil volunteered to buy alcohol production five years in advance to help capitalize the young industry. Some 30 different products were manufactured at Kutno, including a synthetic "Ker" rubber produced from potato-derived ethanol-to-butadiene process. Chemists who developed this technology were smuggled out of Poland in 1939, and set up government-run rubber plants in the U.S. These plants outperformed the petroleum-based synthetic rubber plants and produced 75% of the Allied rubber consumption during WWII.

SWEDEN–Power alcohol was first introduced in 1911, and irregular blends of kerosene and turpentine with alcohol during the first World War were reported damaging to some engines. Most of this alcohol came from the sulfite liquor of the paper industry's 22 plants, which accounted for 36 million liters per year in the 1930s. Alcohol was sold in 25% blends with gasoline called "Lattbentyl." Purchases of power alcohol by gasoline importers were required in 1934, the amount depending on the state-controlled surplus. In 1941, according to one report, 46 million liters were being produced per year. With the beginning of the war, ingenious fuel systems were devised, and there is one report of a baker who ran his delivery trucks on alcohol recovered from bread ovens. By 1952, about 130 million liters of power alcohol were produced, but expanding fuel and chemical industry demands led one U.N. official to comment that "the future for power alcohol does not look bright" in Sweden.

SWITZERLAND–Methanol and ethanol from wood were common fuels in Switzerland before WWII, with some 31 million liters used in 1939. A major wood alcohol plant at Attisholz produced not only alcohols and paper pulp but also developed a yeast process for livestock feed which was considered one of the most important pre-war developments in the distillery business. During the war only vitally necessary vehicles were allowed to operate on liquid fuels, mostly alcohols, and civilians had to be content with gasogen generators. Many wood alcohol plants remained open after the war for the chemical market, and at least one has reopened for the fuel market.

SUDAN–A sisal fiber plant in Same produced 30,000 liters per year from the small amount of sugar byproduct in the sisal production process in the 1940s.

UNITED STATES–Prohibition of beverage alcohol effectively killed any political support for an alcohol fuels industry, but after it was lifted in 1932, widespread interest in the American Midwest led to several power alcohol projects. A 10% blend was marketed by the Agrol Company of Atchison, Kansas as an experiment into broadening markets for surplus crops. The experiment was supported by Henry Ford and the Chemical Foundation, and at its peak in 1938, some 2,000 service stations in 8 different states sold Agrol. But production was indifferent and costs were higher, and Agrol suspended business in 1939. Company official charged that the oil companies had spread malicious rumors, but the economics of the project could not support open competition with blending. Legislation to support an alcohol industry was proposed in most Midwestern state capitals and in Washington, D.C., dozens of times, but was fought by oil interests who claimed "The Alky gas scheme outrages common sense" and "farmers would make motorists pay for farm relief."

With the crucial shortage of crude rubber from Malaya at the outbreak of WWII, the government reopened the Agrol plant and built several other grain and wood alcohol plants in the Midwest and Pacific Northwest. Over 75% of the synthetic rubber for the war was produced from these plants, despite a larger investment in a petroleum-based process. Alcohol was used as a supplementary fuel in some aircraft, and according to one report it was consumed at 24 million gallon per year levels. The plants were closed down at the end of the war, much to the dismay of many farmers and farm state politicians.

UNION OF SOUTH AFRICA–Blends of 40% ethanol made from sugarcane were marketed as "New Union Motor Spirit" and "Natalite" in the 1930s in Natal province and Durban. One report said 45,000 liters were sold in the first few months of 1931.

YUGOSLAVIA–A 1933 decree specified a 20% blend with all gasoline, but was rescinded in 1934 in favor of tax advantages for alcohol blends. One 1936 report said 2.8 million liters of ethanol were blended with gasoline.

Reference Documents

1. "Energy Resources and National Policy," report of an executive committee to Congress, January, 1939, House document #160.
2. Rufus Frost Herrick, *Denatured or Industrial Alcohol*, Chapman Hall & Sons, London, 1907.
3. *The Use of Alcohol in Motor Fuels*, Progress reports #1 & 2, May 1933, Iowa State College, Ames, Iowa.
4. John Geddes McIntosh, *Industrial Alcohol*, Scott, Greenwood & Sons, London, 1907
5. *The Production and Use of Power Alcohol in Asia and the Far East*, report of a seminar held at Lucknow, India, Oct. 23, 1952, organized by the Technical Assistance Administration and the Economic Commission for Asia and the Far East, United Nations, New York, 1954.
6. Gustav Egloff, *Motor Fuel Economy of Europe*, American Petroleum Institute, 1939; also "Substitute Fuels as a War Economy," 1942, paper to the American Chemical Society.
7. American Petroleum Institute "Economic and Technical Aspects of Alcohol-Gasoline Mixtures," October, 1935, New York, N.Y.

For an updated version of the International Historical Survey of Alcohol Fuel Programs, please visit the Web edition of this paper at http://www.forbiddenfuel.com or http://www .radford.edu/~wkovarik/papers/International.History.Ethanol.Fuel.html.

Appendix D
Books
on Making
Alcohol Fuel

Making Fuel In Your Backyard
Written by: Jack Bradley

Available from:
Biomass Resources, P.O. Box 2912, Wenatchee, WA 98801

1979, 63 pages, $10.95

A fine, down-to-earth book by a man who built a still that he's writing about. It is in narrative form with easy instructions and useful drawings and pictures. Although it is not the highest class publication in terms of glossy print, it is one of the most useful small on-farm manuals to come along in a while

Brown's Alcohol Motor Fuel
Written by: Michael H. Brown

Available from:
Desert Publications, Cornville, Arizona 86325

1979, 140 pages, $6.95

Probably the most well photographed, easy-to-understand book on car conversion yet printed. The author includes full page pictures along with technical illustrations and step-by-step instructions. The book covers carburetor modification, increasing the compression, ignition and cold starting. In the cold start chapter, the author skimps on

the newest Volkwagen approach using propane injection but on the whole this auto conversion book covers what any average person needs to know.

The second half of Brown's book covers alcohol production in the most easy and understandable way beginning with moonshine production (5 gallons), batch production (50 gallons) and then covering column design, stripper plates and solar stills. This is followed by 11 pages of photographs interspersed with technical drawings. All in all, this is a great publication and should be read by everyone interested in this field.

How to Make Your Own Alcohol Fuels
Written by: Larry W. Carley

Available from:
Tab Books, Inc., Blue Ridge Summit, PA 17214

1980, 195 pages, $9.95

Of all the books reviewed, Carley's is the most complete and approaches Brown's *Motor Alcohol Fuel* (1979, Desert Publications) as the most easy to follow and use. Although the book could use more pictures, the author correctly leads the reader through a basic course on the reasons for ethanol, fermentation, feedstocks, making mash, distillation, basic still designs, planning, manufacturers and suppliers, some models you can build, regulations and engine conversions. In addition, the author includes a glossary and a list of organizations you can write for further information. The book is accurate and well written and as of this date, is the best book on the market for the alcohol fuels do-it-yourselfer.

The Small Fuel-Alcohol Distillery: General Description and Economic Feasibility Workbook

Written by: Robert S. Chambers

Available from:
ACR Corporation, 808 S. Lincoln Avenue #14, Urbana, IL 61801

21 pages, free

This is an excellent resource for the amateur and professional alike. It systematically walks you through the economics of the technology and marketing.

Making Alcohol Fuel—Recipe and Procedure
Written by: Lance Crombie

Available from:
Rutan Publishing, P.O. Box 3585, Minneapolis, MN 55403

1979, revised, 40 pages, $4.50

Crombie's book was the first on the market with usable data for the modern alcohol producer. His revised version is better particularly in the plant design section. The book has few pictures but some illustrations and is a useful handbook. The book underlines the cautions you should heed and has useful lists of tables and resource people. Crombie has built both a still and has modified an auto to run on 100% pure ethanol.

Converting Gas and Diesel Engines to Use Alcohol Fuel
Compiled by: Gregory James

available from:
VITA, 3706 R.I. Avenue, Mt. Ranier, MD 20822

1980, 5 pages, $1.00

This publication is one of the many fact sheets that the Volunteers in Technical Assistance, Inc. puts out. The fact sheet briefly covers gasoline and diesel conversion. Although there are slight mistakes, the fact sheet is useful for individuals with little skill in automobile technology (precisely for whom the publication was geared).

Food for Thought
Prepared by: P.J. Kolachov and H.F. Willkie

Available from:
Indiana Farm Bureau, Inc. 130 East Washington Street, Indianapolis, IN 46204

1942, reprinted 1978, 209 pages, $2.00

This book, prepared in 1942 by scientists from Seagram and Sons, Inc. for the Indiana Farm Bureau, discusses the technical aspects of ethyl alcohol as a motor fuel and also the availability of raw materials for the production of alcohol. The lengthy appendix contains three papers dealing with the development and design of continuous cooking, washing and fermentation systems. The experimental procedures and data will be helpful to people interested in the cooking and fermentation aspects of alcohol production.

Methanol and Other Ways Around the Gas Pump
Written by: John Ware Lincoln

Available from:
Garden Way Publishers, 530 Ferry Road, Charlotte, VT 05445

Published in 1976 before the gasohol craze, this book describes authomobile conversion using methanol. Many of Lincoln's principles apply to ethanol fuel as well.

Driving Without Gas
Written by: John Ware Lincoln

Available from:
Garden Way Publishing, 530 Ferry Road,

Charlotte, Vermont 05445

1980, 150 pages, $5.95

This book is a sequel to Lincoln's earlier book. It is a political and technical review of alcohol and transportation technology. Electric cars and sterling engines are reviewed along with steam and gas turbines. The book is accurate and provides a general overview of the technology, the arguments both pro and con, resources for further information and some historical data.

Grain Alcohol Study
Prepared by: Fred S. Lindsey

Available from:

Indiana Department of Commerce, State House, Room 336, Indianapolis, IN 46204

1975, 67 pages, free

The book was put together by Rock Island Refining Corp., along with Longardner and Associates, for the Indiana Department of Commerce. It's an indexed handbook on alcohol-blended motor fuels, alcohol production, industrial alcohol, agricultural capabilities, economics and conclusions. This fine resource handbook is set up in an easy-to-use manner, footnoted and well documented.

Makin' It On The Farm
Written by: Micki Nellis

Available from:
American Agriculture Movement, P.O. Box 100, Iredell, Texas 76649

1979, 88 pages, $2.95

A really fine, concise, usable energy primer on alcohol production covering enzyme use, solar stills, methanol and engine conversions. What makes this book different from all others is its low cost, documenting production success stories (Albert Turner, Gene Schroder, Archie Zeitheimer and Lance Crombie, etc.) and thorough coverage of the field. This book is a must for anyone who intends to produce ethanol, methanol or convert their car.

The book includes an excellent list of materials, resource people, and useful data. Lastly, it answers any questions you might have about the actual hands-on production.

The Practical Brewer
Prepared by: Edward H. Vogel and others

Published by: Master Brewers Association of America, Second Edition

Available from: Micro-Tex Laboratories, Inc., Rt. 2, Box 19, Logan, Iowa 51546

1977, 475 pages, $37.50

Written for the practical brewer working in large scale systems, this detailed book covers water, malting, safety, terminology, brewing control and storage. Although the book deals specifically with drinkable alcohol, the information is invaluable to the alcohol fuel producer.

Reference
Notes

Chapter 1

1. Robert Stobaugh and Daniel Yergin, *Energy Future*, rev. ed. (New York: Random House, 1979), p. 60.
2. "Energy: A Breather for Oil Prices," *Business Week*, 25 May, 1981, p. 104.
3. U.S. National Alcohol Fuels Commission, *Fuel Alcohol: An Energy Alternative for the 1980s*, final report (Washington, D.C.: NAFC, 1981), p. 35.
4. Daniel Yergin, "Lulled to Sleep by the Oil Glut Mirage," *New York Times*, 28 June, 1981 III, 2:3.
5. U.S. Department of Energy Information Administration, *Energy Annual 1980* (Washington, D.C.: DOE, 1980).
6. Kenneth F. Weaver, "Our Energy Predicament," *Energy: National Geographic Special Report* (Washington, D.C.: National Geographic, February 1981), p. 17.
7. Stobaugh and Yergin, *Energy Future*, p. 55.
8. Charles Walters, untitled article, *Gasohol U.S.A.*, September 1981.
9. U.S. National Alcohol Fuel's Commission, *Fuel Alcohol*, p. 49.
10. U.S. Congress, Office of Technology Assessment, *Energy from Biological Processes*, vol. 2., July 1981.
11. Ibid.

Chapter 2

1. Reynold Millard Wik, "Henry Ford's Science and Technology for Rural America," *Technology and Culture*, Summer 1963.
2. Rufus Frost Herrick, *Denatured or Industrial Alcohol* (London: Chapman & Hall, 1907), p. 287.
3. Ibid., p. 289.
4. Ibid., p. 287. Also, John Geddes McIntosh, *Industrial Alcohol* (London: Scott, Greenwood & Sons, 1907); Charles Simmonds, *Alcohol: Its Production, Properties and Applications* (London: Macmillan, 1919).
5. Charles Lucke, *Use of Alcohol and Gasoline in Farm Engines* (Washington D.C.: USDA Farmers' Bulletin #227, 1907). Also, H. W. Wiley, *Industrial Alcohol: Uses and Statistics* (Washington, D.C.: USDA Farmers' Bulletin #269, 1906).
6. *New York Times*, 28 November, 1915, VII, 7:8.
7. Conger Reynolds, "The Alcohol Gasoline Proposal," *Proceedings of the American Petroleum Institute*, 9 November, 1939, p. 51.
8. *New York Times*, 2 January 1906.
9. Robert M. Strong, *Commercial Deductions from Comparisons of Gasoline and Alcohol Tests on Internal*

237

Combustion Engines (Washington, D.C.: U.S. Geological Survey Bulletin #392, 1909); idem, *Commercial Deductions*, rev. ed. (Bulletin #32, 1911).

10. Lucke, *Use of Alcohol*, and Wiley, *Industrial Alcohol*.
11. *New York Times*, 12 February 1907.
12. Ibid.
13. Herrick, *Denatured or Industrial Alcohol*.
14. Ibid.
15. Oscar B. Ryder, *Industrial Alcohol* (Washington D.C.: U.S. Tariff Commission Report, "War Changes in Industry" series, 1944), p. 11.
16. T. A. Boyd, "Motor Fuel from Vegetation," *Industrial and Engineering Chemistry*, vol. 13:9, September 1921, p. 836.
17. *National Geographic*, vol. 31, February 1917, p. 131.
18. Thomas Coffey, *The Long Thirst: Prohibition in America 1920–1933* (New York: W. W. Norton, 1975), p. 7.
19. *Encyclopaedia Britannica*, 1963 edition, s.v. "Prohibition."
20. Coffey, *The Long Thirst*, pp. 296–297.
21. Christy Borth, *Modern Chemists and Their Work* (New York: Bobbs-Merrill, 1939), p. 69.
22. John Kenneth Galbraith, *The Great Crash of 1929*, Sentry Edition 10 (Boston: Houghton Mifflin, 1972), p. 173; also, Gerald Gunderson, *A New Economic History of America* (New York: McGraw-Hill, 1976), p. 475.
23. William Manchester, *The Glory and The Dream* (Boston: Little, Brown, 1973), p. 636.
24. Clarence Berger, "Social Adjustments in the Cities," *Social Changes During the Great Depression* (New York: Da Capo Press, 1974), p. 30.
25. John Steinbeck, *The Grapes of Wrath* (New York: Viking Press, 1939), p. 26.
26. Carl Solberg, *Oil Power* (New York: Mason Charter, 1976), p. 71.

27. "Farm-Brewed Fuel: Farmers Would Make Motorists Pay For Farm Relief," *Business Week*, 15 March 1933.
28. August W. Giebelhaus, "Resistance to Long-Term Energy Transition: The Case of Power Alcohol in the 1930s," *Proceedings of the American Association for the Advancement of Science*, 4 Jaunary 1979, p. 9.
29. Gustav Egloff, "Alcohol Gasoline Motor Fuels," paper to the National Petroleum Association (New York: American Petroleum Institute, 1933).
30. Giebelhaus, "Resistance to Transition," p. 11.
31. John A. Simpson, "Quack Remedies," National Broadcasting Co., 22 April 1933, quoted in *Economic and Technical Aspects of Alcohol-Gasoline Mixtures* (New York: American Petroleum Institute, 1935), p. N2042.
32. American Automobile Association, *The Facts About An Alcohol-Gasoline Blend as a Motor Fuel* (Washington, D.C.: AAA, 1933), p. 10.
33. Ibid., p. 12. Also, *Alcohol Gasoline Fuel and the Motorists* (Washington, D.C.: AAA, 1933).
34. "Corn Alcohol Gas Defended," newspaper source unknown, Leo Christensen, alumni file, Iowa State University Archives, Ames, Iowa. Also, E. I. Fulmer, R. M. Hixon and L. M. Christensen, *The Use of Alcohol In Motor Fuels* (Ames, Iowa: Iowa State University, 1933).
35. Oscar C. Bridgeman, "Utilization of Ethanol-Gasoline Blends as Motor Fuels," *Industrial and Engineering Chemistry*, vol. 28:9, September 1936, p. 1102.
36. U.S. Congress, Senate, *Response of U.S. Department of Agriculture to Senate Resolution #65*, 74th Congress, 1st Session, Senate Document #57, 8 May 1933.
37. Borth, *Modern Chemists*, p. 44.
38. *New York Times*, 12 May 1935.
39. Francis P. Garvan, "Scientific Method

of Thought In Our National Problems," *Proceedings of the Second Dearborn Conference on Agriculture, Industry and Science* (New York: The Chemical Foundation, 1936), p. 86.

40. August W. Giebelhaus, personal communication, August 1981.

41. Leo Christensen, alumni file, Iowa State University Archives, Ames, Iowa.

42. "Alky Gas Flops in Sioux City," *Business Week*, 30 July 1938, p. 20.

43. Ibid.

44. Borth, *Modern Chemists*, p. 169.

45. U.S. Department of Agriculture, *Motor Fuels From Farm Products* (Washington, D.C.: USDA Misc. Publication #327, 1938).

46. U.S. Congress, Senate, Committee on Finance, *Hearings on Senate Bill 552*, 76th Congress, 1st Session, May 1939.

47. Reynolds, "The Alcohol-Gasoline Proposal," p. 51. Also, Gustav Egloff, *Power Alcohol: History and Analysis* (Chicago: American Petroleum Institute Committee on Motor Fuels, 1940).

48. U.S. Congress, *Hearings on Senate Bill 522*, p. 128.

49. Gustav Egloff, *Motor Fuel Economy of Europe* (Chicago: American Petroleum Institute Committee on Motor Fuel, 1939); idem, "Substitute Fuels as a War Economy," paper to the American Chemical Society reprinted by the American Petroleum Institute, 1942. Also, U.S. Congress, House, *Report of the President on Energy Resources and National Policy*, 76th Congree, 1st session, House Document #160, 1939, pp. 8–11.

50. American Motorists' Association, *What French Motorists Say About Alcohol Gasoline Motor Fuel Blends* (New York: AMA, 1933), p. 1.

51. Gustav Egloff and P.M. Van Arsdell, "Substitute Fuels as a War Economyu." *Journal of Institute of Petroleum*, LVol. 28, No. 223, July, 1942, p. 118.

52. Ibid. Also, Egloff, *Motor Fuel Economy of Europe*, p. 3.

53. *New York Times*, 28 April 1942.

54. Egloff, "Substitute Fuels," p. 118.

55. U.S. Congress, *President's Report on Energy*, 1939.

56. U.S. Congress, Senate, Committee on Agriculture and Forestry, Select Subcommittee on Utilization of Farm Crops, Industrial Alcohol and Synthetic Rubber, *Hearings on Senate Resolution 224*, 77th Congress, 1st Session, 1942.

57. Bernard Baruch, *The Public Years* (New York: Holt, Rinehart & Winston, 1957).

58. August W. Giebelhaus, personal communication, 1981.

59. U.S. Congress, *Hearings on Senate Resolution 224*, 1942, p. 21.

60. Harold Ickes, *Fighting Oil* (New York: Alfred A. Knopf, 1943), p. 7.

61. Joseph Borkin, *The Crime and Punishment of I.G. Farben* (New York: Free Press, 1978), p. 79.

62. Drew Pearson, *Washington Post*, 14 July 1942 and 17 December 1942.

63. U.S. Congress, Senate, Special Committee Investigating the National Defense Program, *Hearings on Senate Resolution 71*, 77th Congress, 1st Session, 26 March 1942, p. 4313, statement of Attorney General Thurmond Arnold.

64. Charles Wilson, *Trees & Test Tubes* (New York: Henry Holt & Co., 1943), p. 124.

65. Drew Pearson, *Washington Post*, 17 December 1942. Also, U.S. Congress, *Hearings on Senate Resolution 224*, 1943, p. 259.

66. William Stephenson, *A Man Called Intrepid* (New York: Ballantine, 1976), p. 284.

67. U.S. Congress, *Hearings on Senate Resolution 224*, November 1943, testimony of Donald Nelson.

68. Al Frisbie, "The Old Alcohol Plant—Is There A Lesson There?" *Omaha World Herald*, 28 May 1978, p. 55.

69. Ryder, *Industrial Alcohol,* pp. 1–5.
70. James Kerwin, personal communication, October 1978.
71. August W. Giebelhaus, personal communication, 1981.
72. Drew Pearson, "U.S. May Face Another Rubber Shortage," *Washington Post,* 29 November 1950.
73. United Press, "Chemurgy Is For the Farmer: Big Chemical Firms Oppose It," *Omaha World Herald,* 6 September and 9 September 1945.
74. U. S. Congress, Senate, Committee on Agriculture and Forestry, *Report by the Commission on Increased Industrial Use of Agricultural Products,* 85th Congress, 1st Session, Senate Document #45, 1957, p. 99.

Chapter 3
1. Alvin Mavis, "The Challenge is Yours, the Time is Now," State of Illinois Department of Agriculture (Springfield, Ill.: Department of Agriculture pamphlet, 1979).
2. Reprinted by permission of *The Mother Earth News,* copyright © 1979 by the Mother Earth News Inc., from "Lance Crombie: Energy Self-Sufficiency Now!" (Hendersonville, N.C.: February 1979), p. 118.
3. Ibid., p. 22.
4. Reprinted by permission of Micki Nellis, from *Makin' It on the Farm,* copyright © by Micki Nellis, 1979 (Iredell, Tex.: American Agriculture Movement, 1979), pp. 58–60.
5. Cliff Bradley and Ken Runion, *Fuel Alcohol Production* (Butte, Mont.: National Center for Appropriate Technology, 1981), p. 40; idem, *Making More Fuel Alcohol* (NCAT).
6. "Snake Oil Sams Cash In," *Farmland News,* 30 April 1980.
7. Ibid.
8. Ibid.
9. Dan Jantzen and Tom McKinnon, *Preliminary Energy Balance and Economics of a Farm-Scale Ethanol Plant*

(Golden, Colo.: Solar Energy Research Institute, #SERI RR 624 699R, 1980), p. 7.
10. Spokesmen for U.S. Treasury Department, Bureau of Alcohol, Tobacco and Firearms, personal communication, November 1980.
11. Ed Clark, "Dennis Day, A Pioneer in Farm Produced Alcohol," *Farm Energy* (Des Moines, Iowa: Iowa Corn Growers Association, August 1980), p. 10–13.
12. U.S. National Alcohol Fuels Commission, *Fuel Alcohol on the Farm: A Primer on Production and Use* (Washington, D.C.: NAFC, 1980), p. 2.
13. Ed Clark, "Farmer's First Chance to Compare Alcohol Stills: Cookoff," *Farm Energy* (Des Moines, Iowa: Iowa Corn Growers Association, August 1980), pp. 18–19.

Chapter 4
1. Charles Walters, untitled article, *Gasohol U.S.A.,* September 1981, p. 34.
2. U.S. Department of Energy, *First Annual Report to Congress on the Use of Alcohol in Motor Fuel* (Washington, D.C.: DOE, April 1980), p. 102.
3. Associated Press, "Gasohol Craze," *Denver Post,* 19 April 1979, p. 27.
4. Iowa Corn Promotion Board, *Gasohol Acceptance in Established Markets* (Des Moines, Iowa: Iowa Development Commission, April 1979, p. 1.
5. *New York Daily News,* 10 April 1979, p. 24.
6. DOE First Annual Report to Congress on the Use of Alcohol Motor Fuel (Washington, DC, Department of Energy, April 1980.)
7. Irwin Ross, "Dwayne Andreas' Bean Has A Heart of Gold," *Fortune,* October 1973, p. 4 of reprint.
8. Interview with David Susskind, undated, transcript provided by Archer Daniels Midland Corp., Decatur, Ill., p. 32.
9. Ibid., p. 33.

10. Ross, "Andreas' Bean . . . ," p. 1.
11. Michael Iskioff, "Andreas Generous," *Decatur [Ill.] Herald,* 25 July 1978.
12. Reprinted by permission Viking-Penguin, Inc., from Dan Morgan, *Merchants of Grain,* copyright © 1979 by Viking-Penguin Inc. (New York: Viking Press, 1979), p. 310.
13. U.S. National Alcohol Fuels Commission (Washington, D.C.: NAFC, 18 June 1980), testimony of Dwayne Andreas.
14. Susskind transcript, p. 12.
15. "Midwest Solvents," *Atchison [Kansas] Daily Globe,* 28 October 1975.
16. U.S. Congress, House, Committee on Science and Technology, Subcommittee on Advanced Energy Technologies and Energy Conservation (Washington, D.C., 11–13 July 1978), p. 438, testimony of Cloud Cray.
17. Roscoe Born, "Sweet Proposition: How to Turn Surplus Sugar into Gasohol—And A Fast Buck," *Barron's,* 20 August 1979, pp. 9, 22.
18. Ibid.
19. Ibid.
20. Leobard Schroben, "Gasohol Bubble," *Proceedings of the Distillers' Feed Council* (Louisville, Ky.: DFC, March 1978), vol. 33, p. 4.
21. TRW Corp., *Energy Balances in the Production and End-Use of Alcohols Derived from Biomass* (Washington, D.C.: U.S. National Alcohol Fuels Commission, October 1980), p. 1.
22. Robert S. Chambers, *The Small Fuel Alcohol Distillery: General Description and Economic Feasibility Workbook* (Champaign, Ill.: ACR Process Co., 1979), p. 6.
23. Bill Kovarik, "Dramatic Cost Reductions Seen in Membrane Technology," *Energy Resources and Technology,* 3 November 1978.
24. Bernard D. Davis, "Frontiers of the Biological Sciences," *Science,* 4 July 1980, p. 78.
25. Gulf Chemical Co., "The Cellulose Project," *Orange Disc* (Houston, Tex.: Gulf Chemical Co., 1978), p. 2.
26. Leo Spano, personal communications 1979.
27. Gulf, "The Cellulose Project."
28. Leo Spano, personal communication, 1978.
29. Jeremy Rifkin, *Who Shall Play God?,* (New York: Delacorte, 1977)
30. Anthony J. Parisi, "Gene Engineering Industry Hails Court Ruling as Spur to Growth," *New York Times,* 17 June 1980.
31. *Business Week,* 22 October 1979.
32. *Wall Street Journal,* 24 October 1979.
33. Bill Kovarik, "Synfuels Projects Will Be at Mercy of Private Markets for $60 Billion in Capital," *Coal Daily,* 20 November 1980.
34. Leo Spano, "Enzymatic Hydrolysis of Cellulose to Fermentable Sugar for the Production of Ethyl Alcohol," *Proceedings of the First Inter-American Conference on Renewable Sources of Energy,* New Orleans, November 1979.
35. U.S. National Alcohol Fuels Commission, *Fuel Alcohol,* p. 92.
36. "Alcohol Fuels? Ethanol's Good, Methanol's Better," *Chemical Week,* 24 June 1981, pp. 53–54.
37. "Alaska Meeting Will Test Level of Interest in Gas-to-Methanol Option," *Alcohol Week,* 10 August 1981, p. 8.
38. U.S. Congress, Office of Technology Assessment, *Energy From Biological Processes,* vol. 2, p. 140.
39. U.S. National Alcohol Fuels Commission, *Proceedings of Public Hearing June 18–19, 1980* (Washington, D.C.: NAFC, 1980), p. 119.
40. U.S. Congress, Office of Technology Assessment, *Gasohol: A Technical Memorandum* (Washington, D.C.: OTA, 1979), p. 20.
41. U.S. Library of Congress Research Service, Congressional Budget Office, *Memo to Sen. Charles Percy,* November 1980.
42. Marion Anderson and Carl Parisi, *Re-*

port on Job Creation in the Alcohol
Fuels Industry (Washington, D.C.:
Employment Research Associates,
1980).

Chapter 5

1. J. P. Smith, "Gasohol Advocates Lobbying Intensely," Washington Post,
   23 April 1979.
2. Richard Merritt, personal communications, 1978–1981.
3. Ibid.
4. Editorial, Daily Sun, Beatrice, Nebraska, 27 August 1977.
5. Merritt, personal communications.
6. Ernest Smith, Alcohol-Gasoline Fuel
   and the Motorists (Washington, D.C.:
   American Automobile Association,
   1933).
7. U.S. Department of Energy, Consumer Affairs Special Impact Advisory Committee, Spring 1978, hearing
   transcript, p. 52.
8. Ray Daley, personal communications,
   1978–1981.
9. Daniel Boatwright, personal communications, 1979.
10. Memo to Chevron employees, un-
    dated, authenticated in personal com-
    munications with Chevron spokesmen,
    1978.
11. Robert Lindquist, "Alcohol and Motor
    Fuels: The Promise and the Prob-
    lems," briefing document prepared for
    U.S. Senate staff, November 1977, p.
    1.
12. Fred Wahl, personal communication,
    1979.
13. Press release, "Mobil Demonstrates
    Pilot Plan for Making Gasoline from
    Alcohol," Mobil Oil Co., undated
    [1978].
14. Ibid.
15. "Mobil Proves Gasoline-From-Meth-
    anol Process," Chemical and Engi-
    neering News, 30 January 1978, p. 26.
16. Richard Merritt, personal communi-
    cations.
17. Daniel Horowitz, personal communi-

cation, 17 November 1978.
18. Joe Pennick, response to Senate Re-
    publican Conference Report, Mobil
    Oil Co., undated [1978].
19. Sen. Carl Curtis, "Reply to Mobil
    Oil," statement distributed by Senate
    Republican Conference, 1978.
20. Mike Ware, personal communication,
    1979.
21. Jack Anderson, Washington Post, 9
    August 1977.
22. Mike Ware, personal communication.
23. Jack Freeman, personal communica-
    tion.
24. National Energy Plan, White House
    briefing document, 20 April 1977.
25. U.S. Congress, House, Report of the
    President on Energy Resources and
    National Policy, 76th Congress, 1st
    Session, House Document #160, 1939,
    p. 1.
26. Eugene Ecklund, personal communi-
    cations, 1979–1980.
27. U.S. Energy Research and Develop-
    ment Administration, Identification of
    Probable Automotive Fuels Composi-
    tion: 1985–2000 (Washington, D.C.:
    ERDA, 1977).
28. U.S. Congress, Senate, Staff Report
    prepared for the Subcommittee on For-
    eign Policy, August 1977, pp. 5–6.
29. Group letter, various U.S. Senators,
    to James Schlesinger and Bob Berg-
    land, 20 October 1979.
30. Ray Daley, personal communications.
31. U.S. Department of Energy, Report of
    the Alcohol Fuels Policy Review
    (Washington, D.C.: DOE, June, 1979),
    p. 3.
32. "Alcohol Fuels," The Energy Con-
    sumer (Washington, D.C.: U.S. De-
    partment of Energy, January 1980), p.
    6.
33. U.S. Department of Agriculture
    spokesmen, personal communication,
    1978.
34. New York Times 3 May 1980.
35. Jack Anderson, "DOE Ties to Mobil
    Revealed: Gasohol Report Chal-

lenged," *Washington Post,* 24 May 1980.

36. Bill Holmberg, "Memorandum for Dr. John Deutch: Comments on Alcohol Fuels Memorandum," U.S. Department of Energy, 26 December 1979, p. 3.

37. Jack Anderson, *Washington Post,* 24 May 1980.

38. Mobil Oil Co., "Science and Politics Don't Mix," *New York Times* (advertisement), 19 June 1980.

39. U.S. National Alcohol Fuels Commission, *Fuel Alcohol,* p. 48.

40. Reagan Transition Team briefing document, "Report of the Energy Policy Task Force," November 1980, pp. 3, 30.

41. "Ethanol Projects Receive Backing of U. S. Agency," *Wall Street Journal,* 17 August 1981, p. 6.

42. David Haulberg, personal communication, 1981.

Chapter 6

1. U.S. National Alcohol Fuels Commission, *Fuel Alcohol on the Farm* (Washington, D.C.: NAFC, 1980), p. 60.

2. W. C. Lowdermilk, *Conquest of the Land Through Seven Thousand Years* (Washington, D.C.: USDA, reprinted 1975), p. 7.

3. David Pimental, *Biomass Energy,* unpublished discussion paper (Ithaca, N.Y.: Cornell University, September 1980), p. 37.

4. James Risser, "A Renewed Threat of Soil Erosion: It's Worse Than the Dust Bowl," *Smithsonian,* March 1981, p. 121.

5. Benny Martin, U.S. Department of Agriculture, personal communication, 1981.

6. Risser, "Renewed Threat," p. 121.

7. *New York Times,* 26 October 1980.

8. "Is America Running Out of Water?" *Newsweek,* 23 February 1981.

9. Folke Dovring, "Export or Burn? America's Grain and Energy Equations," unpublished paper, March 1980.

10. "Household Wells, Too, Show Nitrate Pollution," *Norfolk [Neb.] Daily News,* 13 September 1977.

11. Spokesman, U.S. Department of Agriculture Extension Office, Pinal Co., Arizona, personal communication, 1981.

12. National Association of Counties Research Foundation, *The Disappearing Farmlands: A Citizen's Guide to Agricultural Lands Preservation* (Washington, D.C.: National Association of Counties, 1980), p. 3.

13. *National Agricultural Lands Study, Interim Report #1* (Washington, D.C.: Council on Environmental Quality and U.S. Department of Agriculture project, March 1980), pp. 14–15.

14. U.S. Department of Agriculture, *A Time To Choose: Summary Report on the Structure of Agriculture* (Washington, D.C.: USDA, January 1980), p. 83.

15. U.S. Congress, House, Agriculture Committee, *Hearings on Agricultural Productivity and Environmental Quality,* 89th Congress, 1st Session, 25 July 1979, testimony of Sylvan Wittwer.

16. Dovring, "Export or Burn?"

17. Claudia Waterloo and Tom Knudson, "An Examination of Iowa's Security Blanket," *Des Moines Register,* April 1978.

18. William Tucker, "The Next American Dust Bowl and How to Avert It," *Atlantic,* July 1979, p. 42.

19. Dan Morgan, *Merchants of Grain* (New York: Viking Press, 1979), p. 128.

20. U.S. Department of Agriculture, *A Time To Choose.*

21. Morgan, *Merchants of Grain,* p. 343.

22. Frances Moore Lappé and Joseph Collins, *Food First: Beyond the Myth of Scarcity* (Boston: Houghton Mifflin Co., 1977), p. 213.

23. From Ellen Goodman's column, 5 June 1981, reprinted with permission of the Boston Globe Newspaper Co. and The Washington Post Writers Group, copyright © 1981, The Washington Post Writers Group.

24. Charles Morris, "Hydroponic Vegetables," *Food Engineering,* March 1981.

25. Lester Brown, *Food or Fuel: New Competition for the World's Cropland* (Washington, D.C.: Worldwatch Institute, 1980), p. 6.

26. Bill Kovarik, "Food or Fuel? Or Food and Fuel?" *A.T. Times* (Butte, Mont.: National Center for Appropriate Technology, May 1980). Also, Berkeley Bedell, "Doomsaying on Synthetic Fuels Will Get Us Nowhere," *Agenda,* December 1980.

27. U.S. National Alcohol Fuels Commission, Public Hearing (Washington, D.C.: NAFC, 18–19 June 1980), pp. 165–185.

28. Robert Rodale, Editorial, *New Farm Magazine,* September 1979, p. 19.

29. "Effects of Tillage and Crop Residue Removal on Erosion, Runoff and Plant Nutrients," *Journal of Soil and Water Conservation,* March–April 1979, pp. 74–98.

30. Pimental, *Biomass Energy,* p. 22.

31. "Curbing the Chemical Fix," *Progressive Magazine,* December 1978.

32. U.S. Department of Agriculture, *Report and Recommendations on Organic Farming* (Washington, D.C.: USDA, July 1980), pp. 10, 24, and 46.

33. U.S. Congress, House, Agriculture Committee, *Hearings on Agricultural Productivity and Environmental Quality,* 89th Congress, 1st Session, 25 July 1979, testimony of W. Lockeretz.

34. U.S. Congress, House, Science and Technology Committee, 89th Congress, 1st Session, 13 July 1979, testimony of G. Ray Sawyer.

35. U.S. Department of Agriculture, *Organic Farming,* p. 13.

36. Ibid., p. 46.

37. Ibid., pp. 45–48.

38. T. W. Ryan, W. Likos, and C. A. Moses, *Hybrid Fuels for Highway Transportation* (San Antonio, Tex.: Southwest Research Institute, 1980).

39. J. J. Bruwer et al., *Sunflower Seed Oil as an Extender for Diesel Fuel in Agricultural Tractors* (Silverton, Republic of South Africa: Department of Agriculture and Fisheries, 1980), p. 6.

40. James D. Ritchie, "Diesels Run on Sunflower Power," *New Farm,* January 1981, p. 52. Also, Benjamin H. Beard, "The Sunflower Crop," *Scientific American,* May 1981, p. 150.

41. Thomas Lukens, "Jerusalem Artichokes: A Special Crop for Alcohol," photocopied (Seattle, Wash.: Brown Bulb Ranch of Washington, Inc., 1980). Also, L.A. Underkofler, W. K. McPherson, and Ellis I. Fulmer, "Alcoholic Fermentation of Jerusalem Artichokes," *Industrial and Engineering Chemistry,* October 1937, pp. 1160–1164.

42. David Hayes, *Fodder Beet Manual,* photocopied (Albany, Oregon: Pacific Seed Co., 1980).

43. E. W. Earl and W. A. Brown, "Alcohol Fuels from Biomass in New Zealand," *Proceedings of the Third International Symposium on Alcohol Fuels Technology,* Asilomar, Calif., May 1979 (hereafter cited as *Third Alcohol Fuels Symposium).*

44. John Gallian, University of Idaho, personal communication, 1981.

45. Mark W. Martin, "Crops for Production of Alcohol," photocopied (Prosser, Wash.: USDA Extension Office, 1980), pp. 1–4.

46. U.S. Department of Agriculture, *A Time To Choose.*

Chapter 7

1. Norman Gall, "Brazil's Alcohol Program," *Common Ground,* journal of the American University field staff

(Washington, D.C.: American University, Winter 1978), p. 34. Also, G. Pischinger and N.L.M. Pinto, "Experiences with the Utilization of Ethanol," *Third Alcohol Fuels Symposium*, p. I-8.

2. Bill Stout et al., "Brazil Promotes Proalcool for Petroleum Independence," *Agricultural Engineering*, April 1978, pp. 30–33.

3. Government of Brazil, "Programa Nacional Do Alcool Regulamento Das Operacões Rurais," Proclamation of President Ernesto Giesel, 14 November 1975, p. 4.

4. Miguel Arrares, "Yes, Brazil is Dominated," *Brazil*, LADOC, Division for Latin American of the United States Congregational Churches, undated photocopied document, p. 1.

5. Dan Griffin, *Washington Post*, 27 May 1973.

6. José de Castro, *Death in the Northeast* (New York: Random House, 1966), p. 13.

7. Celso Furtado, *The Economic Growth of Brazil: A Survey From Colonial to Modern Times*, trans. Ricardo W. D'Aguiar and Eric Charles Crysdale, 3rd. ed. (Berkeley, Calif.: University of California Press, 1968). p. 9.

8. De Castro, *Death in the Northeast*, p. 120.

9. Dedini Corp.: private communications, 1979.

10. Urban Stumpf, personal communication, 1979. Also, Allen Hammond, *Science*, vol. 195, 4 February 1977, pp. 564–567.

11. J.M.F. Miccolis, "Alternative Energy Technologies in Brazil," *Proceedings of the American Association for the Advancement of Science*, Denver, Colo., February 1977, pp. 12–13.

12. Stout, "Brazil Promotes Proalcool."

13. David Welna, "Progress of Brazil's Alcohol Fuel Program Disputed by Leadership," *Latin American Energy Report* (Silver Spring, Md: Business Publishers, Inc., 5 June 1980).

14. Carl Duisberg, "Perspectives on Brazil's Alcohol Program," paper for master's program, Johns Hopkins University, 1980.

15. Sergio Trindade, *Centro de Technologia Promon Newsletter* (São Paulo, Brazil: CTP, February 1979).

16. Duisberg, "Perspectives on Brazil's Alcohol Program."

17. Allen Hammond, *Science*, vol. 506, 11 February 1977, p. 195.

18. Meg Roggensack, "Vegetable oil Use as Fuel," *Latin American Energy Report*, 14 August 1980, p. 127.

19. David Welna, "Head of Brazil's National Alcohol Commission Expresses Optimism Despite Problems," *Latin American Energy Report*, 10 April 1980, p. 58.

20. *Chemical Week*, 14 March 1979, p. 38.

21. *New York Times*, 1 December 1980, p. D2.

22. *Wall Street Journal*, 15 Jaunary 1981.

23. Herbert Heitland et al., *On the Track of New Fuels* (Wolfsburg, Federal Republic of Germany: Bundesministerium Fur Forschung und Technologie, 1974).

24. Herbert Heitland, H. W. Czaschke, and N. Pinto, "Applications of Alcohol from Biomass and their Alternatives as Motor Fuels in Brazil," *Proceedings of the Second International Symposium on Alcohol Fuel Technology*, Wolfsburg, Federal Republic of Germany, November 1977, p. I-3 (hereafter cited as *Second Alcohol Fuels Symposium*).

25. Hal Bernton, "Brazilian Ethanol: The New Fuel Comes of Age," *Latin American Energy Report*, 12 April 1979, p. 1.

26. Welna, "Progress Disputed . . . "

27. Ford do Brazil, advertising campaign, July 1979.

28. *New York Times*, 13 October 1980.

29. *Latin American Energy Report*, 23 October, 1980.

30. Lester Brown, *Food or Fuel: New*

Competition for the World's Cropland (Washington, D.C.: Worldwatch Institute, 1980).

31. New York Times, 13 October 1980.

32. World Bank, Alcohol Production from Biomass in Developing Countries (Washington, D.C.: World Bank, September 1980, p. 53.

33. David Welna, " . . . Optimism Despite Problems."

34. Latin American Economic Report, May 1979.

35. Duisberg, "Perspectives on Brazil's Alcohol Program."

36. Lewis Beman, "The Last Billionaires," Fortune, November 1976, p. 132.

37. Loren MacIntyre, "A Massive Technology Transplant Takes Root in the Amazon Jungle," National Geographic, May 1980, p. 693.

38. George Hawrylyshyn, "Amazon Still An Attraction," Brazil Herald, 4 July 1979; idem, "Foreign Participation in Brazil's Alcohol Fuels Development," Latin American Energy Report, 6 December 1979.

Chapter 8

1. S.J.W. Pleeth, Alcohol: A Fuel For Internal Combustion Engines (London: Chapman & Hall, 1949).

2. U.S. Commission on Agricultural Products, 1957; also, A Study of Ethanol-Gasoline Blends as a Fuel for Automobile Engines (San Antonio, Tex.: Southwest Research Institute, 1963); also, M. D. Gurney and Jerry Alsup, "Gasohol: Laboratory and Fleet Evaluation" (Bartlesville, Okla.: U.S. Department of Energy, Society of American Engineers paper #800892, 1980).

3. William Scheller, "Tests on Unleaded Gasoline Containing 10% Ethanol," Second Alcohol Fuels Symposium, II-I.

4. W. E. Bernhardt, "Economic Approaches to Utilize Alcohol Fuels in Automobiles," Second Alcohol Fuels Symposium, I-5.

5. Carlos Luengo, "Efficient Burning of Concentrated Gasohol of Variable Composition in Spark Ignition Engines," Proceedings, of the Fourth International Symposium on Alcohol Fuel Technology, Guaruja, Brazil, October 1980 (hereafter cited as Fourth Alcohol Fuels Symposium).

6. Victor Codling, "The Development and Testing of An Anti-Corrosive Additive for Alcohol Fuels," Fourth Alcohol Fuels Symposium, p. 94.

7. Herbert Heitland et al., "Applications of Alcohols from Biomass . . . ," Second Alcohol Fuels Symposium.

8. T. B. Reed and R. M. Lerner, "The Methanol Economy—A Practical Version of the Hydrogen Economy," Science, Vol. 182, 28 December 1973, pp. 1299–1304.

9. Allen Hammond, "Methanol at M.I.T.: Industry Influence Charged in Project Cancellation," Science, 21 November 1975.

10. Norman Brinkman, "Vehicle Evaluation of Neat Methanol: Compromises Among Exhaust Emissions, Fuel Economy and Driveability" (Warren, Michigan: General Motors Research Labs paper #GMR 2825 F&L 662, January 1979): also, "General Motors Advanced Emission Control System Development Progress," Report to the Environmental Protection Agency, 15 December 1976, pp. VIII–62 to VIII–77; also, Norman Brinkman, personal communication, March 1981.

11. U.S. Department of Defense, Foreign Science and Technology Center, "Large-Scale West German Test With Gasoline-Methanol Blend (Charlottesville, Va.: FSTC 538–77, June 1977).

12. J. Cruz et al., "Dual Fueling a Diesel Engine with Carbureted Alcohol," p. II-27, and F. Pischinger et al., "A New Way of Direct Injection of Methanol in a Diesel Engine," p. II-28,

Third Alcohol Fuels Symposium.

13. Allen Poole, *The State of the Art of Alcohol Use in Engines,* unpublished photocopied report to Agency for International Development, June 1979; also, E. Holmer, B. Bertilsson, "The Utilization of Different Fuels in a Diesel Engine with Two Separate Injection Systems," *Third Alcohol Fuels Symposium,* II-29.

14. T. W. Ryan, W. Liuos, and C. A. Moses, *Hybrid Fuels For Highway Transportation* (San Antonio, Tex.: Southwest Research Institute, 1980), p. 53. Also, J. J. Bruwer et al., *Sunflower Oil as an Extender for Diesel Fuel in Agricultural Tractors.* (Silverton, Republic of South Africa: Department of Agriculture and Fisheries, 1980).

15. *The Washington Post,* 24 July 1923.

16. "Alternative Fuels at Oshkosh," *Sport Aviation,* August 1979, p. 5.

17. Eugene Ecklund, "Potential Funding of R&D on Use of Alcohol(s) as Aircraft Fuel," memo to Bert Greenglass, Director Office of Alcohol Fuels, U.S. Department of Energy, 4 February 1981, p. 1.

18. C. E. Seglem, "Performance of Combined Cycle Power Plants Fueled by Methanol," *Proceedings of the American Chemical Society, 13th Middle Atlantic Regional Meeting,* W. Long Branch, N.J., March 1979, p. 1; also, idem, preliminary paper to 25th Annual Area Power Conference, Grand Forks, N.D., November 1978.

19. "Status of Alcohol Fuels Utilization Technology for Stationary Gas Turbines," Mueller Associates, Inc., April 1979, prepared for U.S. Department of Energy, Assistant Secretary for Conservation and Solar Applications, pp. 1–20.

20. "Texas Engineer Claims Water Unnecessary for Coal Pipelines," *Coal Daily,* 22 March 1978, p. 1.

21. U.S. Department of Energy Information Administration, *Energy Annual.* (Washington, D.C.: DOE, 1979).

22. Victor Yang, "Agro-Industrial System for Ethanol and Ethylene Production," *Proceedings of the Institute of Gas Technology, Symposium on Energy from Biomass and Wastes,* Washington, D.C., August 1978. Also, V. K. Malik, "Present Status and Future Projections of Alcohol Based Chemicals in India," *Proceedings of the First Inter-American Conference on Renewable Sources of Energy,* New Orleans, November 1979, p. 106.

23. "Special 1980 Summary: Alcohol," *Brazil Energy,* 24 January 1981, p. 6.

24. U.S. Congress, Senate, Committee on the Judiciary and Committee on Small Business, *Joint Hearing on S. 2251,* 96th Congress, 1st Session, 5 March 1980, p. 88.

25. J. Patrick Wright, *On A Clear Day You Can See General Motors* (New York: Avon Books, 1979), p. 4.

26. "News From the World of Ford," press release, Ford Motor Co., 5 February 1981.

27. Jim Yeager, "Is There a Government and Oil Company Conspiracy?" *Big Valley,* May 1979.

28. Herb Michaelson, *Sacramento Bee,* 18 May 1979.

29. Hal Bernton, "Finally On His Own, 'Don Alcoholte' Is Carving a Niche for a Bio-Fuels Market," *A.T. Times,* August 1980.

30. *Farm Energy News,* February 1981, p. 18.

Chapter 9

1. Peter Van der Linde, *Time Bomb* (Garden City, N.Y.: Doubleday, 1978), p. 47.

2. "Methanol Seen as Safe, Economic Alternative to LNG," press release, Davy Powergas Corp., undated [Fall, 1978].

3. Mohammed Al-Mady, "International Joint-Venture Petrochemical Proj-

ects," *Conference on Chemical Feedstocks, Chemical Week,* Houston, Tex., November 1980.

4. Argonne National Laboratory, "The Environmental Implications of the Large-Scale Utilization of Alcohol Gasoline Blends in the United States," *Fourth Alcohol Fuels Symposium,* p. 205.

5. Peter D'Eliscu, "Biological Effects of Methanol Spills on Marine Estuarine and Freshwater Habitats," *Second Alcohol Fuels Symposium,* VIII-3.

6. National Academy of Sciences, *Petroleum In the Marine Environment* (Washington, D.C.: NAS, 1975), p. 107.

7. Noel Mostert, *Supership* (New York: Alfred A. Knopf, 1974).

8. Carl Duisberg, "Perspectives on Brazil's Alcohol Program," paper for master's program, Johns Hopkins University, 1980.

9. Ibid.

10. Ibid.

11. U.S. Department of Energy, *Environmental Assessment: Notice of Proposed Rulemaking for the Allocation and Pricing of Gasohol,* June 1980.

12. Carol Camelio, *Bio-Energy Directory* (Washington, D.C.: Bio-Energy Council, 1980).

13. Rocket Research Corp., *Technical and Economic Feasibility of Ethanol Production in Washington State* (Seattle, Wash.: RRC, February 1980), pp. 6-26 to 6-29.

14. U.S. Department of Energy, Report of the *Alcohol Fuels Policy Review,* (Washington, D.C., DOE, June, 1979).

15. Graham Edgar, "Tetra Ethyl Lead," paper to the American Chemical Society, September 1951, reprinted by the Ethyl Corp., Richmond, Va.

16. *New York Times,* 31 October 1924.

17. Ibid.

18. *New York Times,* 21 May 1925.

19. Ibid.

20. *New York Times,* 22 April 1925.

21. *New York Times,* 20 June 1925.

22. *New York Times,* 20 September 1925.

23. "Study Indicates Strong Correlation Between Lead Levels in Blood, Lead in Gasoline," *Environment Reporter* (Washington, D.C.: Bureau of National Affairs, 27 April 1979), p. 2321. Also, "Get the Lead Out With Gasohol," *Gasohol U.S.A.,* September 1979.

24. B. D. Cohen, "Greater Dangers for Young Children Indicated in Study on Lead Effects," 29 March 1979.

25. Morris Wessel and Anthony Dominski, "Our Children's Daily Lead," Yale University Medical School, 1979.

26. Thomas O'Toole, "Industry Loses on MMT," *Washington Post,* 4 April 1977. Also, *White House Fact Sheet* on delaying lead phase-down, (Washington, D.C., White House, April, 1979).

27. The Petrochemical Energy Group, *Petrochemicals: Their Role in Human Needs* (Washington, D.C.: PEG, June 1978).

28. Testimony of Ralph W. Kienker of Monsanto for The Petrochemical Energy Group to the Environmental Protection Agency, Washington, D.C., 20 June 1979, pp. 3–4.

29. Ibid., p. 1.

30. "Octane Boosters Compete for Gasoline Pool," *Chemical and Engineering News,* 23 October 1978. Also, Stephen C. Stinson, "New Plants, Processes Set for Octane Booster," *Chemical and Engineering News,* 25 June 1979.

31. U.S. Department of Energy, *Environmental Assessment of Gasohol.* Also, "The High Cost of Taking the Lead Out," *Conoco '79* (Stamford, Conn.: Continental Oil Co., 1979). Also, Barry Commoner, "Ethanol's Role in the Current Gasoline Crisis" (St. Louis, Mo.: Center for the Biology of Natural Systems, August 1979).

32. G. H. Unzelman, "Value of Oxygen-

ates as Motor Fuels,'' paper to the American Petroleum Institute meeting, May 1981.

33. Richard Pefley et al., ''Study of Decomposed Methanol as a Low-Emission Fuel,'' *Final Report to the Office of Air Programs, Environmental Protection Agency,* 1971; idem, ''Comparative Automotive Engine Operation When Fueled With Ethanol and Methanol,'' 1978; idem, ''Predicted Methanol Water Fueled SI Engine Performance and Emissions,'' 1977; ''Characterization and Research Investigation of Methanol and Methyl Fuels,'' 1978; and ''Current and Future Prospects for Alcohol Energy Systems,'' 1980 (Santa Clara, Calif.: University of Santa Clara).

34. Argonne National Laboratory, ''The Environmental Implications of Large-Scale Utilization of Alcohol-Gasoline Blends in the United States,'' *Fourth Alcohol Fuels Symposium.*

35. G. Hagey et al., ''Methanol and Ethanol Fuels: Environmental Health and Safety Issues,'' *Second Alcohol Fuels Symposium,* VIII-2.

36. Argonne National Laboratory, ''Environmental Implications . . . ''

37. ''Gasohol Experiments Conducted by Fire Service Extension in Cooperation With Archer Daniels Midland Corn Sweeteners and Iowa Fire Equipment Inc.'' (Ames, Iowa: Iowa State University, 1980, reprinted by ADM).

38. U.S. Department of Energy, *Summary of the Carbon Dioxide Effects Research and Assessment Program* (Washington, D.C.: DOE, February 1979), unpaged edition.

39. National Academy of Sciences, *Energy and Climate* (Washington, D.C.: NAS, 1977).

40. C. D. Keeling et al., ''Atmospheric Carbon Dioxide Variations at Manua Loa Observatory, Hawaii,'' *Tellus,* vol. 28, pp. 538–551.

41. Stanford Research Institute, *Sociopol-itical Impacts of Carbon Dioxide Buildup in the Atmosphere Due to Fossil Fuel Combustion,* report to the U.S. Department of Energy under the ''Inexhaustibles'' research program, never made public.

42. Dewey MacLean, ''A Terminal Mesozoic 'Greenhouse': Lessons From the Past,'' *Science,* vol. 201, 4 August 1978, p. 401.

43. G. M. Woodwell et al., ''The Carbon Dioxide Problem: Implications for Policy in the Management of Energy and Other Resources'' (Washington, D.C.: White House Council on Environmental Quality, 1979). Also, *Global Energy Future and the Carbon Dioxide Problem* (CEQ, 1981).

44. U.S. Congress, Senate, Governmental Affairs Committee, testimony of Gordon MacDonald, 30 July 1979.

Chapter 10

1. C. Scrifes, *Mesquite Research Monograph* (College Station, Tex.: Texas A & M University, 1973), p. 5.

2. Ibid., p. 5.

3. Ibid., p. 5.

4. Felker et al., ''Utilization of Mesquite Pods for Ethanol Production,'' *Tree Crops for Energy Production* (Golden, Colo.: Solar Energy Research Institute, November 1980). pp. 66–69 (hereafter cited as *Tree Crops*).

5. David Freedman, ''Preliminary Analysis of the Potential for Ethanol Production from Honey Locust Pods,'' *Tree Crops,* pp. 25, 125.

6. J. R. Smith, *Tree Crops: A Permanent Agriculture* (New York: Harcourt, Brace, 1929).

7. David Scanlen, ''A Case Study of Honey Locust in the Tennessee Valley Region,'' *Tree Crops,* p. 28.

8. Freedman, ''Ethanol . . . from Honey Locust,'' *Tree Crops,* p. 131.

9. Mitre Corporation, *Silvicultural Biomass Farms,* Vol. I, (Washington, D.C.: Mitre Corp., 1977).

10. Carol Camelio, *Bio-Energy Directory* (Washington, D.C.: Bio-Energy, Council, 1980), p. 83.

11. Kendall Pye, personal communications, 1981.

12. Spokesmen, C. Brewer & Co., personal communications, 1980.

13. David Welna, "Head of Brazil's National Alcohol Commission Expresses Optimism Despite Problems," Also, *Latin American Energy Report*, "Brazil Out to Show Methanol Grows on Trees," *Chemical Week*, 14 March 1979.

14. "Japan Eyes Eucalyptus as Fuel Hope," *Washington Star*, November, 1979.

15. Tommy Crabb, C. Brewer & Co., personal communication.

16. Fege, Inman and Salo, "Energy Farms of the Future," *Journal of Forestry*, June 1979, p. 358.

17. Mitre, *Silviculture Biomass Farms*, pp. 5–6.

18. Earl Barnhart, "Multi-Use Tree Crops in Solar Villages," *Tree Crops*, p. 156.

19. Marion Clawson, "Forests in the Long Sweep of American History," *Science*, vol. 204, pp. 1168.

20. David Pimental, "Biomass Energy," unpublished briefing document, September 1980.

21. U.S. Congress, Office of Technology Assessment, *Energy from Biological Processes*, vol. 2, July 1981, p. 99.

22. Gedaliah Shelef et al., "Algae Bio-mass Production in High-Rate Wastewater Treatment Ponds," *Proceedings of Bio-Energy '80 Conference*, Atlanta, Ga.: April 1980 (Washington, D.C.: Bio-Energy Council, 1980), p. 114 (hereafter cited as *Bio-Energy '80*).

23. Woverton and MacDonald, "Vascular Plants for Water Pollution Control," *Bio-Energy '80*, p. 120.

24. Ibid.

25. D. C. Pratt and N. J. Andrews, "Wetland Energy Crops," *Bio-Energy '80*, pp. 115–117.

26. Armond J. Bruce, "A Review of the Energy From Marine Biomass Program," IGT's Energy From Biomass Conference, pp. 1–8.

27. U.S. Congress, Office of Technology Assessment, *Energy from Biological Processes*, p. 104.

28. David Pimentel, "Biomass Energy," September 1980, unpublished briefing document; also, U.S. Congress, Office of Technology Assessment Report, *Energy From Biological Processes*, vol. 2, (Washington, D.C.: July 1980). p. 94.

29. Kendall Pye, private communications, July 1981.

30. Thomas Reed, "Biomass Energy—A Two Edged Sword," paper to the American Section of the International Solar Energy Society, Denver, Colo., August 1978, p. 1.

Glossary

ABSOLUTE ALCOHOL: completely dehydrated ethyl alcohol of the highest proof obtainable (200° proof); also "neat" alcohol. (See ANHYDROUS.)

ACID HYDROLYSIS: decomposition or alteration of a chemical substance by water in the presence of acid. The use of acid to turn either starch or cellulose to sugar.

ACIDITY: the measure of how many hydrogen ions a solution contains per unit volume; may be expressed in terms of pH.

AFLATOXIN: the substance produced by some strains of the fungus *Aspergillus flavus*; the most potent carcinogen yet discovered; a persistent contaminant of corn that renders crops unsalable.

AFO: Office of Alcohol Fuels.

ALCOHOL: the family name of a group of organic chemical compounds composed of carbon, hydrogen, and oxygen; a series of molecules that vary in chain length and are composed of a hydrocarbon plus a hydroxyl group, CH_3-$(CH_2)n$-OH; includes methanol, ethanol, isopropyl alcohol, and others; see ETHANOL.

ALCOHOL FUEL PLANT: under BATF regulations, a distilled-spirits plant established solely for producing, processing, and using or distributing distilled spirits to be used extensively for fuel use.

ALCOHOL FUEL PRODUCER'S PERMITS: the document issued by BATF pursuant to the Crude Oil Windfall Profits Tax Act (26 USC 5181, P.L. 96-223) authorizing the person named to engage in business as an alcohol fuel production facility.

ALDEHYDES: any of a class of highly reactive organic chemical compounds obtained by controlled oxidation of primary alcohols, characterized by the common group CHO, and used in the manufacture of resins, dyes, and organic acids.

ALKALI: soluble mineral salt of a low-density, low-melting point, highly reactive metal; characteristically "basic" in nature.

AMBIENT: the prevalent surrounding conditions usually expressed as functions of temperature, pressure, and humidity.

AMINO ACIDS: the naturally occurring nitrogen-containing building blocks of proteins.

AMYLASE: any of the enzymes that accelerate the hydrolysis of starch and glycogen.

AMYLODEXTRINS: see DEXTRINS.

ANAEROBIC DIGESTION: a type of bacterial degradation of organic matter that occurs in the absence of air (oxygen) and produces primarily carbon dioxide and methane.

ANHYDROUS: devoid of water; refers to a compound that does not contain water either absorbed on its surface or as water of crystallization.

ANHYDROUS ETHANOL: 100-percent alcohol, neat alcohol, 200-proof alcohol.

APPARENT PROOF: the proof indicated by a hydrometer after correction for temperature but without correction of the obscuration caused by the presence of solids.

ATMOSPHERIC PRESSURE: pressure of the air (and atmosphere surrounding us) which changes from day to day; it is equal to 14.7 pounds per square inch.

AZEOTROPE: the chemical term for two or more liquids that, at a certain concentration, boil as though they are a single substance; alcohol and water cannot be separated further than 194.4 proof because at this concentration, alcohol and water form an azeotrope and vaporize together.

AZEOTROPIC DISTILLATION: distillation in which a substance is added to the mixture to be separated in order to form an azeotropic mixture with one or more of the components of the original mixture; the azeotrope formed will have a boiling point different from the boiling point of the original mixture and will allow separation to occur.

BACKSET (also called set back): the liquid portion of the stillage recycled as part of the process liquid in mash preparation.

BACTERIAL SPOILAGE: occurs when bacterial contaminants take over the fermentation process in competition with the yeast.

BAGASSE: the cellulosic residue left after sugar is extracted from sugar cane.

BALLING HYDROMETER or BRIX HYDROMETER: a triple-scale wine hydrometer designed to record the specific gravity of a solution containing sugar.

BARREL: a liquid measure equal to 42 American gallons or about 306 pounds of crude oil; one barrel equals 5.6 cubic feet or 0.159 cubic meters.

BASIC HYDROLYSIS: decomposition or alteration of a chemical substance by water in the presence of alkali.

BATCH FERMENTATION: fermentation of a specific quantity of material conducted from start to finish in a single vessel. Fermentation in which mash is held in a single container and the entire volume of mash is fermented before the container is refilled. This contrasts with continuous fermentation, where the mash is fed in continuously as fermented beer if drawn off.

BATCH PROCESS: a unit operation where one cycle of feedstock preparation, cooking, fermentation, and distillation is completed before the next cycle is begun.

BATF: Bureau of Alcohol, Tobacco and Firearms, under the U.S. Department of of the Treasury; responsible for the issuance of permits, for both experimental and commercial facilities, for the production of alcohol.

BEER: the product of fermentation by microorganisms; the raw fermented mash, which contains about 7 to 12% alcohol; usually refers to the alcohol solution remaining after yeast fermentation of sugars.

BEER STILL: the stripping section of a distillation column for concentrating ethanol, or the first column of a two (or more) column system, in which the first separation from the mash takes place.

BEER WELL: the surge tank used for storing beer prior to distillation.

BIOMASS: Organic matter, such as trees, crops, manure, and aquatic plants, that is available on a renewable basis.

BOILER: a device which heats water to produce steam for cooking and distillation processes.

BOILING POINT: the temperature at which the transition from the liquid to the gaseous phase occurs in a pure substance at fixed pressure.

BOND: a type of insurance which gives

the government security against possible loss of distilled spirits tax revenue; not required for alcohol fuel plants producing less than 10,000 proof gallons per year.

BRITISH THERMAL UNIT (Btu): the amount of heat required to raise the temperature of one pound of water one degree Fahrenheit under stated conditions of pressure and temperature (equal to 252 calories, 778 foot-pounds, 1055 joules, and 0.293 watt-hours); it is a standard unit for measuring quantity of heat energy.

BULK DENSITY: the mass (weight) of a material divided by the actual volume it displaces as a whole substance, expressed in $1b/ft^3$; kg/m^3; etc.

CALORIE: the amount of heat required to raise the temperature of one gram of water one degree Centigrade.

CARBOHYDRATE: a chemical term describing certain neutral compounds made up of carbon, hydrogen, and oygen; includes all starches and sugars; a general formula is $C_x(H_2O)y$.

CARBON DIOXIDE: a gas produced as a byproduct of fermentation; chemical formula is CO_2.

CASSAVA: a starchy root crop used for tapioca; can be grown on marginal croplands along the southern coast of the United States.

CATALYSIS: the effect produced by a small quantity of a substance (catalyst) on a chemical reaction, after which the substance (catalyst) appears unchanged.

CELL RECYCLE: the process of separating yeast from fully fermented beer and returning it to ferment a new mash; can be done with clear worts in either batch or continuous operations.

CELLULASE: an enzyme capable of decomposing cellulose into simpler carbohydrates.

CELLULOSE: the main polysaccharide in living plants, forms the skeletal structure of the plant cell wall; can be hydrolyzed to glucose. A molecule composed of long chains of glucose molecules bound together. In cellulose the linkage between glucose molecules is different from the linkage in starch. Cellulose molecules are much larger and more structurally complex than starch molecules, which makes the breakdown of cellulose to glucose more difficult, (e.g., wood, plant, fiber, sewage).

CELLULOSIC FEEDSTOCKS: materials, such as wood, crop stalks, and newsprint, containing sugar units linked by bonds that are not easily ruptured.

CELSIUS (Centigrade): a temperature scale commonly used in the sciences; at sea level, water freezes at 0°C and boils at 100°C. $C° = 5/9 (F - 32)$

CENTRIFUGE: a rotating device for separating liquids of different specific gravities or for separating suspended colloidal particles according to particle-size fractions by centrifugal force.

CETANE NUMBER (cetane rating): measure of a fuel's ease of self-ignition; the higher the number, the better the fuel for a diesel engine.

COD: Chemical Oxygen Demand; a measure of water pollution.

COLUMN: vertical, cylindrical vessel containing a series of perforated plates or packed with materials through which vapors may pass, used to increase the degree of separation of liquid mixtures by distillation or extraction.

COMPLETELY DENATURED ALCOHOL (CDA): ethyl alcohol which is at least 160 proof blended, pursuant to formulas prescribed by BATF, with sufficient quantities of various denaturants to make it unfit for and not readily recoverable for beverage use; this may then be distributed through retail outlets without permits. (Compare to SPECIALLY DENATURED ALCOHOL.)

COMPOUND: a chemical term denoting a specific combination of two or more distinct elements.

CONCENTRATION: the quantity of a substance (e.g., ethanol or sugar) in a known quantity of water. Weight percent is the weight of alcohol or sugar per volume of water. Volume percent is the volume of alcohol or sugar per volume of water.

CONDENSER: a heat-transfer device that reduces a fluid substance from its vapor phase to its liquid phase by reducing its temperature as it contacts cooling surfaces in its path.

CONTINUOUS FERMENTATION: a steady-state fermentation system that operates without interruption; each stage of fermentation occurs in a separate section of the fermenter, and flow rates are set to correspond with required residence times.

COOKER: a tank or vessel designed to cook a liquid or extract or digest solids in suspension; the cooker usually contains a source of heat, and is fitted with an agitator; its purpose is to aid in breaking down starches into fermentable sugars.

COPRODUCTS: the resulting substances and materials that accompany the production of ethanol by fermentation processes.

DDG: see Distillers Dried Grains.

DDGS: see Distillers Dried Grains with Solubles.

DDS: see Dried Grains with Solubles.

DEHYDRATION: the process of removing water from any substance by exposure to high temperature or by chemical means.

DENATURANT: a substance added to ethanol to make it unfit for human consumption so that it is not subject to alcohol beverage taxes.

DENATURE: the process of adding a substance to ethyl alcohol to make it unfit for human consumption; the denaturing agent may be gasoline or other substances specified by the Bureau of Alcohol, Tobacco, and Firearms.

DESICCANT: a substance having an affinity for water; used for drying purposes.

DEWATERING: removal of the free water from a solid substance.

DEXTRINS: a polymer of D-Glucose which is intermediate in complexity between starch and maltose and formed by partial hydrolysis of starches.

DEXTROSE: the same as glucose.

DISACCHARIDES: the class of compound sugars which yield two monosaccharide units upon hydrolysis; examples are sucrose, maltose, and lactose.

DISTILLATE: that portion of a liquid which is removed as a vapor and condensed during a distillation process.

DISTILLATION: the process of separating the components of a mixture by differences in boiling point; a vapor is formed from the liquid by heating the liquid in a vessel and the vapor is successively collected and condensed into liquids. The process by which the components of a mixture are separated by boiling and recondensing the resultant vapors.

DISTILLERS DARK GRAINS: see DISTILLERS DRIED GRAINS WITH SOLUBLES (DDGS).

DISTILLERS DRIED GRAINS (DDG): the water-insoluble, dried distillers grains coproduct of the grain fermentation process which may be used as a high-protein (28 percent) animal feed. (see DISTILLERS GRAINS.)

DISTILLERS DRIED GRAINS WITH SOLUBLES (DDGS): a grain mixture obtained by mixing distillers dried grains and distillers dried solubles.

DISTILLERS DRIED SOLUBLES (DDS): a mixture of water-soluble oils and hydrocarbons obtained by condensing the thin stillage fraction of the solids obtained from fermentation and distillation processes.

DISTILLERS FEEDS: coproducts resulting from the fermentation of cereal grains by the yeast *Saccharomyces cerevisiae;* the nonfermentable portion of grain mash.

DISTILLERS GRAIN: the nonfermentable portion of a grain mash comprised of protein, unconverted carbohydrates and sugars, and mineral material.

DOE: Department of Energy.

DRY MILLING: a process of separating various components of grains, such as germ, bran, and starch without using water.

DSP (DISTILLED SPIRITS PLANT): a plant, including fuel alcohol plants, authorized by the Bureau of Alcohol, Tobacco and Firearms to produce, store, or process ethyl alcohol in any of its forms.

ENERGY CROPS: includes such agricultural crops as corn and sugarcane; also nonfood crops such as poplar trees. (See BIOMASS.)

ENGINE: a machine for converting any of various forms of energy into mechanical force and motion.

ENZYMES: the group of catalytic proteins that are produced by living microorganisms; enzymes mediate and promote the chemical processes of life without themselves being altered or destroyed. Organic substances (proteins) produced in the cells of living organisms that create specific chemical changes. Enzymes used in the liquefaction and saccharification steps of starch conversion are produced by bacteria or fungi.

EPA: Environmental Protection Agency.

ETHANOL: chemical formula C_2H_5OH; the alcohol product of fermentation that is used in alcoholic beverages, a pure fuel, and for industrial purposes; blended with gasoline to make gasohol; also known as ethyl alcohol or grain alcohol.

ETHYL ALCOHOL: see ETHANOL.

EVAPORATION: conversion of liquid to the vapor state by the addition of latent heat of vaporization; usually refers to vaporization into the atmosphere.

EXCISE TAX, GASOLINE: a tax collected at the pump to support the construction and maintenance of highways. Gasohol is exempt through 1992 from the $.04 federal excise tax and in some states from the state excise tax.

FAHRENHEIT SCALE: a temperature scale in which the boiling point of water is 212 and its freezing point 32°; to convert °F to °C, subtract 32, multiply by 5, and divide the product by 9 (at sea level). $C° = (F° - 32) \times 5/9$.

FEED PLATE: the theoretical position in a distillation column above which enrichment occurs and below which stripping occurs.

FEEDSTOCK: the base raw material that is the source of sugar for fermentation.

FERMENTABLE SUGAR: sugar (usually glucose) derived from starch or cellulose that can be converted to ethanol (also known as reducing sugar or monosaccharide).

FERMENTATION: a microorganically mediated enzymatic transformation of organic substances, especially carbohydrates, generally accompanied by the evolution of a gas; the conversion of simple sugars to ethanol with the aid of enzymes.

FERMENTATION ETHANOL: ethyl alcohol produced from the enzymatic transformation of organic substances.

FOSSIL FUEL: any naturally occurring fuel of an organic nature which originated in a past geologic age (such as coal, crude oil, or natural gas).

FRACTIONAL DISTILLATION: a process of separating alcohol and water (or other mixtures) by boiling and drawing off vapors from different levels of the distilling column.

FRUCTOSE: a fermentable monosaccharide (simple) sugar of chemical formula $C_6H_2O_6$; fructose is a ketohexose.

FUEL GRADE ALCOHOL: usually refers to ethanol of 160 to 200 proof, although 100 to 200 proof can be used with diesel fuel.

FUSEL OIL: a clear, colorless, poisonous liquid mixture of alcohols obtained as a byproduct of grain fermentation; major constituents are amyl, isoamyl, propyl, isopropyl, butyl, and isobutyl alcohols.

GASOHOL (GASAHOL): a registered trademark held by the State of Nebraska for a fuel mixture of agriculturally derived 10 percent anhydrous fermentation ethanol and 90 percent unleaded gasoline; it is often incorrectly used to mean any mixture of alcohol and gasoline to be used for motor fuel.

GASOLINE: a volatile, flammable liquid obtained from petroleum that has a boiling range of approximately 200° to 216°C and is used as fuel for spark-ignition internal-combustion engines.

GLUCOSE: a monosaccharide; occurs free or combined and is the most common sugar; chemical formula $C_6H_2O_6$; glucose is an aldohexose.

GRAIN ALCOHOL: See ETHANOL.

HEAT EXCHANGER: a device that transfers heat from one fluid (liquid or gas) to another, or to the environment.

HEAT OF CONDENSATION: the same as the heat of vaporization, except that the heat is given up as the vapor condenses to a liquid at its boiling point.

HEAT OF VAPORIZATION: the heat input required to change a liquid at its boiling point to a vapor at the same temperature (e.g., water at 212°F to steam at 212°F).

HEATING VALUE: the amount of heat obtainable from a fuel and expressed, for example, in Btu/1b.

HYDRATED: Chemically combined with water.

HYDROCARBON: a chemical compound containing hydrogen and carbon.

HYDROLYSIS: the decomposition or alteration of a substance by chemically adding a water molecule to the unit at the point of bonding.

HYDROMETER: a long-stemmed glass tube with a weighted bottom; it floats at different levels depending on the relative weight (specific gravity) of the liquid; the specific gravity or other information is read where the calibrated stem emerges from the liquid.

INDOLENE: a standard mixture of chemicals used in comparative tests of automotive fuels.

INDUSTRIAL ALCOHOL: ethyl alcohol produced and sold for other than beverage purposes; depending on the use, may or may not be denatured.

INULIN: a polymeric carbohydrate comprised of fructose monomers found in the roots of many plants, particularly jerusalem artichokes.

LACTOSE: a crystalline disaccharide made from whey and used in pharmaceuticals, infant foods, bakery products, and confections; also called "milk sugar." A sugar with the chemical formula $C_{12}H_{22}O_{11}$ most commonly found in milk and cheese whey. Lactose is fermentable only by some strains of yeast.

LEADED GASOLINE: gasoline containing tetraethyl lead to raise octane value.

LEAN FUEL MIXTURE: an excess of air in the air-fuel ratio; gasohol has leaning effect over gasoline because the alcohol adds oxygen to the system.

LIGNIFIED CELLULOSE: cellulose polymer wrapped in a polymeric sheath and extremely resistant to hydrolysis because of the strength of its linkages.

LIGNIN: a polymeric, noncarbohydrate constituent of wood that functions as a binder and support for cellulose fibers.

LINKAGE: the bond or chemical connection between constituents of a molecule.

LIQUEFACTION: the change in the phase of a substance to the liquid state; in the case of fermentation; the conversion of water-insoluble carbohydrate to water-soluble carbohydrate.

MALT: barley softened by steeping in water, allowed to germinate, and used especially in brewing and distilling as a source of amylase. Sprouted grain, usually special varieties of barley, that contains enzymes which convert starch to sugar.

MALTOSE: a disaccharide of glucose.

MASH: a mixture of grain and other ingredients with water to prepare wort for brewing operations.

MEAL: a granular substance produced by grinding.

MEMBRANE: a sheet polymer which separates components of solutions by permitting passage of certain substances but preventing passage of others.

METHANOL: a light volatile, flammable,

poisonous, liquid alcohol, CH_3OH, formed in the destructive distillation of wood or made synthetically and used especially as a fuel, a solvent, an antifreeze, or a denaturant for ethyl alcohol, and in the synthesis of other chemicals; methanol can be used as fuel for motor vehicles; also known as methyl alcohol for wood alcohol.

METHYL ALCOHOL: also known as methanol or wood alcohol; see METHANOL.

MOLECULAR SIEVE: a compound which separates molecules by selective penetration into the sieve space on the basis of size, charge, or both.

MOLECULE: the chemical term for the smallest particle of matter that is the same chemically as the whole mass.

MONOMER: a simple molecule which is capable of combining with a number of like or unlike molecules to form a polymer.

MONOSACCHARIDES: see FERMENTABLE SUGAR.

MULTIPLE-EFFECT EVAPORATOR: a series of evaporators in which the vapor removed from each unit is used to supply heat to the next unit in the series.

NET ENERGY BALANCE: the amount of energy available from fuel when it is burned, less the amount of energy it takes to produce the fuel.

OCTANE NUMBER: a rating which indicates the tendency to knock when a fuel is used in a standard internal combustion engine under standard conditions.

OFFERING STATEMENT: a prospectus of an investment in the form of an offering of stock or partnership in the investment in return for cash invested into the company.

OFFICE OF ALCOHOL FUELS: an independent office within the Department of Energy responsible for administration of all alcohol fuels programs in the Department and coordination of related programs in other federal agencies.

OVERHEAD: the relatively low-boiling-point liquids removed from the top of a distillation unit.

PACKED DISTILLATION COLUMN: a column or tube constructed with a packing of ceramics, steel, copper, or fiberglass-type material to increase surface area through which vapors or liquid may pass.

pH: a scale that measures the acidity or alkalinity of a solution. An acidic solution has a pH of 0–7, a neutral solution has a pH of 7, and an alkaline solution has a pH of 7–14.

PLATE DISTILLATION COLUMN (sieve tray column): a distillation column constructed with perforated plates or screens.

POLYMER: a substance made of molecules comprised of long chains or cross-linked simple molecules.

POUND OF STEAM: one pound (mass) of water in the vapor phase, not to be confused with the steam pressure which is expressed in pounds per square inch.

POUNDS PER SQUARE INCH ABSOLUTE (psia): the measurement of pressure referred to a complete vacuum of 0 pressure.

PRACTICAL YIELD: the amount of product that can actually be derived under normal operating conditions; i.e., the amount of sugar that normally can be obtained from a given amount of starch or the amount of alcohol that normally can be obtained is usually less than theoretical yield.

PROCESS GUARANTEE: refers to the financial ability of an engineering company to successfully make its process perform, within given tolerance levels, in the event that the process does not meet production levels orginally agreed to in the contract.

PROOF: a measurement of the alcohol concentration in an alcohol-water mixture, equal to twice the percentage by volume of the alcohol; e.g., 80 percent alcohol equals 160; proof, 100 percent alcohol equals 200 proof.

PROOF GALLON: a U.S. gallon of liquid

which is 50 percent ethyl alcohol by volume or the alcohol equivalent thereof; also one tax gallon.

PROTEIN: high-molecular-weight compounds composed of amino acids. Proteins are essential to the animal diet.

PURE ETHYL ALCOHOL: ethyl alcohol that has not been denatured and is usually sold as 190 proof and 200 proof (absolute or anhydrous).

QUAD: one quadrillion (10^{15} or 1,000,000,000,000,000) Btu (British Thermal Units).

RECTIFICATION: with regard to distillation, the selective increase of the concentration of a component in a mixture by successive evaporation and condensation.

RECTIFYING COLUMN: the portion of a distillation column above the feed tray in which rising vapor is enriched by interaction with a countercurrent falling stream of condensed vapor.

REFLUX: the condensate return to a rectifying column to maintain the liquid-vapor equilibrium.

RENEWABLE RESOURCES: renewable energy; resources that can be replaced after use through natural means; examples: solar energy, wind energy, energy from growing plants.

ROAD OCTANE: a numerical value for automotive antiknock properties of a gasoline; determined by operating a car over a stretch of level road.

SACCHARIDE: a simple sugar or a compound that can be hydrolyzed to simple sugar units.

SACCHARIFICATION: a conversion process using acids, bases, or enzymes in which carbohydrates are broken down into fermentable sugars. Saccharification converts short glucose chains formed during liquefaction to single molecules of glucose.

SACCHAROMYCES: a class of single-cell yeasts which selectively consume simple sugars.

SBA: Small Business Administration.

SCRUBBING EQUIPMENT: equipment for counter-current liquid-vapor contact of flue gases to remove chemical contaminants and particulates.

SIMPLE SUGARS: see FERMENTABLE SUGARS.

SPECIALLY DENATURED ALCOHOL (SDA): ethyl alcohol to which sufficient quantities of various denaturants have been added, pursuant to formulas prescribed by federal regulations, to render it unfit for beverage purposes without impairing its usefulness for other purposes; specially denatured alcohol may be distributed only to persons holding BATF permits.

SPECIFIC GRAVITY: the ratio of the mass of a solid or liquid to the mass of an equal volume of distilled water at 4°C.

SPENT GRAINS: the nonfermentable solids remaining after fermentation of a grain mash.

STARCH: a molecule composed of long chains of glucose molecules bound together. Starch is a major component of many agricultural crops such as potatoes and grains and can be broken down into glucose. Starch is found in most plants and is a principal energy storage product of photosynthesis; starch hydrolyzes to several forms of dextrin and glucose.

STILL: An alcohol distillation unit that consists of a container for heating a beer solution, a distillation column for separating the alcohol and water, and a condenser for capturing the alcohol vapors. An apparatus for distilling liquids, particularly alcohols; it consists of a vessel in which the liquid is vaporized by heat, and a cooling device in which the vapor is condensed.

STILLAGE: the mixture of nonfermentable solids and water that remains after removal of the alcohol by distillation; also called "spent beer."

STOICHIOMETRIC RATIO: the ratio

of chemical substances necessary for a reaction to occur completely.

STOVER: the dried stalks and leaves of a crop remaining after the grain has been harvested.

STRIPPING SECTION: the section of a distillation column below the feed in which the condensed water is progressively decreased in the next higher component by stripping.

SUCROSE: a crystalling disaccharide carbohydrate found in many plants, mainly sugarcane, sugar beets, and maple trees; $C_{12}H_{22}O_{11}$.

SULFURIC ACID: a strong acid with the chemical formula H_2SO_4 used to lower the pH (increase the acidity) of a solution; also known as "battery acid."

SURETY BOND: a type of insurance which satisfies the government's bonding requirements on distilled spirits production (see BOND); obtainable from U.S. Treasury-authorized insurance companies, surety bonds usually carry an annual premium of 1 to 2 percent of face value.

SURFACTANT: surface-active agent, a substance that alters the properties, especially the surface tension, at the point of contact between phases; e.g., detergents and wetting agents are typical surfactants.

TAX-FREE ALCOHOL: pure ethyl alcohol withdrawn free of tax for government, for hospital use, for science, or for humanitarian reasons; it cannot be used in foods or beverages; all purchasers must obtain BATF permits, post bonds, and exert controls upon storage and use of Tax-Free Alcohol.

TAX-PAID ALCOHOL: pure ethyl alcohol which has been released from federal bond by payment of the federal tax of $21.00 per gallon at 200 proof or $19.95 per gallon at 190 proof.

TETRAETHYL LEAD (TEL): an octane enhancer for gasoline now under environmental restriction.

THERMAL EFFICIENCY: energy heating value; the ratio of energy output to energy input.

THERMOPHILIC: capable of growing and surviving at high temperatures.

THIN STILLAGE: the water-soluble fraction of a fermented mash plus the mashing water.

TRAY: one of several types of horizontal pieces in a distillation column.

TURBINE: a rotary engine actuated by the reaction or impulse or both of a current of fluid (as water or steam) subject to pressure and usually made with a series of curved vanes on a central rotating spindle.

UDAG: U.S. Department of Agriculture

VACUUM DISTILLATION: the separation of two or more liquids under reduced vapor pressure (a vacuum) reduces the boiling points of the liquids being separated.

VAPORIZE: to change from a liquid or a solid to a vapor, as in heating water to steam.

VAPOR PRESSURE: the pressure at any given temperature of a vapor in equilibrium with its liquid or solid form.

VOLUMETRIC FUEL ECONOMY: miles per gallon.

WET MILLING: a process similar to dry milling except that the various components of grain are separated in water.

WHEY: the watery part of milk separated from the curd in the process of making cheese; it is produced commercially in large quantities and can be used as a fertilizer, animal feed, or feedstock in the production of ethanol.

WHOLE STILLAGE: the undried "bottoms" from the beer comprised of nonfermentable solids, distillers solubles, and the mashing water.

WINE GALLON: a United States gallon of liquid measure equivalent to the volume of 231 cubic inches.

WOOD ALCOHOL: see METHANOL

WORT: the liquid remaining from a brewing mash preparation following the filtration of fermentable beer.

YEAST: single-cell microorganisms (fungi) that produce alcohol and carbon dioxide under anaerobic conditions and acetic acid and carbon dioxide under aerobic conditions; the microorganism that is capable of changing sugar to alcohol by fermentation; single-celled microscopic plants that ferment simple sugars into ethyl alcohol and carbon dioxide; a type of fungus.

ZYMOSIS: see FERMENTATION

Selected Bibliography

"Alcohol: a Brazilian Answer to the Energy Crisis," *Science*. February 11, 1977.

"Alcohol from Babacu," *Amazon News Letter*. Belem P.A. Brazil. March/April 1977.

"Alcohol Fuels," *The Energy Consumer*, U.S. Department of Energy publication, Washington, D.C. January, 1980.

"Alcohol Fuels? Ethanol's Good, Methanol's Better," *Chemical Week*. June 24, 1981.

Altsheller, William B. et al. "Design of a Two-Bushel Per Day Continuous Alcohol Unit," *Chemical Engineering Progress*. Vol. 43, No. 9.

A Marvelous Chicken Burning Motor Car, *Mother Earth News*. Hendersonville, N.C., Issue #10. July 1971.

American Petroleum Institute, Alcohols: a technical assessment of their applications as fuels, Paper #4261. Washington, D.C. July 1976.

Anderson, Jack. "DOE Ties to Mobil Revealed: Gasohol Report Challenged," *Washington Post*, Washington, D.C. May 24, 1980.

Anderson, Marion and Carl Parisi. *American Jobs from Alcohol Fuel*, Employment Research Associates. Minnesota 1980.

A Time to Choose: Summary Report on the Structure of Agriculture, U.S.D.A. Washington, D.C. January 1980.

Baratz, B., et al. *Survey of Alcohol Fuel Technology*, Prepared for the National Science Foundation, Office of Energy Research Foundation. McLean, Va. The Mitre Corporation, V.I., November 1976.

Barnhart, Earle. "Multi-Use Tree Crops in Solar Villages," *Tree Crops for Energy Production*, published proceedings of conference sponsored by U.S. Department of Energy, November 12-14, 1980.

Barr, William J. and Frank A. Parker. *The Introduction of Methanol as a New Fuel into the United States Economy*, American Energy Research Co., McLean, Va., 22101. Published by the Foundation for Ocean Research, San Diego, Ca. 92121. March 1976.

Berger, Clarence. *Social Adjustments in the Cities*, printed in Social Changes During the Great Depression. De Capo Press, 1974.

Bernhardt, W. E. and W. Lee. *Engine Performance and Exhaust Emission Characteristics of a Methanol-Fueled Automobile*, Reprinted from Future Automotive Fuels-Prospects, Performance, Perspective. Plenum Publishing Corp. New York 1977.

Blaser, Richard F. *Heat Balanced Cycle*, U.S. Naval Academy, Division of Engineering and Weapons, Annapolis, Md. 1974.

Bolt, Jay A. *A Survey of Alcohol as a Motor Fuel*, Society of Automotive Engineers, Special Publication 254, 1964.

Borkin, Joseph. *The Crime and Punishment of I.G. Farben*, New York, The Free Press, 1978.

Born, Roscoe. "Sweet Proposition: How to Turn Surplus Sugar into Gasohol—And a Fast Buck," *Barron's* August 20, 1979.

Borth, Christy. *Modern Chemists and Their Work*, New York, Bobbs-Merrill, 1939.

Boyd, T.A. "Motor Fuel from Vegetation," *Industrial and Engineering Chemistry*, 13:9. September 1921.

Brackett, A.T. and others. *Indiana Grain Fermentation Alcohol Plant*, House. Indianapolis, Indiana 46204. Indiana Department of Commerce. No. 336. 1976.

Bradley, Cliff and Ken Runion. "Fuel Alcohol Production," Butte, Montana, National Center for Appropriate Technology, 1981; also, "Making More Fuel Alcohol," NCAT.

Bridgeman, Oscar C. "Utilization of Ethanol-Gasoline Blends as Motor Fuels," *Industrial and Engineering Chemistry;* 28.9 September 1936.

Brinkman, Norman D. "Vehicle Evaluation of Neat Methanol: Compromises Among Exhaust Emissions, Fuel Economy and Drivability," General Motors Research Labs (GMR 2825 F&L 662). Warren, Michigan. January 1979.

Brown, Lester. "Food or Fuel: New Competition for the World's Cropland," Worldwatch Institute. Washington, D.C. March 1980.

Bruwer, J.J. "Sunflower Seed Oil as an Extender for Diesel Fuel in Agriculture Tractors," Division of Agricultural Engineering, Department of Agriculture. Silverton, South Africa. 1980.

Christensen, Leo. "Testimony to Hearings on Senate Bill 552, Committee on Finance," U.S. Senate, May 23-29, 1939, Washington, D.C. Government Printing Office. 1939.

Clark, Ed. "Denis Day, A Pioneer in Farm Produced Alcohol," *Farm Energy*. August 1980. Des Moines, Iowa, Iowa Corn Growers' Association.

——— "Farmers' First Chance to Compare Alcohol Stills: Cookoff," *Farm Energy*. Iowa, 1980.

Clawson, Marion. "Forests in the Long Sweep of American History," *Science Magazine*. Volume 204. June 1979.

Coffey, Thomas. *The Long Thirst: Prohibition in America 1920-1933*, W.W. Norton and Company. 1975.

Conway, Hugh. "Alcohol Fuel," *Bugantics*. England. Vol. 40. No. 2.

Cowper, William. "The Alcohol Fuels," *Energy and Earth Machine Magazine*. Norton & Co. 1976.

Cruz, J., W. Chancellor, J. Gross. "Dual Fueling a Diesel Engine with Carbureted Alcohol," F. Pischinger, C. Havenith, "A New Way to Direct Injection of Methanol in a Diesel Engine," Proceedings of the Third International Symposium on Alcohol Fuel Technology. Asilomar, California. May 1979.

Curry, Richard F. "Future may be now for Alcohol Fuels," Falls Church, Va. American Automobile Association. May 1977.

Curtis, Carl. "Commission on Increased Industrial Use of Agricultural Products," Doc. #45. Senate PL-540 report. Washington, D.C. U.S. Government Printing Office. 1957.

Davis, Bernard D. "Frontiers of the Biological Sciences," *Science*. July 4, 1980.

de Castro, Jose. *Death in the Northeast*, Random House. New York, N.Y. 1966.

"The Disappearing Farmlands: A Citizen's Guide to Agricultural Lands Preservation," National Association of Counties Research Foundation. August 1980.

Doubleday, Henry. *Fuel Plus Fertility*, Andrews Singer Publishers. Bottisham, England. 1954.

Earl, W.B., and W.A. Brown. "Alcohol Fuels from Biomass in New Zealand," Proceedings of the Third International Symposium on Alcohol Fuels Technology. Asilomar, California. May 29, 1979.

Egloff, Gustav. "Alcohol-Gasoline Motor Fuels," New York. American Petroleum Institute. 1933.

Egloff, Gustav. "Motor Fuel Economy of Europe," Industrial and Engineering Chemistry. Vol. 30. 1937.

———— "Motor Fuel Economy of Europe," American Petroleum Institute 1939; "Substitute Fuels in a War Economy," paper to the American Chemical Society. 1942. Reprinted by A.P.I.

"Energy Balances in Production and End-Use of Alcohols Derived from Biomass," Report to the NAFC by TRW Corp. October 1980.

"Energy for Rural Development," Renewable resources and alternative technologies for developing countries. N.A.S. Washington, D.C. 1976.

"Energy from Biological Processes—Volumes I and II," U.S. Congress, Office of Technology Assessment. Washington, D.C. July 1980.

"Ethanol Projects Receive Backing of U.S. Agency," Wall Street Journal. August 17, 1981.

"The Facts About an Alcohol-Gasoline Blend as Motor Fuel," American Automobile Association. Washington, D.C. 1933.

"Farm Brewed Fuel: Farmers Would Make Motorists Pay for Farm Relief," The Business Week. March 15, 1933.

Feger, Inman, and Salo. "Energy Farms for the Future," Journal of Forestry. June 1979.

"First Annual Report to Congress on the Use of Alcohol in Motor Fuel," U.S. Department of Energy. Washington, D.C. April 1980.

"Five Grain Alcohol Plants May Spring Up in State," Evening Journal. Lincoln, Nebraska. April 14, 1978.

Furtado, Celso. "The Economic Growth of Brazil: A Survey from Colonial to Modern Times," 3rd Edition. Trans. Ricardo W. D'Aguiar, Eric Charles Crysdale. University of California Press. Berkeley, CA. 1968.

Gaden, Elmer, ed., "Enzymatic Conversion of Cellulosic Materials, Technology and Applications," John Wiley and Sons. New York, New York 1976.

Galbraith, John Kenneth. The Great Crash 1929, Sentry Edition 10. Boston, Houghton Mifflin. 1972.

Gall, Norman "Brazil's Alcohol Program," Common Ground. American University Field Staff Journal. IV:4. Washington, D.C. Winter 1978. p. 34.

Garvan, Francis P. "Scientific Method of Thought in our National Problems," Proceedings of the Second Dearborn Conference on Agriculture, Industry and Science. The Chemical Foundation. New York. 1936.

"Gasohol Acceptance in Established Markets," Iowa Corn Promotion Board. The Iowa Development Commission. Des Moines. April 1979.

"Gasohol Backer Sees Economic Benefits," Sunday Journal and Star. Lincoln, Nebraska. May 29, 1977.

"Gasohol Craze," Associated Press. The Denver Post. April 19, 1979. p. 27.

"Gasohol Sold Like Hotcakes at Illinois Pump," San Francisco Sunday Examiner. From Chicago Daily News. December 11, 1977.

Gilbert, M.M., et al., "High Compression Tractor Engines with Alcohol-Gasoline Blends," Philippine Agr. Engin. Jour. Vol. 90. No. 2. 1961. pp. 239-248.

Goldstein, I.S. "The potential for converting wood into plastics and polymers or into chemicals for the production of these materials," NSI-Rann Report. Raleigh, N.C. North Carolina State, School of Forest Resources. Department of Wood and Paper Science. N.C. 1974.

Green, Farno L. Energy potential from agricultural field residues, New Orleans,

LA. Manufacturing Development G.M. Technical Center. Warren, Michigan. June 1975.

Gunderson, Gerald. *A New Economic History of America*, McGraw-Hill, Inc. New York. 1976.

Hagen, David L. *Methanol as a Fuel*, A Review with Bibliography. Society of Automobile Engineers. 770292. 1977.

Hagey, Graham and others. *Methanol and Ethanol Fuels—Environmental, Health and Safety Issues*, U.S. D.O.E and Mueller Associates. Baltimore, Md. 1977.

Hammond, Allen. "Methanol at MIT: Industry Influence Charged in Project Cancellation," *Science*. November 21, 1975.

Hammond, Dean C., Jr., Sidney G. Liddle and William L. Huellmantel. "Combustion of Methanol in an Automotive Gas Turbine," *Resource Recovery and Energy Review*. Wakeman-Walworth, Inc. Summer 1977.

Harney, Brian M. "Methanol from Coal—A Step Toward Energy Self Sufficiency. *Energy Sources*. Vol. 2. No. 3. 1975.

Havemann, H.A., et al., "Alcohol in Diesel Engines," *Automobile Engineer*. June 1954. p. 256.

——— "Alcohol With Normal Diesel Fuels," *Gas and Oil Power*. Vol. 50. January 1955. pp. 15-19 and Vol. 50. February 1955.

Hearings before the Subcommittee on Antitrust, Monopoly and Business Rights of the Committee on the Judiciary and the Committee on Small Business, U.S. Senate, S.2251 No. 96:58. March 5, 1980.

Heid, Waller G., Jr. The Performance and Economic Feasibility of Solar Grain Drying Systems, Economics, Statistics, and Cooperative Service of the U.S. Department of Agriculture. Agricultural Economic Report #396. Washington, D.C. February 1978.

Heitland, Herbert, H.W. Czaschke, N. Pinto. "Applications of Alcohol from Biomass and their Alternatives as Motor Fuels in Brazil," *Proceedings of the Second International Symposium on Alcohol Fuel Technology*. Wolfsburg, W. Germany. November 21, 1977.

Herrick, Rufus Frost. *Denatured or Industrial Alcohol*, Chapman and Hall. London. 1907.

Hildenbrand, Barb. *Methanol: An Automobile Fuel in our Future?* G.M. Research Laboratory. January-February 1977.

Hieronymus, William. "Brazil tries mixing Alcohol from Sugar with Gasoline to reduce its Oil Imports," *The Wall Street Journal*. November 28, 1977.

Hixon, R.M., LM. Christensen. *The Use of Alcohol in Motor Fuels*, Iowa State College. May 1, 1933.

Holsendolph, Ernest. "Farmers Produce High-Test Brew as Car Fuel," *The New York Times*. New York. May 9, 1978.

Holtzman, David. "Biomass Fuel: Fermenting Self-Sufficiency," *People and Energy*. Washington, D.C. February 1978.

Ickes, Harold. *Fighting Oil*, Knopf. New York. 1943.

"Industrial Alcohol," *War Changes in Industry Series*. U.S. Tariff Commission. Washington, D.C. January 1944.

"Is Alcohol Next Candidate for Fuel Pumps?" *Chemical Week*. January 30, 1974.

Jacobs, P. Burke and Harry P. Newton. *Motor Fuels from Farm Products*, U.S.D.A. miscellaneous publication #327. Washington, D.C. 1938.

Jaffe, Bernard. *New World of Chemistry*, Silver Burdett Co. New York. 1952.

Janeway, Eliot. "Out of the Prairie into the Capital," *Washington Star*. Washington, D.C. October 23, 1977.

Jantzen, Dan and Tom McKinnon. *Preliminary Energy Balance and Economics of a Farm-Scale Ethanol Plant*, Golden, Colo., Solar Energy Research Institute, 1980. SEPI RR 624 699R.

"Japan Eyes Eucalyptus as Fuel Hope," *Washington Star*. Associated Press. November 1979.

Jonchere, J.P. "Methanol Seen as a Hydrogen Source," *Oil & Gas Journal*.

Petroleum Publishing Co. June 14, 1976.

Katzen, Raphael *Chemicals from Wood Waste,* USDA. Forest Service, Forest Products Laboratory. Madison, Wisc. December 24, 1975.

Lappé, Francis Moore and Joseph Collins. *Food First: Beyond the Myth of Scarcity,* Houghton Mifflin Co. Boston, Mass. 1977.

Lincoln, John W. *Methanol and Other Ways Around the Gas Pump,* Gateway Publishing. Charlotte, Vermont. 1976.

Lindsey, Dred S., Project Manager. *Grain Alcohol Study,* Prepared for the Indiana Department of Commerce. Long Rock J.V. Study Group. Indianapolis, Indiana. July 2, 1975.

Lindsley, E.F. "Alcohol Power: Can it help you meet the soaring cost of gasoline?" *Popular Science.* Vol. 206. April 1975.

Longwell, J.P. and W.J. Most. *Single-Cylinder Engine Evaluation of Methanol Improved Energy Economy and Reduced NOx,* Society of Automotive Engineers. New York. 1975.

Lovins, Amory B. *Soft Energy Paths: Toward a Durable Peace,* Ballinger Publishing Co. Cambridge, Mass. 1977.

Lowdermilk, W.C. *Conquest of the Land Through Seven Thousand Years,* U.S.D.A. Washington, D.C. reprinted 1975.

Lucke, Charles. *Use of Alcohol and Gasoline in Farm Engines,* U.S. Department of Agriculture Farmer's Bulletin. #227. Washington, D.C. 1907.

Ludvigsen, Karl. "Alcohol Comes Back to Power your Car," *Mechanics Illustrated.* January 1978.

Lukens, Thomas. *Jerusalem Artichokes: A Special Crop for Alcohol,* Unpublished, Brown Bulb Ranch of Washington, Inc. Seattle, Washington. 1980.

Lyles, Thomas. "Alcohol in the Gas Tank: A Renewable Energy Source," *The Washington Star.* Washington, D.C. May 21. 1977.

MacIntyre, Loren. "A Massive Technology Transplant Takes Root in the Amazon Jungle," *National Geographic.* May 1980.

MacLean, Dewey. "A Terminal Mesozoic 'Greenhouse': Lessons From the Past," *Science.* Vol. 201. No. 4354. August 4, 1978.

Malik, V.K. *Present Status and Future Projections of Alcohol Based Chemicals in India,* Proceedings of the First InterAmerican Conference on Renewable Sources of Energy. New Orleans, La. November 1979.

Manchester, William. *The Glory and the Dream,* Little Brown & Co. Boston, Mass. 1973.

Marvin, Murphy, ed. "Methanol Output Seen Topping 100,000 tons/year by 2000," *The Oil Daily.* Washington, D.C. February 24, 1978.

McElheny, Victor K. "Multi-fuel Car Engine Gets New Tests," *New York Times.* New York. September 12, 1977.

McIntosh, John Geddes. *Industrial Alcohol,* Scott Greenwood & Sons. London. 1907.

Menrad, Holger, W. Lee and W. Bernhardt. *Development of a Pure Methanol Fuel Car,* Society of Automobile Engineers Passenger Car Meeting. #770790. Detroit, Michigan. September 1977.

Methane Generation from Human, Animal and Agricultural Wastes, National Academy of Sciences. Washington, D.C. 1977.

"Methyl Gasoline Fuel Tested by Volkswagen," *The Detroit News.* Detroit, Mich. April 17, 1975.

Meyer, A.J. *Development of a Practical Method for Burning Alcohol in a Gasoline Tractor—Calculation and Charting of Thermodynamic Properties of Ethyl Alcohol—Air Mixture and its Combustion Products,* Kentucky Univ. of Eng. Exp. Station. Bul. No. 3, June 1948.

Miccolis, J.M.F. "Alternative Energy Technologies in Brazil," *Proceedings of the American Association for the Advancement of Sciences.* Denver, Colo. February 1977.

Michelson, Herb. "A Hero on Ninth Street,"

The Sacramento Bee. Sacremento, Ca. January 24/25, 1978.

"Midwest Solvents," *Atchison Daily Globe.* Atchison, Kansas. October 28, 1975.

Miller, Dwight "Fermentation of Ethyl Alcohol," Biotechnology and Bioengineering Symposium. No. 6307-312. *USDA proceedings of the 7th National Conference on Wheat Utilization Research* (Prepared by USDA, Manhattan, Kansas. November 1971.) John Wiley & Sons, Inc. 1976.

"Mobil Proves Gasoline-From-Methanol Process," *Chemical and Engineering News.* January 30, 1978.

Mohr, B.J. and W.A. Scheller. *Net Energy Analysis of Ethanol Production,* American Chemical Society. Fuel Chemicals Division. Reprints, 29, 21, Vol. 2. 1976.

Moll, A.J. and C.F. Clark. An Overview of Alcohol Production Routes Presented at Alcohols as Alternative Fuels, Stanford Research Institute. Toronto. November 19, 1976.

Morgan, Dan. *Merchants of Grain,* Viking Press. New York, N.Y. 1979.

Motor Fuels From Farm Products, USDA. Publication No. 327. Washington, D.C. December 1938.

Myers, P.S. *Automobile Emissions—A Study in Environmental Benefits Versus Technological Costs,* SAE Paper 700132. November 1969.

Nellis, Micki. *Makin' It on the Farm,* Iredell, Texas, American Agriculture Movement. Colorado. 1979.

Nelson, Diane. "Grain Fields-Oil Wells of the Future?" *Minnesota Motorist.* A.A.A. March 1978.

Nyberg, Bartell. "Gasohol: Fuel for the Future?," *The Denver Post. Empire Magazine.* Washington, D.C. February 13, 1977.

Omang, Joanne. "Mr. Bell had a Solution to the Energy Crisis," *The Washington Post.* Washington, D.C. December 15, 1977.

O'Toole, Thomas. "Industry Loses on MMT," *The Washington Post.* Washington, D.C. April 4, 1977.

Parisi, Anthony J. "Gene Engineering Industry Hails Court Ruling as Spur to Growth," *New York Times.* New York. June 17, 1980.

Pearson, Drew. "U.S. May Face Another Rubber Shortage," *Washington Merry Go Round.* Bell Syndicate. Washington, D.C. November 29, 1950.

Pefley, R.K., et al., *Performance and Emission Characteristics Using Blends of Methanol and Dissociated Methanol as an Automotive Fuel,*" Paper 719008. Intersociety Energy Conversion Engineering Conf., Boston, Massachusetts. August 3-5, 1971.

Petroleum In the Marine Environment, National Academy of Sciences. Washington, D.C. 1975.

Pimental, David and others. "Food Production and the Energy Crisis," *Science.* Washington, D.C. November 1973.

Pischinger, G. and N.L.M. Pinto. "Experiences with the Utilization of Ethanol," *Proceedings of the Third International Symposium on Alcohol Fuels Technology.* Asilomar, Calif. May 29, 1979.

Planting Oil, Banco Real. Brazil. April 11, 1977.

Pleeth, S.J.W. "Alcohol: A Fuel For Internal Combustion Engines," Chapman Hall. London. 1949.

——— "Alcohol Motor Fuels," *Automobile Engineer.* Vol. 42. April 1952.

Porteous, Dr. A. "The Recovery of Industrial Ethanol from Paper in Waste," *Chemistry and Industry.* 1967.

Posner, Herbert S. "Biohazards of Methanol in Proposed New Uses," *Journal of Toxicology and Environmental Health.* 1975.

"Potato Alcohol: A Solution to the Energy Crises and Higher Prices for Spuds?," *Potato Grower of Idaho.* April 1978.

Pouring, A.A., R.F. Blaser, E.L. Keating and B.H. Renkin. "Influence of Combustion with Pressure Exchange on the Performance of Heat Balanced Internal Combustion Engines," *Society of Au-*

tomotive Engineers. 400 Commonwealth Drive. Warrendale, Pa. 15096. February 28, 1977.

Povich, M.J. *Some Limitations of Fuel Farming*, General Electric Co. Schenectady, New York. 1978.

Reed, Thomas. *Biomass Energy—A Two Edged Sword*, Paper presented to the American Section of the International Solar Energy Society. Denver, Colorado. August 28-31, 1978.

Reed, T.B. and R.M. Lerner. "Methanol-A Versatile Fuel for Immediate Use," *Science*. Vol. 182. No. 4119. 1973.

Reynolds, Conger. "The Alcohol Gasoline Proposal," *Proceedings of the American Petroleum Institute*. Chicago, Illinois. November 9-17, 1939.

Risser, James. "A Renewed Threat of Soil Erosion: It's Worse than the Dust Bowl," *Smithsonian*. Washington, D.C. March 1981.

Ritchie, James D. "Diesels Run on Sunflower Power," *New Farm*. January 1981. Also, Benjamin H. Beard, "The Sunflower Corp," *Scientific American*. May 1981.

Rodale, Robert. Editorial. *New Farm Magazine*. Emaus, Pa. September-October 1979.

Roggensack, Meg. "Vegetable Oil Use as Fuel," *Latin American Energy Report*. Washington, D.C. August 14, 1980.

Rosenbluth, R.F. and C.R. Wilke. *Comprehensive Studies of Solid Wastes Management: Enzymatic Hydrolysis of Cellulose*, Sanitary Engineering Research Laboratory. Berkeley, California. December 1970.

Ross, Irwin. "Bean Has A Heart of Gold," *Fortune Magazine*. New York. October 1973.

Rothberg, Paul. *LNG: Hazards, Safety Requirements and Policy Issues*, Library of Congress. Science Policy Research Division. U.S. Government Printing Office. Washington, D.C. June 10, 1977.

Ryder, Oscar B. *Industrial Alcohol*, U.S. Tariff Commission Report. Government

Printing Office. Washington, D.C. 1944.

Saeman, Jerome F. *Energy and Materials From the Forest Biomass*, Madison, Wisconsin. U.S. Forest Products Laboratory. Washington, D.C. January 1977.

Scanlen, David. "A Case Study of Honey Locust in the Tennessee Valley Region," *Tree Crops for Energy Production*, published proceedings of the conference sponsored by U.S. Department of Energy. November 12-14, 1980.

Schinto, Jeanne. "Alcohol for Gasoline," *The Progressive*. Wisconsin. November 1977.

Scheller, William A. *Energy and Ethanol*, Department of Chemical Engineering. University of Nebraska. Lincoln, Ne. 68588. April 6, 1978.

Schruben, Leonard. *Gasohol Bubble*, Proceedings of Distillers Feed Council. Vol. 33. Louisville, Kentucky. March 30, 1978.

Seameth: Study Summary—A Marine Methanol Plant. CONOCO. Ponca City, Oklahoma. 1977.

Sheppard, Nathaniel, Jr. "An Energy Saving Device Brews Trouble," *The New York Times*. New York. May 20, 1978.

Silber, Howard. "Exxon Questions Objectivity of State Gasohol Committee," *Omaha-World Herald*. Omaha, Ne. June 8, 1977.

Simmonds, Charles. *Alcohol: Its Production, Properties and Applications*, MacMillan & Co., London. 1919.

Smith, J.P. "Gasohol Advocates Lobbying Intensely," *Washington Post*. Washington, D.C. April 23, 1979.

Smith, J.R. *Tree Crops: A Permanent Agriculture*, Harcourt, Brace & Co., New York. 1929.

Smothers, Ronald. "When Waste Becomes Energy," *The Long Island Weekly. New York Times*. New York. April 30, 1978.

Solar Energy: A Comparative Analysis to the Year 2020, MITRE Corporation technical report MTR-7579. McLean, Virginia. March 1978. p. 65.

Solberg, Carl. *Oil Power*, Mason Charter,

New York. 1976.

Solomon, Burt. "Moonshine and Motor Cars: Alcohol Fuels Come of Age," *The Energy Daily.* Washington, D.C. October 28, 1977. Vol. 5. No. 209.

Spano, Leo. "Enzymatic Hydrolysis of Cellulose to Fermentable Sugar for the Production of Ethyl Alcohol," *Proceedings of the First Inter-American Conference on Renewable Sources of Energy.* New Orleans, La. November 25-29, 1979.

Steinbeck, John. *Grapes of Wrath,* Viking Press. New York. 1939.

Stobaugh, Robert and Daniel Yergin. *Energy Future,* Random House, 1979.

Stout, Bill, Wesley Buchele, Robert Peart et al. "Brazil Promotes Proalcool for Petroleum Independence," *Agricultural Engineering.* April 1978.

Strong, Robert M. *Commercial Deductions form Comparisons of Gasoline and Alcohol Tests on Internal Combustion Engines,* U.S. Geological Survey Bulletin #392. Dept. of the Interior. Washington, D.C. Government Printing Office. 1909.

Teodoro, A.L. "Fifty Thousand Kilometers on Alcohol as a Motor Fuel," *Philippine Agriculture.* Vol. 28. 1939.

ter Horst, J.F. "Gasohol: No Fuel Like an Old Fuel," *Los Angeles Times.* Los Angeles, Ca. November 8, 1977.

Tucker, William. "The Next American Dust Bowl and How to Avert It," *Atlantic Magazine.* July 1979.

Underkofler, L.A., W.K. McPherson, Ellis I. Fulmer. "Alcoholic Fermentation of Jerusalem Artichokes," *Industrial and Engineering Chemistry.* October 1937.

U.S. National Alcohol Fuels Commission. *Fuel Alcohol An Energy Alternative for the 1980's,* Final Report. 1981.

Vetter, R.L. *Cornstalk Grazing and Harvested Crop Residues for Beef Cows,* A.S. Leaflet R186. Iowa State University Cooperative Extension Service. Iowa. July 1973.

Ware, Sylvia. *Fuel and Energy Production by Bioconversion of Waste Materials—State-of-the-art,* Contract RFQ CI75-0786. Washington, D.C. August 1975.

Weaver, Kenneth F. "Our Energy Predicament," *Energy: National Geographic Special Report,* National Geographic, February, 1981.

Welna, David. "Progress of Brazil's Alcohol Fuel Program Disputed by Leadership," *Latin American Energy Report.* Washington, D.C. June 5, 1980.

Wiebe, R. and J.D. Hummell. "Practical Experiences with Alcohol-Water Injection in Trucks and Farm Tractors," *Agricultural Engineering,* Vol. 35, No. 5. May 1954. pp. 319-326.

Wiebe, R. and J. Nowakowska. *The Technical Literature of Agricultural Motor Fuels,* U.S. Department of Agriculture Bibliography. Bul. No. 10. Washington, D.C. 1949.

Wiley, H.W. "Industrial Alcohol: Uses and Statistics," U.S.D.A. Bulletin #269. Washington, D.C. 1906.

Wilke, C.R., ed. *Cellulose as a Chemical and Energy Resource,* John Wiley & Sons. New York. 1975.

Wilkie, Herman Frederick and Dr. Paul J. Kolachov. *Food for Thought,* Indiana Farm Bureau. Indianapolis. 1942.

Wittwer, Sylvan. *Agricultural Technology, Productivity and Environment,* Testimony before subcommittees of the House Agriculture Committee and House Science and Technology Committee. U.S. Congress. July 25, 1979.

Woodwell, G.M., G.J. MacDonald, R. Revelle and C.D. Keeling. *The Carbon Dioxide Problem: Implications for Policy in the Management of Energy and Other Resources,* Prepared for the Council on Environmental Quality. Washington, D.C. 1979.

Wright, J. Patrick. *On a Clear Day You Can See General Motors,* Avon Books. New York. 1979.

Wyss, Al "Major New Fuel Seen for Industries," *The Journal of Commerce*. October 17, 1977.

Yang, Victor. "Agro-Industrial System for Ethanol and Ethylene Production," *Proceedings of the Institute of Gas Technology's Symposium on Energy from Biomass and Wastes*. Washington, D.C. August 1978.

Yeager, Jim "Is There a Government and Oil Company Conspiracy?," *Big Valley Magazine*. Los Angeles, Ca. May 1979.

Index

271